Molecular Imaging: Computer Reconstruction and Practice

NATO Science for Peace and Security Series

This Series presents the results of scientific meetings supported under the NATO Programme: Science for Peace and Security (SPS).

The NATO SPS Programme supports meetings in the following Key Priority areas: (1) Defence Against Terrorism; (2) Countering other Threats to Security and (3) NATO, Partner and Mediterranean Dialogue Country Priorities. The types of meeting supported are generally "Advanced Study Institutes" and "Advanced Research Workshops". The NATO SPS Series collects together the results of these meetings. The meetings are coorganized by scientists from NATO countries and scientists from NATO's "Partner" or "Mediterranean Dialogue" countries. The observations and recommendations made at the meetings, as well as the contents of the volumes in the Series, reflect those of participants and contributors only; they should not necessarily be regarded as reflecting NATO views or policy.

Advanced Study Institutes (ASI) are high-level tutorial courses intended to convey the latest developments in a subject to an advanced-level audience

Advanced Research Workshops (ARW) are expert meetings where an intense but informal exchange of views at the frontiers of a subject aims at identifying directions for future action

Following a transformation of the programme in 2006 the Series has been re-named and re-organised. Recent volumes on topics not related to security, which result from meetings supported under the programme earlier, may be found in the NATO Science Series.

The Series is published by IOS Press, Amsterdam, and Springer, Dordrecht, in conjunction with the NATO Public Diplomacy Division.

Sub-Series

A.	Chemistry and Biology	Springer
B.	Physics and Biophysics	Springer
C.	Environmental Security	Springer
D.	Information and Communication Security	IOS Press
E.	Human and Societal Dynamics	IOS Press

http://www.nato.int/science
http://www.springer.com
http://www.iospress.nl

Series B: Physics and Biophysics

Molecular Imaging: Computer Reconstruction and Practice

edited by

Yves Lemoigne

European Scientific Institute
Site d'Archamps
Archamps
France

Alessandra Caner

European Scientific Institute
Site d'Archamps
Archamps
France

 Springer

Published in cooperation with NATO Public Diplomacy Division

Proceedings of the NATO Advanced Study Institute on
Molecular Imaging from Physical Principles to Computer Reconstruction
and Practice
Archamps, France
9–21 November 2006

Library of Congress Control Number: 2008930998

ISBN 978-1-4020-8751-6 (PB)
ISBN 978-1-4020-8750-9 (HB)
ISBN 978-1-4020-8752-3 (e-book)

Published by Springer,
P.O. Box 17, 3300 AA Dordrecht, The Netherlands.

www.springer.com

Printed on acid-free paper

Preface

This book reports the majority of lectures given during the NATO Advanced Study Institute ASI-982440, which was held at the European Scientific Institute of Archamps (ESI, Archamps – France) from November 9 to November 21, 2006. The ASI course was structured in two parts, the first was dedicated to individual imaging techniques while the second is the object of this volume and focused on data modelling and processing and on image archiving and distribution. Courses devoted to nuclear medicine and digital imaging techniques are collected in a complementary volume of NATO Science Series entitled "Physics for Medical Imaging Applications" (ISBN 978-1-4020-5650-5).

Every year in autumn ESI organises the European School of Medical Physics, which covers a large spectrum of topics ranging from Medical Imaging to Radiotherapy, over a period of five weeks. Thanks to the Cooperative Science and Technology sub-programme of the NATO Science Division, weeks two and three were replaced this year by the ASI course dedicated to "Molecular Imaging from Physical Principles to Computer Reconstruction and Practice". This allowed the participation of experts and students from 20 different countries, with diverse cultural background and professional experience (Africa, America, Asia, and Europe). A further positive outcome of NATO ASI participation is the publication of this book, which contains the lectures series contributed by speakers during the second week of the ASI. We hope it will be a reference book in medical imaging, addressing an audience of young medical physicists everywhere in the world, who are wishing to review the physics foundations of the relevant technologies, catch up with the state of the art and look ahead into future developments in their field.

This volume includes two introductory lectures highlighting the role of imaging applications to the medical field and develops further into an in-depths discussion of digital imaging modelling and data handling for medical applications.

The role of medical imaging in clinical diagnostic as well as a set of medical doctors' requirements was presented by Susanne Albrecht (HCUGE, Geneva). An illustration of a new technological trend in X-ray digital imaging is illustrated by a publication contributed by of D. R. Dance (Royal Marsden Hospital, London). The following lectures focused on analysing in depths the digital imaging modelling and data handling.

Simone Giani (CERN, Geneva) reviewed medical simulation tools, related methodology and functionality. An introduction to Monte-Carlo techniques was presented by Michel Maire (LAPP, Annecy-le-Vieux), naturally followed by John Apostolakis (CERN, Geneva), who introduced the GEANT 4 Toolkit and its applications. Two lectures by Michel Defrise (University Hospital, Brussels) were devoted to image reconstruction algorithms. A full afternoon was allocated to hands-on PC practice with GATE and GEANT, each student equipped with a PC and the relevant software (O. Nix, DKFZ, Heidelberg). Examples of simulation of semi-conductor detectors for biomedical imaging were presented by Paolo Russo (Napoli University/INFN, Napoli) whereas Adele Lauria (Napoli University/INFN, Napoli) discussed CAD and GRID for medical imaging applications. An introduction to computer networks, more and more present in hospital structures, was given by Ulrich Fuchs (CERN, Geneva). About one day was devoted to pharmacokinetics modelling for PET. The theoretical foundations were explained by Wolfgang Müller-Schaünburg (University Clinic of Tübingen, Tübingen) whilst their application and the results obtained in a small animal experiment were presented by Philippe Millet (HCUGE, Geneva). Four hours were dedicated to multimodality techniques and images fusion. They were shared between two lecturers: Pierre Jannin (Rennes University, Rennes) and Luc Bidaut (University of Texas – ACC, Houston). David Bandon (HCUGE, Geneva) explained the PACS system for medical images archiving and sharing and Bernard Gibaud (CHU, Rennes) illustrated the DICOM standard framework, state of the art and future applications.

Lectures were not the only activity proposed to participants of the ASI: the possibility was offered to visit the equipment at the neighbouring Geneva hospital, hosting PET/SPECT cameras as well as cyclotron for radio tracers production. Production of radioisotopes and preparation for use in PET imaging were explained in-situ by Yan Seimbille (Geneva Hospital). To add to their scientific education, students were invited to CERN, located only 13 km away from Archamps. At the renowned European experimental research centre, they visited and studied a huge crystal electromagnetic calorimeter, i.e. a detector representing a very large scale version of the main constituent of modern PET cameras. Note that lab courses were organised at the Geneva University, for the students who wished to practise their skills in dealing with experimental equipment for beta and gamma radiation detection.

We wish to thank all the participants, who allowed the ASI at Archamps to be a success within an excellent international atmosphere: lecturers, students (who participated actively) and all the ESI team (Manfred Buhler-Broglin, Alessandra Caner, Severine Guilland, Violaine Presset and Julien Lollierou).

Many thanks to The Physics Department of the Geneva University which allowed us to propose practical exercises in its laboratory directed by Jean Divic Rapin. We deeply thank also the CERN team in Geneva for opening the laboratory and the Hôpital Cantonal de Genève, which hosted us twice and gave us access to its radiotracers and imaging equipment (Cyclotron, PET and SPECT camera).

Finally, we wish to thank and express the gratitude of all participants to the Cooperative Science and Technology sub-programme of the NATO Science Division, lead by Professor Fausto Pedrazzini, without whom this Advance Study Institute would have not been possible.

A few days before the beginning of the ASI, we were informed of the sudden death of Professor Vitali KAFTANOV from ITEP Moscow. Vitali was one the co-director of the ASI, as representative of "Partner Countries". He had spent time and used much energy in the preparation of the courses.

We would like this book to be dedicated to his memory.

Co-Director of ASI-982440   *Yves Lemoigne*[1]

---

[1] European School of Medical Physics, European Scientific Institute, Bâtiment Le Salève, Site d'Archamps F-74166 Archamps (France).

Contents

List of Participants

Lecturers' surnames are in bold

Surname	Name	Country	City	E-mail address
Abo-Elmagd	Nevien	Egypt	Mansora	vivamagd@yahoo.com
Acar	Hilal	Turkey	Istanbul	hilalacar@hotmail.com
Acun	Hediye	Turkey	Istanbul	acunhediye@yahoo.com
Albrecht	Susanne	Switzerland	Geneva	susanne.albrecht@hcuge.ch
Al Faraj	Achraf	France	Lyon	achraf_f1@hotmail.com
Apostolakis	John	Switzerland	Geneva	john.apostolakis@cern.ch
Bandon	David	Switzerland	Geneva	bandon@sim.hcuge.ch
Ben Abbas	Wassila	Algeria	Batna	bwassila@hotmail.com
Bidaut	Luc	USA	Houston	luc.bidaut@di.mdacc.tmc.edu
Caner	Alessandra	France	Archamps	alessandracaner@freesurf.fr
Carp-Rusu	Adela	Romania	Magurele	adell_23@yahoo.com
Defrise	Michel	Belgium	Brussels	mdefrise@minf.vub.ac.be
Djarova	Anna	Bulgaria	Sofia	anna_djarova@abv.bg
Dominietto	Marco	Italy	Novarra	marco.dominietto@cern.ch
Dordai	Dan	Romania	Cluj-Napoca	dan@nat.vu.nl
Dyakov	Iliya	Bulgaria	Sofia	i_diakov@yahoo.com
Florescu	Maria Gabriela	Romania	Baia Mare	ghilac@gmail.com
Fuchs	Ulrich	Switzerland	Geneva	t.fuchs@dkfz.de
Georgiev	Ivaylo	Bulgaria	Sofia	iv3georg@yahoo.com
Giani	Simone	Switzerland	Geneva	Simone.Giani@Cern.Ch
Gibaud	Bernard	France	Rennes	bernard.gibaud@chu-rennes.fr
Ghitulescu	Zoe	Romania	Bucharest	zoe.ghitulescu@cncan.ro
Giuliacci	Arianna	Italy	Novarra	arianna.giuliacci@cern.ch
Hamal	Mohammed	Morocco	Oujda	hama_m@yahoo.com
Jannin	Pierre	France	Rennes	pierre.jannin@irisa.fr
Kaftanov[†]	Vitali	Russian Federation	Moscow	valeri.kaftanov
Kadjo	Aziz	France	Lyon	aziz-kad@hotmail.fr
Karadag	Nazli	Turkey	Istanbul	nazlikaradag@yahoo.com
Kostova-Lefterova	Desislava	Bulgaria	Sofia	dessi.zvkl@gmail.com
Kryvenko	Sergiy	Ukraine	Donetsk	kryvserg@rambler.ru

Lauria	Adele	Italy	Napoli	alauria@na.infn.it
Lemoigne	Yves	France	Archamps	psf1@compuserve.com
Lollierou	Julien	France	Lyon	jlollierou@yahoo.fr
Luchian	Alina Mihaela	Romania	Iasi	luchian_alinutza@yahoo.co.uk
Madsen	Esben	Denmark	Aarhus	esym@mta.aaa.dk
Maire	Michel	France	Annecy-le-Vieux	michel.maire@lapp.in2p3.fr
Marzeddu	Roberto	Italy	Cagliari	roberto.marzeddu@ca.infn.it
Müller-Schauenburg	Wolfgang	Germany	Tübingen	wolfgang.mueller-schauenburg@uni-tuebingen.de
Neamtu	Daniela	Romania	Craiova	daniela.neamtu@rdslink.ro
Nix	Oliver	Germany	Heidelberg	o.nix@dkfz-heidelberg.de
Pasternak	Vladislav	Ukraine	Donetsk	v_pasternak@ukr.net
Russo	Paolo	Italy	Napoli	paolo.russo@na.infn.it
Solevi	Paola	Italy	Milano	Paola.Solevi@mib.infn.it
Sumova	Andrea	Czech Republic	Prague	asumova@bluewin.ch
Sunjic	Svetlana	Croatia	Sarajevo	ssc@lsinter.net
Udovyk	Oleg	Ukraine	Kyiv	oleg_udovyk@hotmail.com
Younes	Mahmood	Egypt	Cairo	melmokh@yahoo.com
Zajacova	Zuzana	Slovakia	Bratislava	zuzana.zajacova@cern.ch

Photographs

Exercises

Computer room

Group photo

Classroom

Image reconstruction exercises

Pharmakokinetics

Farewell party

Clinical PET and PET-CT

Susanne Albrecht

Abstract The aim of this report is to provide general information about the use of PET and PET–CT in clinical routine, mainly in oncology and to explain with some selected examples what one can see and what one cannot see in PET and PET-CT.

Keywords: PET · PET-CT

1 Introduction

Functional imaging using Positron Emission Tomography (PET) is a powerful imaging technique that is used to study metabolic processes in a large clinical spectrum. It gives complementary information to other so called structural imaging techniques such as CT, MRI and ultrasound. Modern PET-CT scanners permit the acquisition of co-registered PET and CT and thus combine functional and structural information.

In nuclear medicine, radiolabeled tracers are used to image organ function. The type of tracer being used depends on the proximity of a cyclotron unit and the pathology to be explored.

In clinical routine, oncological patients represent a large number of the daily routine examinations. The most commonly used PET tracer is F-18-fluoro-2-deoxy-D-glucose (18F-FDG), the radiolabeled glucose analog, a sensitive diagnostic tool to image tumors based on increased uptake of glucose. This tracer accumulation can be quantified and support the visual interpretation.

However, depending on the tumor and patients characteristics, FGD uptake can be low, leading to false negative results. On the other hand, uptake can be increased by e.g. infection or inflammation, mimicking tumor localization. This effect has more recently been exploited, by using 18F-FDG PET for imaging of infection as

S. Albrecht
Hôpitaux universitaires de Genève, rue Micheli-du-Crest 24, CH-1211 Genève 14, Switzerland
e-mail: susanne.albrecht@hcuge.ch

Y. Lemoigne, A. Caner (eds.) *Molecular Imaging: Computer Reconstruction and Practice, and Experiments,*
© Springer Science+Business Media B.V., 2008.

1

well as of inflammation. Besides 18F-FDG, a large spectrum of other tracers is also being used in daily routine practice as well as in research in non oncological indications such as neurology, cardiology and psychiatry.

2 Background

Functional imaging by PET-CT has become a powerful non-invasive medical imaging tool, which provides tomographic images of metabolic processes in the human body. The most common tracer in clinical routine is 18F-FDG. The rational for its use in oncology is that FDG is accumulated in the cell due to a transformation processes in cancer cells leading to a high tumor to background ratio. FDG will be eliminated from the body by the renal system and will not be reabsorbed.

2.1 Physiological FDG Distribution

FDG is a glucose analogue that is transported into the viable cell by normal glucose transport mechanisms without being metabolized. Under physiological conditions FDG will be observed in organs like heart, cerebral cortex, kidney and urinary tract. Concerning the digestive system, a variable metabolic activity can be found in the intestinal tract and gastric mucosa can show discrete diffuse tracer accumulation. In lymphatic tissue metabolism can be increased e.g. in the caecum and in muscle tissue can also show increased activity. In children, physiologic uptake in the thymus is found. Knowledge of this physiological distribution is important for image interpretation.

2.2 Patients History

Not only the knowledge of the physiological FDG uptake is important for the physician who is interpreting the images, but also the knowledge of patient's recent history concerning surgery, radiation therapy, chemotherapy, infection and inflammation in order to avoid false positive results by non neoplastic lesions mimicking a cancer or false negative results by missing or masking cancer tissue. Information concerning medication is also an important point in order to correctly plan the exam. Drugs that induce elevated blood glucose levels should be avoided prior to tracer injection to guarantee a good quality study. In a patient who received chemotherapy 24 h prior to PET scanning, cancer cells can temporarily loose their ability to incorporate tracer leading to falsely negative results due to the stunning of the cancer cell.

2.3 Blood Glucose Levels

Blood glucose levels should not be higher than 130 mg/dl in order to obtain good images. Higher levels of blood glucose lead to a competition with FDG that is an analogue of glucose.

For the diabetic patient there are no explicit guidelines. In many centres nowadays insulin will not be injected to decrease blood glucose levels in case a patient presents with blood glucose level above 130 mg/dl. This attitude is chosen in order to avoid the consecutive insulin induced glucose shift into extra vascular tissue, for example the muscle. This would decrease tumor to background ratios. However, values should not exceed 200 mg/dl.

2.4 Administration

The tracer is injected intravenously. Extravasate by mistake into the surrounding soft tissue leads to tracer accumulation locally in the tissue but also to drainage of the tracer into the lymphatic system and lymph nodes and can mimic pathological lesions. Therefore a documentation of injection site and injection procedure can be helpful for image interpretation.

After injection, a delay of about 60 min between tracer injection into the human body and the beginning of imaging is usually recommended to obtain diagnostic images.

2.5 Patients Preparation

For administration of FDG the patient should be installed in a quiet room and for studies of the brain, the room should be darkened. Patients that are investigated in the head and neck region should not talk or chew immediately after tracer injection in order to avoid non specific hypermetabolism that could mimic tumoral tissue. Some patients may require pre-medication for muscular relaxation of the neck muscles in order to avoid that muscular activity is masking cancer tissue. Therefore the patient receives e.g. Valium^R before tracer injection. Some patient may also need pre-medication with anxiolytic drugs for claustrophobic reasons in order to pass the study in the scanner.

3 Whole-Body PET

3.1 18F-FDG PET in Oncology

In oncology, 18F-FDG represents the all round tracer. 18F-FDG PET is a functional imaging method that is part of the clinical work-up of patients in different settings.

One objective is the diagnosis of malignant lesions and 18F-FDG PET is commonly used for the investigation of malignant tumors of the aero-digestive pathways, malignant lymphoma and malignant melanoma.

3.2 Diagnosis

In the lung a common incidental diagnosis is a solitary pulmonary nodule (SPN). SPN is defined as a solitary lesion within the lung parenchyma that can be functionally characterized by 18F-FDG PET. In case of malignancy this lesion will show increased metabolism due to increased glycolysis in cancer cells. In the context of SPN, PET-CT contributes important information concerning the evaluation of malignancy of that lesion. In addition PET-CT permits the evaluation of nodal involvement of mediastinal and hilar lymph nodes and of distant metastasis.[1] A PET negative lymph node result has in this context a high negative predictive value permitting to avoid further invasive investigations with surgical lymph node exploration. However, false positive results can be caused by inflammation such as chronic inflammation in granulomatosis in the lung parenchyma as well as in lymph nodes but can also result from chronic inflammation in a smoker or by infection. However, depending on the histology of the underlying cancer, cells can be less avid for FDG and cause false negative results. Interpretation has to be done carefully by integrating different clinical aspects and patient conditions. Thus whole body PET-CT can help to select patients suitable for invasive diagnosis like mediastinoscopy for lymph node staging or invasive therapy with surgical resection.

3.3 Evaluation of Disease Extent (Staging)

When a cancer is diagnosed, the extent of the disease has to be determined in order to choose adequate treatment. 18F-FDG PET is a sensitive imaging tool used frequently in staging of tumors of aero-digestive pathways,[2] malignant lymphoma and malignant melanoma.

With the help of this powerful imaging tool with high special resolution and sensitivity, small lesions can be detected. Thus patients can be selected in early stages of disease when risk of metastasis is very low and patients profit from local therapy.

3.4 Evaluation of Response to Therapy

In malignant lymphoma, the response to chemotherapy can be evaluated during therapy and thus treatment can eventually be modified according to the pattern of response.[3,4] One should be aware that in young adults the activity of the thymus gland can be reactivated by chemotherapy and must not be interpreted as tumoral uptake18F-FDG PET.

3.5 Differentiation between Residual Tumor or Recurrence and Therapy Induced Tissue Alterations

In a patient who underwent radiotherapy for a head and neck cancer with or without adjuvant surgery, therapy induced tissue alteration can make image interpretation difficult. During follow-up, 18F-FDG PET can be helpful to distinguish between viable cancer tissue and post therapeutic alterations. However, therapy induces inflammation in the tissue and therefore metabolic imaging should not be performed 6 weeks or better 3 months after end of therapy.

3.6 Biochemical Evidence of Recurrence Without Morphological Findings

In prostate cancer PET can be used for early diagnosis of cancer recurrence in case of biochemical evidence of recurrence or residual disease. Imaging of the prostate is limited for 18F-FDG due to low avidity of the prostate cancer to this tracer and its elimination by the urinary tract and consecutively accumulation in the bladder, and tumor lesions will be missed.

Therefore other PET tracers have been developed; among these tracers 18F-Fluoro-Choline is increasingly being used in clinical routine.

3.7 Search for Primary Tumor

In clinical practice patients can present initially with metastasis located in lymph node or other organs without indication for the primary tumor. In this case whole body PET can be beneficial for the identification of the primary tumor.

3.8 Planning of Radiotherapy

Metabolic imaging with PET in combination with CT and MRI can be used for target volume definition for therapy planning in radio-oncology in order to optimize localization and extent of the irradiated volume.[5] In addition it can be used for staging and characterize the metabolic distribution within the tumor giving the possibility for 2.9.

3.9 Determination of Most Aggressive Tumor Component

The detection of the most active region within a tumor can guide biopsy to the most aggressive cancer component. With the histo-pathological results the adequate therapy can be selected.

4 Infection and Inflammation

The knowledge of false positive imaging results in oncological PET imagining during the last years, has lead to the concept of using 18F-FDG PET for diagnosis in these indications. At present there are not enough data to express specificity, though results are promising.

5 Brain PET

Brain PET has broad clinical indications including oncology, neurology, psychiatry and vascular diseases, for example. In all these indications, the choice of tracer being used depends on the pathology being studied and the availability of a cyclotron unit in order to use tracers with very short half-life.

5.1 Neuro-Oncology

Cell proliferation markers like Fluoro-ethyl-tyrosine (FET) are increasingly being used in the workup of glioma and high grade astrocytoma. FET-PET can visualize the extent of a known tumor and eventually spot additional lesions. In addition it can help to visualize the most metabolic region of the tumor in order to guide neurosurgical stereotactic biopsy for histo-pathological analysis and to eventually describe additional tumor lesions.

5.2 Neurology

18F-FDG PET is being used to evaluate cerebral metabolic activity in patients with suspected dementia. The pattern of regional or diffuse hypometabolism will be evaluated. These alterations can precede structural changes seen on CT or MRI.

In the preoperative evaluation of epilepsy in children and adults that are resistant to medication, PET is being performed in the seizure-free interval to visualize regions of hypometabolism corresponding to the epileptic focus. However the occurrence of seizures during tracer injection and fixation will influence the study by regions of hypermetabolism and should therefore be documented and taken into consideration during interpretation.

F-DOPA PET is increasingly being used to study Parkinson disease and additional information by CT and MRI will increase performance in interpretation.

6 Conclusions

Functional imaging by Positron Emission Tomography has become a powerful tool for imaging of metabolic processes. Modern PET-CT scanners allow the acquisition of co-registered PET and CT and thus combine functional and structural information. Clinical PET is performed in a large clinical spectrum of indications. The optimal requirements for clinical interpretation of PET studies include good knowledge of patient and tumor characteristic[6] as well as knowledge of physiologic tracer distribution.

References

1. R. M. Pieterman et al. Preoperative staging of non-small-cell lung cancer with positron-emission tomography. N Engl J Med. 2000 Jul 27; 343(4):254–61.
2. T. Stuckensen et al. Staging of the neck in patients with oral cavity squamous cell carcinomas: a prospective comparison of PET, ultrasound, CT and MRI. J Craniomaxillofac Surg. 2000 Dec; 28(6):319–24.
3. M. G. Mikhaeel et al. Use of FDG-PET to monitor response to chemotherapy and radiotherapy in patients with lymphomas. Eur J Nucl Med Mol Imaging. 2006 Jul; 33 Suppl 1:22–6.
4. P. L. Zinzani et al. Early positron emission tomography (PET) restaging: a predictive final response in Hodgkin's disease patients. Ann Oncol. 2006 Aug; 17(8):1296–300. Epub 2006 Jun 9.
5. O. Vrieze et al. Is there a role for FGD-PET in radiotherapy planning in esophageal carcinoma? Radiother Oncol. 2004 Dec; 73(3):269–75.
6. P. Seam et al. The role of FDG-PET scans in patients with lymphoma Blood. 2007 Nov 15; 110(10):3507–16. Epub 2007 Aug 20.

Digital X-Ray Imaging

David R. Dance

Abstract The use of X-ray image receptors that produce a digital image is becoming increasingly important. Possible benefits include improved dynamic range and detective quantum efficiency, improved detectability for objects of low intrinsic contrast, and reduced radiation dose. The image can be available quickly. The display is separated from the image capture so that processing and contrast adjustment are possible before the image is viewed. The availability of a digital image means ready input into PACS and opens up the possibility of computer-aided detection and classification of abnormality. Possible drawbacks of digital systems include high cost, limited high contrast resolution and the fact that their clinical value is sometimes not proven in comparison with conventional, analogue techniques. The high contrast resolution attainable with such systems is discussed and the problem of sampling limitations and aliasing considered. The properties and limitations of digital systems using computed radiography, caesium iodide plus CCDs and active matrix arrays with either caesium iodide or selenium detectors are demonstrated. Examples are given of digital systems for mammography and general radiography and their performance is demonstrated in terms of clinical assessment and measurements of the modulation transfer function and detective quantum efficiency.

Keywords: Digital radiology · mammography · chest radiology

1 Introduction

The use of digital detectors for medical X-ray imaging is now well-established. The digital capture of projection images has become the norm in many hospitals, where

D.R. Dance
Joint Dep. of Physics, Institute of Cancer Research
and The Royal Marsden NHS Foundation Trust, London SW3 6JJ, UK
e-mail: DrDance@f2s.com

Y. Lemoigne, A. Caner (eds.) *Molecular Imaging: Computer Reconstruction and Practice,*   9
and Experiments,

screen/film receptors and image intensifiers have been or are being replaced with digital receptors. Digital capture is essential if images are to be stored and accessed using a PACS.

This lecture provides a brief introduction to digital receptors, and gives examples of their performance taken from mammography and general radiography. Further details can be found in the list of references provided. The treatment is limited to consideration of the image receptor; the image display and the merits of soft-copy reporting of images are not considered here.

2 Benefits and Drawbacks of Digital Systems

Digital systems have a number of important advantages over conventional analogue imaging systems. Probably the most important are those connected with the limited dynamic range of film. This is illustrated in Figs. 6 and 7 of the companion lecture on mammography, which show a typical characteristic curve, film gamma and detective quantum efficiency (DQE) for a screen/film system. The contrast achievable in the screen/film image is proportional to the slope of the characteristic curve, or film gamma, which is only high over a limited range of optical densities. This means that it is critical to get the exposure right and even then parts of the image will show reduced contrast. Moreover, the detective quantum efficiency (DQE) of the system only has its highest value in the regions where the film gamma is high. In the high and low density regions of the image the DQE falls to low values. In addition, since the highest DQE is typically only 0.3 (at 1 line pair/mm), there is considerable potential for improvement and hence in dose reduction. Digital systems in general have a wide dynamic range and images can be captured over a wide range of exposures. These are important advantages, which, as well as avoiding the need for retakes (important for mobile radiography), means that the exposure can be reduced in some situations where a noisier image can be accepted. The rapid availability of the image can also be helpful. A good example is stereotactic localization, where films must be reviewed before the procedure can be completed.

Another important limitation of screen/film imaging is that the image cannot be manipulated before it is displayed. This of course is not true if the image is captured digitally, which opens up the further possibility of achieving a dose saving by using a higher energy X-ray spectrum and manipulating the image before it is displayed. In addition, the digital image is directly available for computer aided detection and classification of abnormalities.

A particular problem associated with digital systems is that it may be difficult to match the high contrast resolution achievable with screen/film systems because of the pixel sizes which are available and the limitation imposed by the Nyquist sampling theorem (see below). In practice, however, small objects do not present with very high contrast and image noise has an important effect on detectability. In such a situation it is the NEQ (noise equivalent quanta, or signal-to-noise ratio) which is important, which in turn is determined by the exposure used and the DQE. In this respect the digital system *may or may not* have a performance superior to that of the screen/film system.

3 Performance SPECIFICATION and Evaluation

The performance requirement for a digital radiology system can be specified in terms of parameters such as NEQ, DQE, MTF, patient dose and dynamic range. For evaluation purposes, these parameters for digital systems are often compared against those for analogue systems. However, some of the parameters can be difficult to measure and an alternative is to compare images of test phantoms and the detectability of details contained within the phantoms. Whilst this can be of great utility, it does not replace the need for clinical evaluation. Such evaluation should be based on the clinical objective of the examination, for example using ROC curves, and may be difficult to make with adequate statistics.

There is not space here for a detailed discussion of performance requirements, and we consider just three aspects, using mammography as an example.

3.1 Sampling

The Nyquist sampling theorem tells us that the sampling frequency must be at least twice the highest spatial frequency that is contained within the image. For mammography, we would like to be able to detect microcalcifications of around 100 μm in diameter, leading to a sampling interval of 50 μm, and a pixel size of similar magnitude. However, the satisfying of this constraint does not necessarily mean that such a small object will be visualized as it takes no account of the intrinsic resolution (or the pre-sampling MTF) of the image receptor or the noise. It should also be realised that information in the image above the Nyquist frequency will lead to aliasing of both the imaged information and the noise.[1]

3.2 Dynamic Range

For a typical mammographic X-ray spectrum, the attenuation through a 6 cm breast is about a factor 40. The contrast of a 3 mm carcinoma viewed against fatty tissue is about 4%, thus giving a noise level requirement of say 1%. To see 1% noise over a 40:1 transmission range gives a dynamic range requirement of 4,000:1 or 12 bits (for a detailed discussion of dynamic range requirements in digital mammography, see Maidment et al.[2]). For a receptor area of 18×24 cm^2, a pixel depth of 12 bits and a pixel size of 50 μm, the storage requirement for an uncompressed image is 25 Mbyte.

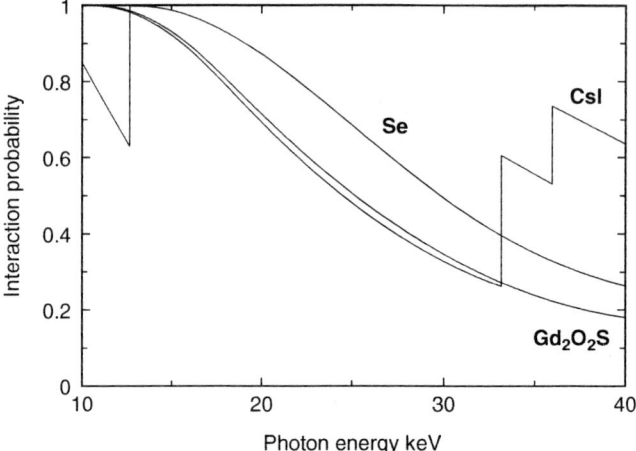

Fig. 1 Variation of interaction probability with photon energy. Data are given for caesium iodide, gadolinium oxysulphide and selenium receptors 100 μm thick

3.3 Quantum Sensitivity and DQE

In order to achieve a high DQE, it is important that the X-ray photons incident on the image receptor have a high probability of interacting. Figure 1 shows a calculation of the interaction probability for three different receptor materials in an extended mammographic energy range. All three receptors are 100 μm thick, but the gadolinium oxysulphide receptor has a packing density of just 50%. It will be seen that in this energy range, the efficiencies of the cesium iodide and gadolinium oxysulphide are similar up to the region of the K-edges for the former. Selenium, however, has a K-edge just above 12 keV, and as a consequence has a high interaction probability for much of this energy range.

The interaction of the X-ray photon, however, is only the start of the story. The energy absorbed from the X-ray has to be converted into an electrical signal, which is then digitized. This whole process may involve a number of steps, each of which may increase the noise in the image and hence decrease the DQE.[3]

4 DIGITAL RECEPTORS using Photostimulable Phosphors

Computed radiography (CR) systems have been available since the 1980s and are widely used. They comprise a photostimulable phosphor, (usually barium fluorobromide with an Eu^{2+} dopant deposited on a plastic substrate). When the phosphor is irradiated, energy absorbed from the incident X-rays produces electron-hole pairs. Many of these pairs recombine promptly, giving rise to the emission of fluorescent light. However, some electrons are trapped at colour or F-centres within the

Fig. 2 Single-side read-out
of a CR image plate (Figure
based on IPEM (2005). Copy-
right Institute of Physics
and Engineering in Medicine
2005. Reproduced with per-
mission)

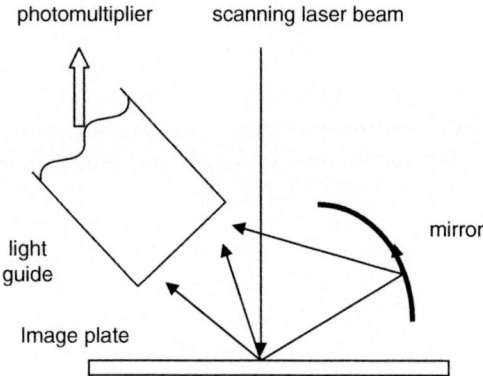

material, caused by the presence of the dopant. Such traps are metastable so that a latent image is built up consisting of electrons trapped at the F-centres. Once the exposure is complete, the plate is stimulated by irradiating with laser light of the right frequency. This frees the electrons from the traps and the energy released leads to the emission of light photons, which can be detected to create an image.

The CR image plate is built into a radiographic cassette, which is exposed in the same way as a screen/film cassette. For image read-out, the image plate is scanned using a fine laser beam which traverses the plate in a in a raster pattern. The light emitted is collected by a light guide and detected by a photomultiplier (Fig. 2).

The resolution of the CR system is limited by the spread of the read-out light within the image plate and the spot size of the laser beam. Some improvement of the resolution can be achieved by reading the plate out from both sides, and systems are available for mammography which use this approach.

The dynamic range of a CR system can be in excess of 10^4, which is an important advantage compared with screen/film systems, but, depending upon the particular systems chosen and the operating conditions, the resolution and DQE may not be as good. A very full account of CR systems, their limitations and potential is given in Rowlands.[4]

5 Direct Digital Systems using Phosphors

The images produced by the CR systems described above are only available after the exposed digital cassette is taken to the digital processor for read-out. For some digital systems, however, the electrical signal generated following the interaction of the X-rays can be read-out in a short period immediately following the expo-sure, and the image is thus available for viewing straight away. Such systems are sometimes called direct digital systems, and may use a light emitting phosphor or a photoconductor to absorb the incident X-rays.

5.1 Direct Digital Systems using Phosphors

Direct digital systems using phosphors rely on the detection of the light fluorescent photons emitted by the phosphor. Possible phosphors include caesium iodide
and gadolinium oxysulphide. Cesium iodide will give a better resolution at the
same phosphor thickness because its 'cracked' structure of parallel crystals channels
the fluorescent photons emitted in the phosphor towards the light photon detector.
This reduces the lateral spread which is possible with a fluorescent screen such as
gadolinium oxysulphide. On the other hand, the number of light photons produced
per X-ray absorbed in gadolinium oxysulphide is greater than that for cesium iodide
as they require on average 13 and 19 eV[5] respectively per light photon emitted.

5.1.1 Systems using Charge Coupled Devices

One option for detecting the light fluorescent photons emitted by the phosphor is to
use a charge-coupled device or CCD. This comprises a series of electrodes or gates
on a semi-conductor substrate. An array of metal-on-semiconductor capacitors is
formed, which act as storage wells for the charge generated within the CCD by the
photoelectric absorption of optical quanta. The 'charge image' is built up in lines of
capacitors and is read out by passing the charge from capacitor to capacitor along
each line. A high charge transfer efficiency is required.

Because the size of CCDs is limited, it is necessary to employ some means of demagnification as the light photons pass from phosphor to CCD. This can be achieved
for example by use of a lens or a fibre optic taper. The light loss associated with coupling using a lens is much greater than that for coupling with a fibre optic taper at
the same demagnification. The DQE of systems using the latter approach can therefore be expected to be better. Figure 3 shows the light collection efficiencies for the
two approaches. Even with demagnification, an array of CCDs (e.g. a 3×4 array)
may be used to achieve a large enough image matrix. An alternative approach is to
use a line or slit CCD array which is scanned across the image whilst the image is
read-out in 'time delay integration mode'.[7]

5.1.2 Systems using Active Matrix Arrays

Active matrix LCDs constructed from amorphous silicon are widely used in note-
book computer displays, but active matrix devices can also be configured as an
alternative to the CCD for the detection of light emitted by a phosphor.[5] A layer of
cesium iodide can be evaporated directly onto the active matrix. Each pixel is configured as a photodiode, which converts the light fluorescent photons to electrical
charge. For each pixel there is an associated region of the device which is configured as a thin-film field-effect transistor (TFT) and is used for image read-out.
As the area of the device accommodates both the photodiode and read-out control
circuitry, there is some loss of efficiency, usually expressed as the 'fill-factor'. For

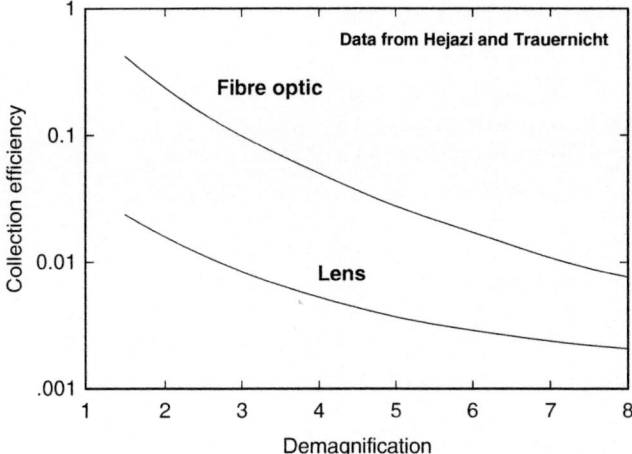

Fig. 3 Light collection efficiency for fibreoptic and lens systems for coupling to the CCD (Data taken from Hejazi and Trauernicht[6])

example, a particular digital system using an amorphous silicon active matrix read-out has a sampling interval of 100 μm, but the pixel size is only 87 μm, leading to a 'fill factor' of 0.75, and a reduction of quantum sensitivity and DQE by this factor. This problem becomes more important as the size of the pixel decreases.

5.2 Direct Digital Systems using Selenium Photoconductors

It can be deduced from Fig. 1 that amorphous selenium can be deposited in thick enough layers to have a reasonably high X-ray absorption efficiency for moderate photon energies. This and its property of photoconduction makes it attractive as part of an imaging device that can directly produce electrical charge following the interaction of an X-ray photon (and without the need to produce fluorescent light photons). The efficiency with which charge is produced depends upon the electric field strength in the selenium, but a typical value is 50 eV per electron/hole pair. Because the movement of the electrical charge within the selenium is along the direction of the electric field, there is very little lateral spread of image information as the charge is moved from its production point to its measurement point. There is thus potential for excellent spatial resolution.

For read-out purposes, the amorphous selenium can be evaporated onto an silicon active matrix in which the pixel elements are electrodes and the read-out can again be controlled via TFTs.

6 Performance of Digital Systems

As noted in section 3, the performance of digital and conventional X-ray imaging systems may be compared in many ways. We give two examples: a clinical comparison using cancer detection rates and a physical comparison using DQE and MTF.

The use of screen/film systems for screening mammography is well established, but following the introduction of digital mammographic systems, there has been strong need to compare the performance of both modalities in the screening situation. The low incidence of breast cancer necessitated a very large clinical trial. Pisano et al.[8] looked at breast cancer detection for 42,760 paired screening examinations, each woman receiving both digital mammographic (DM) and screen-film (SFM) exposures. Cancer detection was measured in terms of the area under the ROC curve. Results are listed below for three groupings: all women; women under 50 years; and women with radiographically dense breasts.

All:	DM 0.78 ± 0.02;	SFM 0.74 ± 0.02
<50years:	DM 0.84 ± 0.02;	SFM 0.69 ± 0.02
Dense:	DM 0.78 ± 0.04;	SFM 0.68 ± 0.03

The differences observed between DM and SFM for women aged less then 50 years and for dense breasts are both statistically significant at the 95% confidence interval.

Samei and Flynn[9,10] have compared different digital receptors in terms of MTF and DQE. Some of their results are shown in Figs. 4 and 5 for: DR1000, a system using a 500 μm thick selenium photoconductor and a pixel size of 139 μm; X/Qi and

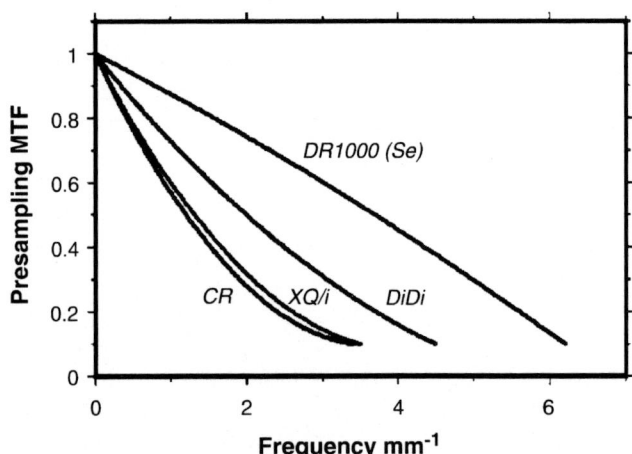

Fig. 4 Digital (pre-sampling) MTF for particular CR, caesium iodide flat panel (XQ/i and DiDi), and selenium flat panel systems (Data taken from Samei and Flynn[9,10])

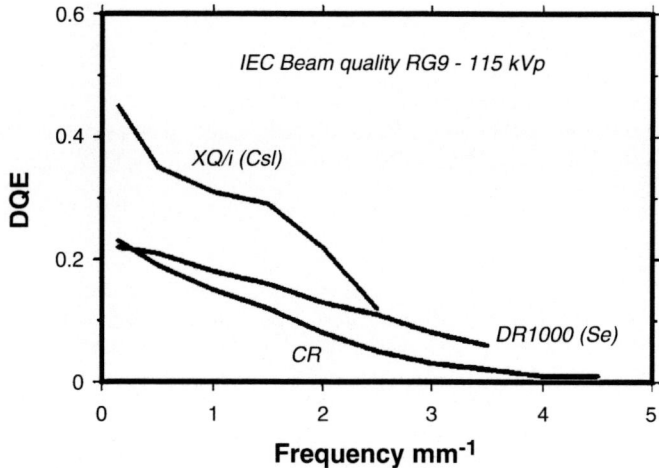

Fig. 5 DQE for particular CR, caesium iodide flat panel (XQ/i), and selenium flat panel systems (Data taken from Samei and Flynn[10])

DiDi, two systems using a CsI phosphor and a 200 µm pixel size; and a CR system with a pixel size of 100 µm. The MTF shown in Fig. 4 for one of the CsI-based systems has been enhanced artificially by the manufacturer, but in spite of this the selenium-based system has the best MTF.

This advantage, however, does not carry over to the DQE when measured at IEC beam quality RG9 at 115 kVp, where the XQ/i CsI system performs best up to the Nyquist frequency of 2.5 cycles per mm. This result is to be expected from the X-ray absorption properties shown in Fig. 1. For this particular example CR has both the worst MTF and DQE.

References

1. Dobbins, J.T., 1995, Effects of under-sampling on the proper interpretation of modulation transfer function, noise power spectra and noise equivalent quanta of digital imaging systems, *Med. Phys.* **22** 171–181
2. Maidment, A.D.A., Fahrig, R., and Yaffe, M.J., 1993, Dynamic range requirements in digital mammography, *Med. Phys.* **20** 1621–1633
3. Cunningham, I.A., Westmore, M.S., and Fenster, A., 1994, A spatial-frequency dependent quantum accounting diagram and detective quantum efficiency model of signal and noise propagation in cascaded imaging systems, *Med. Phys.* **21** 417–427
4. Rowlands, J.A., 2002, The physics of computed radiography, *Phys. Med. Biol.* **47** R123–R166
5. Yaffe, M.J. and Rowlands, J.A., 1997, X-ray detectors for digital radiography, *Phys. Med. Biol.* **42** 1–39
6. Hejazi, S. and Trauernicht, D.P, 1996, Potential image quality in scintillator CCD-based systems for digital radiography and digital mammography, *SPIE* **2708** 440–449
7. Tesic, M.M., Fisher Piccaro, M., and Munder, B., 1999, Full field digital mammography scanner. *Eur. J. Radiol.* **31** 2–17

8. Pisano, E.D., Gatsonis, C., Hendrick, E., et al., 2005, Diagnostic performance of digital versus film mammography for breast-cancer screening, *NEJM* **353** 1773–1783
9. Samei, E. and Flynn, M.J., 2002, An experimental comparison of detector performance for computed radiography systems, *Med. Phys.* **29** 447–459
10. Samei, E. and Flynn, M.J., 2003, An experimental comparison of detector performance for direct and indirect digital radiography systems, *Med. Phys.* **30** 608–622
11. Kengyelics, S.M., Cowen, A.R., and Davies, A.G., 1999, Image quality evaluation of a direct digital radiography system in a UK radiology department, *SPIE* **3659** 124–135
12. Muller, S., 1999, Full-field digital mammography designed as a complete system, *Eur. J. Radiol.* **31** 25–34
13. Siebert, J.A., Filipow, L.J., and Andriole, K.R., (Eds) 1999, *Practical Digital Imaging and PACS*. AAPM monograph no 25. Medical Physics Publishing, Madison, WI
14. Vedantham, S., Karallas, A., Suryanarayanan, S., et al., 2000, Full breast digital mammography with an amorphous silicon-based flat panel detector: physical characteristics for a clinical prototype, *Med. Phys.* **27** 558–567

Physics Simulation Software Foundations, Methodology and Functionality
Simulation Software Methods and Functionality

Simone Giani

Abstract The key methodological aspects of physics simulation software are outlined using highlights from relevant applications. Rigorous understandings of the Monte-Carlo numerical foundations and of the validation and verification processes are necessary to guarantee the reliability of simulation software and results. Basic and advanced functionality of simulation software are explained by analyzing the main subsystems of a Monte-Carlo software toolkit.

Keywords: Simulation · random numbers · Monte-Carlo · Geant4 · physics

1 Foundations and Methods

1.1 Numerical Foundations

Monte-Carlo simulation is based on the calculation of particles' trajectories in bodies of any chemical composition and geometrical shapes, reproducing microscopic physics interactions. Macroscopic observables are constructed from high statistics (~millions) of simulated events.

The correct propagation of radiation crossing boundaries between volumes with different materials, density, etc. (implying discontinuity of cross-sections) is the specific requirement characterizing Monte-Carlo simulation with respect to numerical integration methods. Moreover, the rare interactions between the particles themselves composing the radiation flux, as it is typical in high energy physics (HEP), differentiates HEP simulation from standard fluids dynamics Monte-Carlo (where the complexity of the simulation of a system of N interacting particles evolving in time dominates over the modelling of their individual interactions with solid bodies).

S. Giani
European Organization for Nuclear Research, CERN, CH-1211 Genève 23, Switzerland
e-mail: Simone.Giani@cern.ch

Y. Lemoigne, A. Caner (eds.) *Molecular Imaging: Computer Reconstruction and Practice,* 19
and Experiments,

Applications of physics Monte-Carlo simulation include:

- **HEP and nuclear** physics, with the simulation of accelerators, detectors, and astrophysics experiments
- **Space** physics, addressing the simulation of satellites, of instrumentation, of most ESA and NASA space missions, and the simulation of the space radiation environment
- **Industry**, for the design of high-tech instruments for example at General Electrics, Siemens...
- **Medicine**, ranging from the design and optimization of PET and CAT scanners for medical imaging, to radiotherapy treatment planning (including conventional electron and photons beams, brachy-therapy, and hadron-therapy)

Monte-Carlo simulation software must be based on thorough mathematical and computer science foundations. At the mathematics level, algorithms stability analysis is needed to guarantee the convergence of iterative numerical algorithms, the absence of bias in fitting algorithms, etc. Numerical stability, as well as portability, are requirements to be fulfilled at the hardware and software level in order to guarantee control and reproducibility of results. Pseudo-randomness must be ensured at the level of random numbers generation, for a correct sampling of cross-sections, interaction lengths, probability distributions, etc., and to avoid correlations onto the simulation results from the periodicity of random number sequences.

Pseudo-randomness is achieved when sequences of random numbers generated via algorithms are undistinguishable from true stochastic distributions in nature. Random numbers generators are based on algorithms defined as RandomEngines, which generate uniform random sequences with measured stochastic properties. A summary about some of the most frequently used random generator engines follows below:

- *HepJames*
- *Ranecu*, a Multiplicative Congruential Generator
- *Ranlux*, with a MCG initialisation complemented by skipping numbers logic provoking an effective convolution of two pseudo-random algorithms; Ranlux can be used at different levels of complexity, trading off pseudo-randomness with speed performance:

  - Level-0: Rcarry generator, fails "gap test"
  - Level-1: fails "spectral test"
  - Level-2: passes known tests
  - Level-3: 'no' correlations
  - Level-4: 24 bits chaotic

- *Rand*, a simple and fast generator, but with easily observable biases
- *Drand48*, an extremely reliable double precision generator engine, using just 16 bits for swapping and making 48 bits stochastic

Random Generators use RandomEngines to generate numbers sampling standard probability distributions, such as Exponential, Gauss, BreitWigner, Poisson, Binomial, Chi2, Gamma, Tstudent, and "User-defined" distributions.

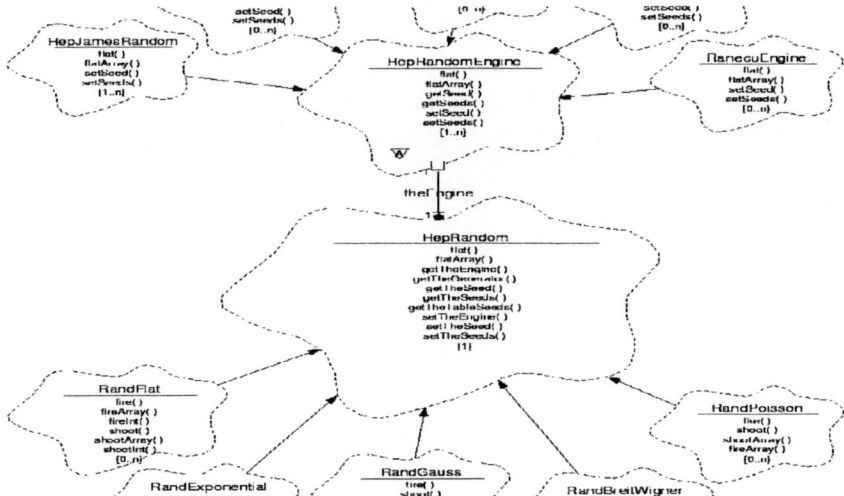

Fig. 1 CLHEP class diagram for random engines and generators

Figures 1 and 2 show the design relation between RandomEngines and Random-Generators in the CLHEP library, both of them implementing abstract interfaces to allow transparent switching and comparison of different generators, as well as extensions of additional statistical distributions samplers.

A key concept in the formalism of random numbers generation is given by the coding of "seeds": seeds represent the parameters needed by a pseudo-random algorithm to (re)produce a given sequence of random numbers, and can be implemented by a simple pair of numbers up to a complex combination of tables and formulae. Seeds thus allow to set or save the status of specific generator algorithms, and hence they allow the control of the initial conditions of random generation for parallel computing (avoiding correlations and repetitions), as well as to reproduce saved run-time conditions of a program (for debugging purposes).

1.2 Software Methodology

1.2.1 A Mission-Critical Application

In order to study the fundamental methodological aspects of Monte Carlo simulation software and produced results, a case-study is selected here from a 'mission-critical' space application. In the second half of 1999, ESA was to operate the first production-phase launch of the Ariane5 generation, with the XMM X-ray telescope as payload (Fig. 3). XMM represented about 15 years of investment for the

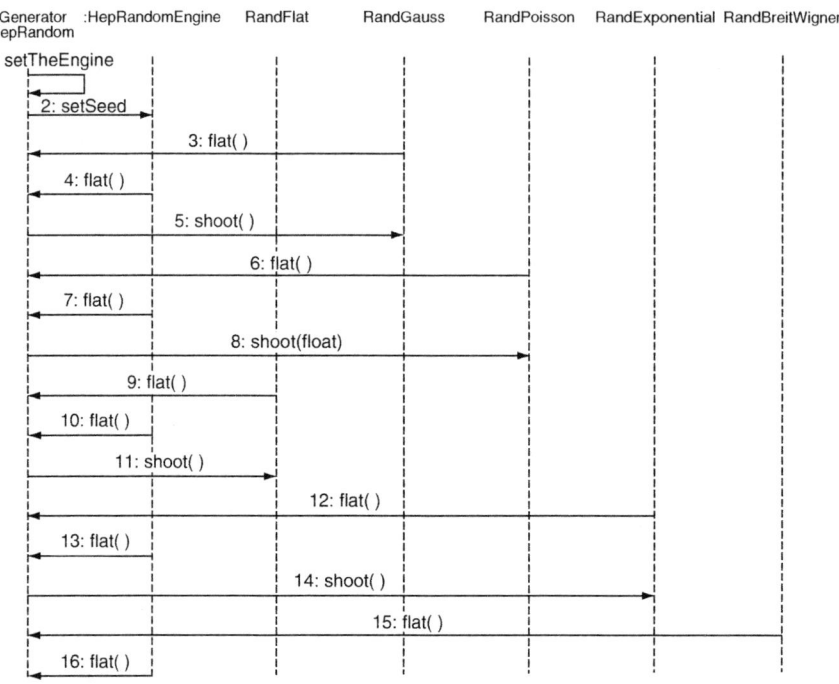

Fig. 2 CLHEP object/interaction diagram for random numbers generation

Fig. 3 XMM-Newton X-ray telescope

astrophysics community, priced at 700 million euros in 1999. Objectives of software simulation were to:

- Discover the sources of radiation-induced degradation experienced by the CCD-based detectors of the NASA Chandra X-ray telescope launched in summer '99
- Compute radiation levels for XMM's CCD-based detectors and determine protection procedures before launch

For those purposes, the simulation software was challenged by the need to correctly predict effects without the prior support of measured data. The transparency of the simulation modelling turned out to be a decisive requirement, thus improving the scientific validation process for the software.

The interactions of different particles, present in the Earth magnetosphere, with the telescope had to be simulated. An open-source analysis of multiple scattering models in the limit of corrected Rutherford scattering[1] was conducted. Tests on integrators in magnetic field to certify ~100 micron accuracy on ~10 m trajectories[2] were passed successfully. An open-source analysis of GEANT4 energy loss parameters, in order to guarantee tracking accuracy on ~10s nm paths in the mirrors coating, was performed. Comparisons to specific qualification data by ESA were checked.

Particles' back-tracking by steps-history recording allowed a full reconstruction of the possible particles' trajectories in the Chandra and XMM geometric models, showing the cause of the radiation damage on the CCDs in the focal planes. In facts, protons trapped in the magnetosphere, and just four times less abundant than electrons, can mimic by scattering processes onto the coating of the telescope mirrors the "grazing incidence" and optical-scattering-guided paths of the X-rays (Fig. 4). In specific angular-momentum acceptance windows, protons can thus be effective in reaching, without being deviated by standard fix-rigidity magnets, the telescopes' focal planes where the pixel detectors are placed.

The simulation computations estimated the following fluxes for XMM:

- Flux (at 1.3 MeV) on XMM RGS spectrometer: = 0.7 million protons/cm^2/month
- Flux (at 200 keV) on XMM EPIC camera: = 60 million protons/cm^2/month

The energy loss of such low-energetic protons in the semiconductor, especially for NIEL processes, would quickly degrade by radiation damage the performance of the silicon devices, particularly for the front-mounted-pixels configuration. Hence it was evident that the EPIC camera needed to be protected and, before the XMM launch, the decision was taken to shut on-board Aluminium doors on EPIC when orbit would cross the Earth radiation belts.

XMM is correctly operating in orbit with no damage since over 7 years.[3] This represents a key example of simulation predictive power, on the basis of which the operational conditions of a space mission were changed before launch.

1.2.2 Validation of Software and Reliability of Simulation Results

These results bring to a review of the key simulation validation concepts. In software engineering, validation of software is achieved via testing with respect to

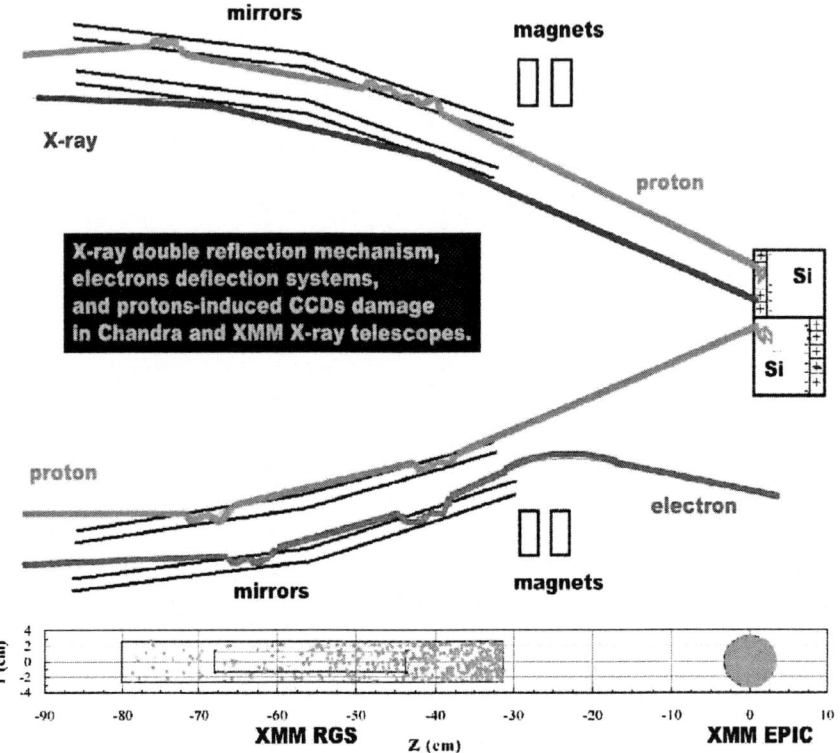

Fig. 4 Proton-induced radiation damage in X-ray space telescopes

user requirements, design deliverables and implementation units. On the contrary, usually the validation/verification of physics/medical simulation software is simply with respect to available experimental data. However the following arguments clearly show the weakness of such a simplistic approach:

- In facts, even assuming simulation results are consistent with the experimental benchmark data, how many software parameters were used to allow the program to fit the data?
- Moreover, how to proceed if there is the need to validate simulation results before knowing the experimental physics data (see XMM case)?

In order to address the question of software validation in general, Software Engineering standards define words such as "validation" and "verification" in terms of the activities to be performed. For example, in the ISO 12207 standard[4] one can find the following definitions:

6.4 Verification process
The Verification Process is a process for determining whether the software products of an activity fulfil the requirements or conditions imposed on them in the previous activities.

Fig. 5 Validation and verification process in the PSS-05 space standards

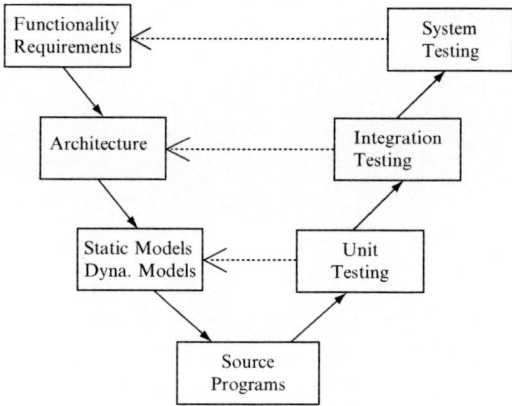

6.5 Validation process
The Validation Process is a process for determining whether the requirements and the final, as-built system or software product fulfils its specific intended use.

However, from the definitions above it is evident that it cannot be effective to just extract "terminology" from a standard, without understanding the overall framework of its software development processes. For this purpose, a schematic view (Fig. 5) of the software engineering processes and their relations, from the ESA PSS-05 space standards, can be very explicative. In facts, it shows the dependency (full arrows) between the various phases of software development and related deliverables, and the traceability (dotted open arrows) between them and the various required levels of testing.

The net result is that validation is performed and achieved incrementally (and not just by comparing end results), thus remarkably reducing the large redundancy of degrees of freedom that one has available to fit results when developing code.

For cases where a fully engineered validation software process cannot be followed for reasons of resources, key criteria can nevertheless be derived from such standard established processes in order to pragmatically assess both the reliability of scientific simulation software, and the reliability of the results produced by simulation software.

The Software Reliability Model (SRM)[5] is defined to quantify the reliability and validability of delivered simulation software with respect to the fulfilment of certain conditions. Increasing levels of reliability and validability correspond to the fulfilment of more stringent conditions.

SRM Levels:

- *Level 0*: Different object code is provided and used for different use-cases.
- *Level 1*: The same publicly distributed object code is used for different use-cases.
- *Level 2*: Different source code is provided and used for different use-cases.
- *Level 3*: The same publicly distributed source code is used for different use-cases.
- *Level 4*: The same public source code also exposes all the parameters which influence the results.

The Results Validability Model (RVM)[5] is defined to quantify the reliability and validability of produced simulation results with respect to the fulfilment of certain conditions. Increasing levels of reliability and validability correspond to the fulfilment of more stringent conditions.

RVM Levels:

- *Level 0*: Simulation results match known experimental results using parameterisations based on those data.
- *Level 1*: Simulation results match known, but independent, experimental results.
- *Level 2*: Simulation results match known experimental results and are (re)produced by random users.
- *Level 3*: Simulation results match known experimental results with a controlled number of parameters.
- *Level 4*: Simulation results correctly anticipate unknown experimental results.

On the basis of what discussed above, one can now investigate the cases exposed in Fig. 6:

Let us assume that the black circles in Fig. 6 represent correct experimental results, and that the drawn error bars are correct error estimations for those data. Let

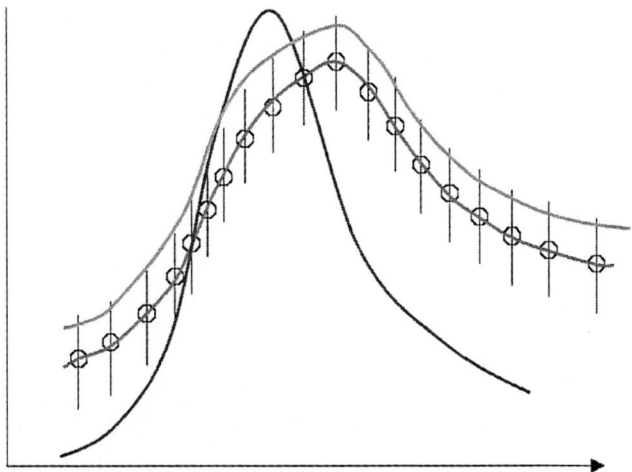

Fig. 6 Validation and verification process in the PSS-05 space standards

us also assume that the green, red, and blue graphs in Fig. 6 represent results of different Monte-Carlo simulations trying to reproduce those experimental data.

What are the Best Simulation Results?

The question is left to be answered by the reader for both the case when the experimental errors are dominated by statistical uncertainty and for the case when the experimental errors are dominated by systematic uncertainty.

In conclusion, we will review here an example of the scientific impact that rigorous simulation software procedures can bring. Figure 7 shows one of first XMM images analysing the Large Magellanic Cloud; incidentally, the data were collected by the EPIC camera CCDs. It has to be noted that the protective measures decided on the basis of the simulation results meant 25% less observation time for the astrophysics community, but the operation of the detector has been optimal since the launch, and results have been excellent. Actually, those XMM results seems to have an important cosmological impact since their interpretation suggested that Universe is hotter than foreseen due to discovered energy in X-ray range.

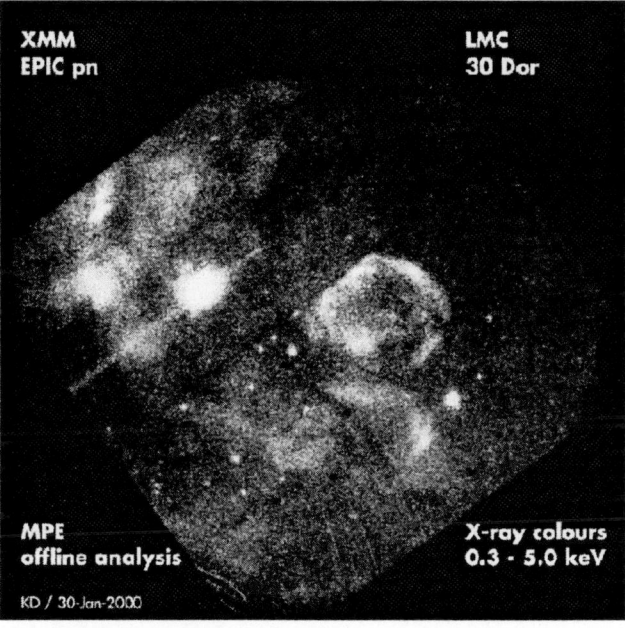

Fig. 7 The Large Magellanic Cloud in the X-ray range observed by the XMM EPIC. See Appendix I for color version of the picture

2 Simulation Functionality

2.1 Specifications

The Monte-Carlo simulation domain extends itself onto two main conceptual directions: in facts, on one side simulation software allows to design and optimise physics detectors and their response before commissioning and construction, and on the other side it helps to evaluate experimental results in the data-analysis phase (efficiencies, backgrounds, deviations...).

As mentioned in section 1, Monte-Carlo simulation software finds its natural applications for physics experiments, accelerators, nuclear physics, heavy-ions, radioprotection computations, space radiation environment, radiation shielding for satellites, cosmic rays applications, astrophysics, medical imaging, dosimetry, radiotherapy treatment planning.

High Energy Physics functional requirements demand already a large physics coverage; indeed, the requirements coming from the physics just of the LHC experiments range from a few kiloelectron volts to teraelectron volts in energy, from heavy ions to CP violation physics, from luminosities provoking event pile-up to non-uniformities of magnetic fields in the order of Tesla, and to the solid modelling of order one million objects per detector. Full shower and/or fast MonteCarlo and/or event biasing (variance reduction) are also useful for simulation of detectors or experimental set-ups (which could also be the human body or a satellite) in specific conditions. Current cosmic rays studies extend the need for physics modelling at energy ranges beyond the picoelectron volts limit, while radiation studies require propagating neutrons down to thermal and cold energies.

Additional requirements come from Space Science and Medicine. Space applications requirements for spacecrafts and instrumentation design, mainly for radiation shielding and human safety, imply low energy extensions of electromagnetic interactions down to 1 KeV and to 250 eV, and below down to the range of 100 eV for some biomedical applications (including the molecular interactions of radiation with DNA sequences). Space science studies also require simulating induced radioactivity with consequent decay products including line X-gamma emissions spectra and hadronic spallation reactions.

Advanced radiotherapy requires extremely accurate simulation of short-range and stopping hadrons, as well as the capability to model the space-time movements of the experimental apparatus for the simulation of imaging devices. Further functional requirements have to be taken into account for the simulation of radiation damage on electronics and circuitry, such as single event upsets on chips in space, including interfaces to electronics design tools.

In the attempt to summarize those specifications for the system functionality of a Monte-Carlo simulation software, one could deduce the following mission statement:

From events and particles generated by the beam collision or any other sources, the software simulates their propagation into, and interaction with, the detector (or human body) matter (including vacuum), and then reconstructs the detector electronic signals and measurements.

2.2 Basic Functionality

In the following, we will analyze functionality, design, and implementation features of Monte-Carlo software at a basic level. A break down of the basic elements for the functionality of simulation software follows below:

- Faithful and precise description of the *Geometry* and *Chemical* composition of the detector (or human body) elements should be modelled.
- The *Kinematics* of the physics events generated in the beam collider or any other instrument should be reproduced.
- *Particles* should be tracked through the "detector", simulating their *Physics* interactions in matter and the effect of *Fields* and boundaries on their trajectories.
- The *Electronics* response of the sensitive "detector" elements should be reproduced.
- *Graphics* visualisation (imaging), user interfaces, object storage,.. should be available.

Figure 8 shows a picture drawing simulated tracks trajectories in a complex detector geometry, highlighting the sensitive detector elements for which the digitized electronic response is modelled.

The multi-disciplinary aspects built-into the functionality of a typical simulation software require a good modularity already at a basic design level. In facts, the Monte-Carlo multi-disciplinary nature, providing functionality in a set of different scientific domains, is to be reflected into dedicated software components (such as digitisation, event-generators, solid geometry, GUI, hadronics, e.m. physics, medical-physics, tracking, visualisation, particles-data, chemistry, data-parameterisations, space, parallel-computing...).

A modular design of the software should allow the user to load and use only the components he needs for each application. At the same time, the (Object-Oriented) design should allow the user to understand, extend, or customise the simulation software toolkit in all the sub-domains. Moreover, abstract interfaces should avoid dependencies on any software, commercial or free, providing functionality or tools external to the simulation domain (such as data storage, graphics rendering...).

Basic implementation criteria and procedures can guarantee and enhance the quality of simulation software and of scientific software in general:

- The physics modelling should be transparent to the user, so that he/she can understand **how the results are produced**, hence improving the scientific validation process.

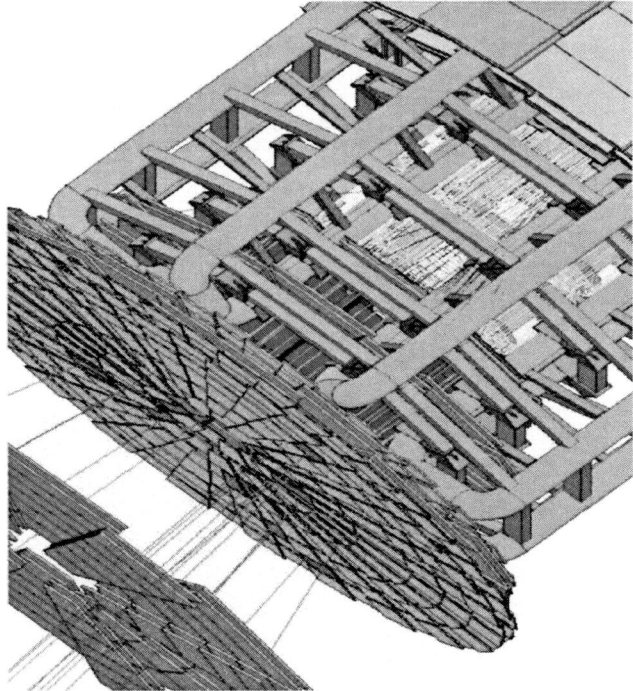

Fig. 8 Visualization of detector solid model,particle tracks,and detector signals

- **No numbers must be hard-coded** in formulae and algorithms, but only variables and constants should be used. Constants should then be initialised to their numerical value followed by the most suitable physical units.
- An extensive set of **units should be defined** (e.g. CLHEP), and all the numerical quantities in the simulation software should be expressed through the most convenient units explicitly. Consequently, the physics code is made independent from the units chosen by the user.

2.3 Advanced Functionality

Let us now concentrate on the specifications of advanced Monte-Carlo functionality, focused to improve the physics performance of the simulation software. As highlights:

- For a full exploitation of the validity ranges of the physics models, only production thresholds should exist, i.e. no tracking cuts, and all particles should be tracked down to 0 range.

- Consistent and material-independent accuracy requires production cuts defined in range (plus automatic correct treatment of near-boundary regions via production of secondary particles below threshold).

These kinds of features guarantee also consistency between the physics modelling for particle production and the physics modelling for the interactions of particles with matter and fields. In all cases, user-defined cuts should anyway be allowed in energy, total path length, time-of-flight, etc. for special treatment of selected areas in the experimental setup. This guarantees user-specified optimizations for specific cases.

A growing sophistication of the functionality demands, and benefits from, an advanced design of the software. Not only a clean subdivision of the functionality into software component, but also a strict and unidirectional flow of dependency relations between them, contributes to improve the quality and the comprehension of the software. As an example, Fig. 9 shows the architectural design of the GEANT4 simulation toolkit: the notation should be read as "the component-box touched by the circle depends/uses the component-box touched by the connected segment".

Advanced design considerations can also be applied to the modelling of the physics. Actually, the way cross sections are calculated (via formulas, data files, etc. and using different data-sets with applicability by particle, energy, material) should clearly be exposed and separated from the way they are accessed and used in the algorithms. Similarly, the way the final state is computed should be separated from the tracking and split into alternative or complementary models, according to the energy range, the particle type, the material. Multiple implementations of physics processes and models should be available/possible within a simulation toolkit.

Special control should also be applied to the interfaces to external physics libraries. In order to guarantee a consistent level of quality, validation and transparency for a simulation software, it is essential that no interfaces are created to

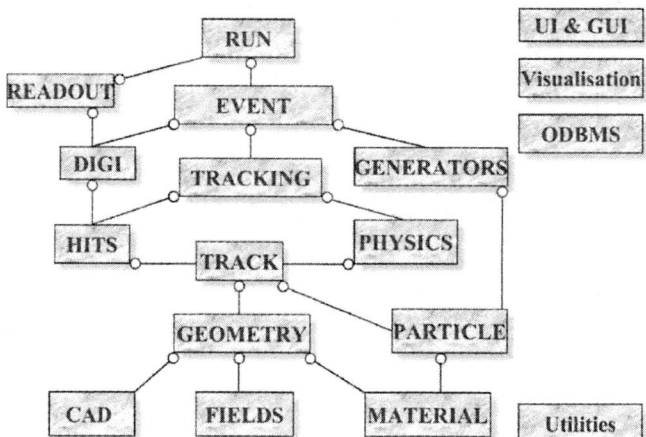

Fig. 9 GEANT4 architectural design diagram

external black-box physics packages: the physicist-user must be able to see and eventually extend the physics modelled into the whole Monte-Carlo software which produces physics results.

Finally, it should be mandatory that simulation software openly interfaces to published standard data libraries for the data sets used by the Monte-Carlo (as opposed to copying the numerical values in the software code). Explicit references and integrity should be assured to specific versions of standard data libraries and evaluations, such as ENDF/B, JENDL, FENDL, CENDL, ENSDF, JEF, BROND, EFF, MENDL, IRDF, SAID for neutron data, EPDL, EEDL, EADL for the Livermore evaluated data sets, SANDIA for QED processes data-tabulations, ICRU for medical physics reference-term published data.

Advanced implementation and algorithmic features in the simulation software improve and extend the physics modelling capabilities. In this context, let us recall that physics processes' implementation mandate is to use cross-sections to determine the tracking steps in the simulation, and hence to trigger the occurrence of physics interactions (determining the update of the particle's final state). During such 'integration' of the equation of motion, processes can compete to determine the particle status by occurring:

- In time ($\sim$regardless the space location)
- In space ($\sim$regardless the time location)
- In space-time along the step

Each modelled process can perform any of these actions or any combination of them, leading to 2^3 - 1 standard types of processes that should be implemented in a Monte-Carlo software. This represents a closed generalisation[6] of the original classification of simulation processes[7] into 'continuous' and 'discrete', which are found back just as two special cases.

2.4 Physics

A review of standard electromagnetic and hadronic physics processes to be modelled in Monte-Carlo simulation software is exposed in the following paragraphs.

Electro-magnetic processes:

- Ionisation and Bremstrahalung energy loss for charged leptons:
  - 'Differential' (standard), 'Sandia-tables' (dE/dx in thin layers & gases[8]), and 'integral' (of cross-sections and ToF when high derivatives in function of energy) approaches
  - low-energy extensions,[9] delta-rays and photon production also below threshold, LPM effect for Bremstrahalung

- Hadrons and ions Energy loss:

    o Low energy extensions[10]
    o Bragg peak and long range tail

- Multiple scattering modelling:

    o Angular and lateral displacement
    o True path length conversion and no step size restrictions

- Annihilation and Decays:

    o In flight and at rest

- Cerenkov, Synchroton, Transition radiation:

    o Including coherent effects

- Gamma physics: Photoelectric, Compton, Pair-production:

    o Photoelectric with shells fluorescence, low energy
    o Compton with polarisation & low energy corrections, form factors

- Photon: Refraction, Reflection, Absorption, Rayleigh, Scintillation:

    o Unified model[11] for surface roughness simulation

- Muons: Ionisation, Bremstrahalung, PairProduction, Muon-nuclear:

    o Theoretical models for validity range up to PeV for cosmic rays applications

- Shower parameterisations for fast MonteCarlo (for e.m and hadronics):

    o Full integration with standard tracking, treatment near cracks
    o Detector geometries independent from standard tracking geometry

Hadronic Processes:
A parameterisation-driven modelling approach is useful when sufficient data are available, particularly from test beams, or from multiple sources in literature.[8] Parameterisation-driven models can include:

- High energy and low energy Inelastic scattering
- Elastic scattering
- Fission and Capture
- Dedicated processes for Stopping kaons, pions physics
- Special Pion-production of particular experiments (e.g. HARP data)

A theory-driven modelling approach is useful for physics beyond test-beams energies and for microscopic reproduction of processes leading to final states. Theory-driven models can include:

- String-parton models (diffractive, dual parton, QGCSM...) in the high energy regime (but interface to event-generators for hard-scattering[12]
- Intra-nuclear transport models & Pre-equilibrium models

- De-excitation models, including Evaporation, Photo-evaporation, Fission, Fermi break-up, Multi-fragmentation
- Lepton-hadron interactions, such as Muon-nuclear interactions, Photo-fission and general Conversion gamma-meson

A data-driven modelling approach is useful for resonances-dominated physics and for discrete spectra treatment (for example low energy neutron transport for radiation background studies). Data-driven models can include:

- Low energy neutrons down to thermal and cold energies; cross-sections must be based on evaluated data-sets (ENDF, JEF, JENDL, CENDL, ENSDF, BROND, IRDF, MENDL....), complementary for isotopes versus light elements data; file-system should ensure granular+transparent access/usage of data (ENDF-B VI format largely adopted for sampling codes)
- Discrete photo-evaporation emission [13]

2.5 Geometry and Mathematics

A key issue in Monte-Carlo simulation is the accuracy of the geometric and material modelling of the experimental setup. Physics and geometric modelling are complementary in order to achieve correct simulation results, as demonstrated in a large variety of challenging simulation applications.

The first aspect of geometric modelling concerns the 3-dimensional solid modelling of detectors, satellites, human body/organs or accelerator elements. Solid modelling functionality should include multiple solid representations, such as CSG Constructive Solid Geometry, BREPS Boundary REPresented Solids (including NonUniformRationalBSplines), SWEPT Solids, BOOLEAN Operations. Ideally, standards such as ISO STEP/EXPRESS [14] for solid modelling in engineering CAD/CAM, space and car industry should be used also for Monte-Carlo; this would allow to perform physics simulation directly into the engineering models of the experimental devices. Standards such as XML can be used as exchange and I/O storage format as in data-base computing.

Effective computations of particle trajectories require fast navigation in the geometrical data-base describing the apparatus, as it is often the case for ray-tracing software and computer graphics in general. This is achieved via the implementation of functionality allowing the optimisation of flat or hierarchical volumes structures (such as voxelization and volumes parameterisations, including by material, position, size). Incidentally, these techniques allow also interfacing the simulation code to the output formats of medical imaging. Additional functionality is required to model the movement of the "setups" during simulation runs, as it is the case for medical scanners or for modelling the breathing of patients.

The second, and equally relevant, key aspect of geometric modelling concerns the trajectory followed by particles during their propagation in the experimental apparatus. The Monte-Carlo software should include mathematically-rigorous equation of motion solvers for different fields (e.g. electric, magnetic, gravitational...) and geometrical boundaries' conditions. The following functionality is to be considered essential to guarantee accuracy of the physics results:

- Different integrators, beyond classical uniform and Runge-Kutta, and including multi-turn perturbative methods, Cash-Karp, etc. are needed for a correct propagation in e.g. E.M. **fields of various non-uniformity in space, variability in time, and differentiability**.
- **Physical thickness as tolerance for geometric boundaries** is important for numerical stability and should be handled in the logic of the intersection algorithms.
- The handling of the uncertainty on the step's end-point returned by the integrators should be used in the Monte-Carlo algorithms, to perform error propagation from the user **required precision on the total path length to the constraining of the accuracy per-step**.
- The **intersection of** (or approximation of) **curvilinear tracks with geometric boundaries** should be based on sound mathematics.
- A proper **integration** should also be performed to update the particles' **time of flight** during transportation.

The geometric modelling and the physics modelling are steered by a dedicated subsystem of the simulation software usually called '*tracking*' system. The Tracking system is the simulation engine managing the evolution of the track's status determined by all the physics interaction occurring at a given time, at a given location, or distributed in space-time. The Tracking system is ultimately responsible for setting the simulation *step*, conceptually corresponding to the increment in the integration of the overall equation of motion. This depends on the probability of occurrence in space-time of the various physics processes weighted with respect to geometry boundaries and fields. The physics status of the particle, after the final state computation of each physics process and propagation in field by a step, is moved to the Tracking kernel, which is responsible to integrate it with the contribution of the other processes involved in the step. Of course the correlations and ordering logic of processes should be visible to the user.

2.6 General Functionality

A quick review of the functionality to be expected in additional typical subsystems of Monte-Carlo software follows below.

The Run, Event and Track management allow the simulation of the event kinematics, together with primary and secondary tracks. Interfaces to event generators are included here. Functionality should enable to perform studies of anything from 'pile-up', to trigger, and to 'loopers'. Multiple stacking mechanisms can be useful to assign tracks priorities for specific physics studies and channels.

A fast parameterisation framework can be triggered on particle type, volume, etc. and should be integrated with the full simulation, allowing independent and simplified detector descriptions and at the same time a correct treatment near cracks. Fast parameterisations should allow the direct production of hits corresponding to a full shower development for several detector types.

Hits and Digi sub-domains provide the functionality to reproduce the read-out structure of the detector and its electronic response, independently from the geometry used for the tracking.

Particle definitions, including hundreds of baryonic and mesonic resonances and ions, decay processes and branching ratios, can be implemented and made compliant to the Particle Data Group coding.[15]

Materials definitions should include isotopes, elements, compounds, compounds of compounds... Automatic code generation for detector description and materials definition can be useful for engineering and medicine applications.

A large variety of *user-actions* can be designed as plug-ins to be implemented by the user and executed automatically in the Monte-Carlo software:

- Primary tracks generation (from event generators to single particle)
- Detector construction (geometry, materials)
- Physics selections (particles and processes)
- Moreover: user code to be executed per Run, per Event, per Track, per Stacking, per Step...

Random number generators can be used as external utilities for the simulation software.[16]

Computer graphics functionality can include engineering quality drawings, 3D rendering, automatic detection of volumes overlaps, user interfaces (batch, interactive by command-lines, GUIs). It should exploit external libraries such as X11, PostScript, OpenGL, Inventor, VRML, TCL/TK, Java, Python... Interactive picking of physics objects, such as tracks and hits, visualising in real time the associated physics information, can also be functional.

2.7 Conclusions

In conclusion, let us consider an example for which advanced Monte-Carlo simulation features allowed to improve the scientific results, having a useful impact for clinical applications. Figure 10 shows the dosimetry calculation in a radiotherapy treatment plan[17] in presence of sharp discontinuities of cross-sections, as for example between the bones, air cavities, and soft tissue in the head of a patient. The exploitation of concepts such as production thresholds and range allow computing 3-dimensional energy release with consistent accuracy in presence of strong density discontinuities and thin layers of different materials.[18] The advanced Monte-Carlo simulation functionalities discussed in this chapter are correctly taking into account the air cavities and bone structures, and allow predicting accurately the dose to tissue that is surrounded by air. Improvements with respect to traditional software are up to 25–30%.

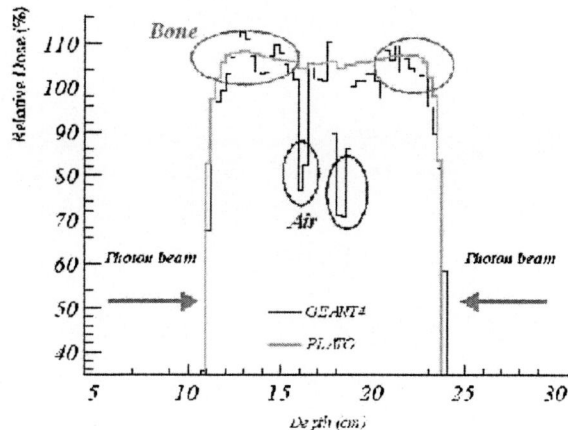

Fig. 10 Monte-Carlo application to radio-therapy treatment planning [17]

References

1. J. Ziegler, www.srim.org, 2008
2. S. Agostinelli, et al., *NIM* A 506 (2003), 250–303
3. sci.esa.int/content/doc/b3/14515_.htm
4. ISO/IEC 12207 : 1994 (E)
5. 8th International Conference on Calorimetry in HEP, Calor99 Proceedings, Lisboa, 1999
6. S. Giani et al., LHCC/RD44, CERN-LHCC 98-44, 1998
7. M. J. Berger, Methods in Computational Physics, Vol. I, Academic, New York, 1963
8. V. Grichine et al., *Nucl. Instr. Meth. A* 453 (2000), 597
9. D. Cullen et al., EPDL97, the Evaluated Photon Data Library, 97 version, UCRL–50400, Vol. 6, Rev. 5, 1997
10. S. T. Perkins et al., Tables derived from the LLNL Evaluated Electron Data Library (EEDL), UCRL-50400 Vol. 31, 1997. International Commision on Radiation Units and Measurements, Journal of the ICRU, Oxford Univercity Press, Oxford
11. F. Cayouette et al., DETECT2000, International Society for Optical Engineering
12. T. Sjöstrand, www.thep.lu.se/~torbjorn/Pythia.html, Lund
13. Sandia National Laboratories, www.sandia.gov
14. ISO 10303 STEP Standards
15. Particle Data Book, pdg.lbl.gov
16. See Section 1
17. LIP-Coimbra Laboratorio de Instrumentaço e Física Experimental, www.lip.fis.uc.pt
18. 9th ICATPP Conference Proceedings, Como 2005, World Scientific

Electromagnetic Interactions of Particles with Matter

Michel Maire

Abstract This document is a brief review to the main mechanisms of electromagnetic interactions of charged particles and photons with matter, pertinent in Bio-Medical applications.

1 'Standard' Em Physics: The Model

The projectile is assumed to have an energy ≥ 1 keV.

- The atomic electrons are quasi-free: their binding energy is neglected (except for photoelectric effect).
- The atomic nucleus is fixe: the recoil momentum is neglected.

The matter is described as homogeneous, isotropic, amorphous.

1. Common to all charged particles

   - Ionization   *(~keV →)*
   - Coulomb scattering from nuclei   *(~keV →)*
   - Cerenkov effect
   - Scintillation
   - Transition radiation

2. Muons

   - (e+,e-) pair production   *(~100GeV →)*
   - Bremsstrahlung   *(~100GeV →)*
   - Nuclear interaction   *(~1TeV →)*

M. Maire
Laboratoire d'Annecy de Physique des Particules (LAPP)
F-74941 Annecy-le-vieux, P.O. Box 110, France
e-mail: michel.maire@lapp.in2p3.fr

Y. Lemoigne, A. Caner (eds.) *Molecular Imaging: Computer Reconstruction and Practice,*   39
and Experiments,

3. Electrons and positrons

- Bremsstrahlung $(\sim 10 MeV \longrightarrow)$
- e+ annihilation

4. Photons

- Gamma conversion $(\sim 10 MeV \longrightarrow)$
- Incoherent scattering $(\sim 100 keV \longrightarrow \sim 10 MeV)$
- Photo electric effect $(\longleftarrow \sim 100 keV)$
- Coherent scattering $(\longleftarrow \sim 100 keV)$

5. Optical photons

- Reflection and refraction
- Absorption
- Rayleigh scattering

1.1 Glossary

σ cross section per atom $(cm^2/atom)$

$n_{at} = \mathcal{N}\rho/A$ number of atoms per unit of volume $(atoms/cm^3)$

$$n_{at} = n_1 + n_2 + \cdots = \frac{\mathcal{N}\rho w_1}{A_1} + \frac{\mathcal{N}\rho w_2}{A_2} + \cdots$$

$\Sigma = n_{at}\,\sigma$ cross section per volume (cm^2/cm^3)

Φ: number of interactions per unit of length $(1/cm)$

μ: absorption, attenuation coefficients, etc.

$\lambda = 1/\Phi$ mean free path, interaction length, etc. (cm)

$t = x\rho$ mass-thickness, mass/surface (g/cm^2)

Φ/ρ cross section per mass (cm^2/g)

μ/ρ: **mass** attenuation coefficient ..etc.

$X_0\rho$ radiation length, expressed in mass/surface (g/cm^2)

dE/dt energy loss per (mass/surface) $(MeV/(g/cm^2))$

2 Compton Scattering

The Compton effect describes the scattering off quasi-free atomic electrons:

$$\gamma + e \rightarrow \gamma' + e'$$

Each atomic electron acts as an independent cible; Compton effect is called incoherent scattering. Thus:

$$\text{cross section per atom} = Z \times \text{cross section per electron}$$

The inverse Compton scattering also exists: an energetic electron collides with a low energy photon which is blue-shifted to higher energy. This process is of importance in astrophysics. Compton scattering is related to (e^+, e^-) annihilation by crossing symmetry.

2.1 Kinematic

Assuming the initial electron free and at rest, the kinematic is given by energy-momentum conservation of two-boby scattering.

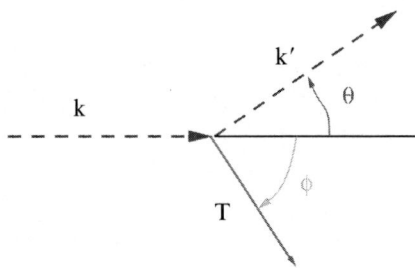

$$k' = \frac{k}{1 + \kappa(1 - \cos\theta)} \ \text{where} \ \kappa = \frac{k}{mc^2}$$

$$T = k - k'$$

$$\cot\phi = (1 + \kappa)\tan(\theta/2)$$

limits $\theta = 0$: $\quad k'_{max} = k \qquad T_{min} = 0 \qquad \phi_{max} = \frac{\pi}{2}$

$\qquad\quad \theta = \pi$: $\quad k'_{min} = k\frac{1}{2\kappa+1} \quad T_{max} = k\frac{2\kappa}{2\kappa+1} \quad \phi_{min} = 0$

2.2 Energy Spectrum

Under the same assumption, the unpolarized differential cross section per atom is given by the Klein-Nishina formula [Klein29]:

$$\frac{d\sigma}{dk'} = \frac{\pi r_e^2}{mc^2} \frac{Z}{\kappa^2} \left[\varepsilon + \frac{1}{\varepsilon} - \frac{2}{\kappa} \left(\frac{1-\varepsilon}{\varepsilon} \right) + \frac{1}{\kappa^2} \left(\frac{1-\varepsilon}{\varepsilon} \right)^2 \right] \qquad (1)$$

where

k' energy of the scattered photon ; $\varepsilon = k'/k$
r_e classical electron radius
κ k/mc^2

2.3 Total Cross Section per Atom

$$\sigma(k) = \int_{k'_{min}=k/(2\kappa+1)}^{k'_{max}=k} \frac{d\sigma}{dk'} \, dk'$$

$$\sigma(k) = 2\pi r_e^2 Z \left[\left(\frac{\kappa^2 - 2\kappa - 2}{2\kappa^3} \right) \ln(2\kappa + 1) + \frac{\kappa^3 + 9\kappa^2 + 8\kappa + 2}{4\kappa^4 + 4\kappa^3 + \kappa^2} \right]$$

limits:

$$k \to \infty: \quad \sigma \text{ goes to 0: } \sigma(k) \sim \pi r_e^2 Z \frac{\ln 2\kappa}{\kappa}$$

$$k \to 0: \quad \sigma \to \frac{8\pi}{3} r_e^2 Z \text{ (classical Thomson cross section)}$$

In fact, when $k \le 100 \, keV$ the binding energy of the atomic electron must be taken into account by a corrective factor to the Klein-Nishina cross section:

$$\frac{d\sigma}{dk'} = \left[\frac{d\sigma}{dk'} \right]_{KN} \times S(k, k')$$

See for instance [Cullen97] or [Salvat96] for derivation(s) and discussion of the *scattering function* S(k,k'). As a consequence, at very low energy, the total cross section goes to 0 like k^2. It also suppresses the forward scattering. At X-rays energies the scattering function has little effect on the Klein-Nishina energy spectrum formula 9. In addition the Compton scattering is not the dominant process in this energy region.

2.4 Pictures

Number of interactions per centimeter in Aluminium

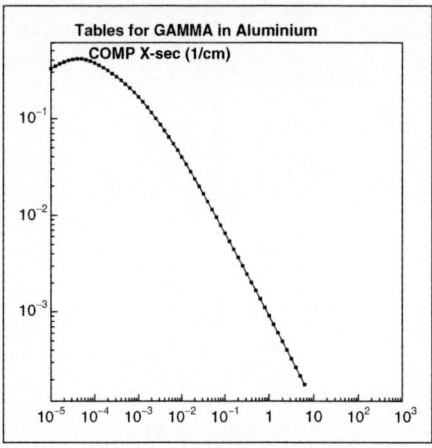

γ 10 MeV in 10 cm Aluminium: Compton scattering

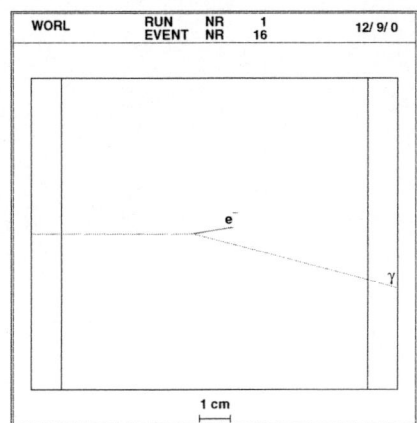
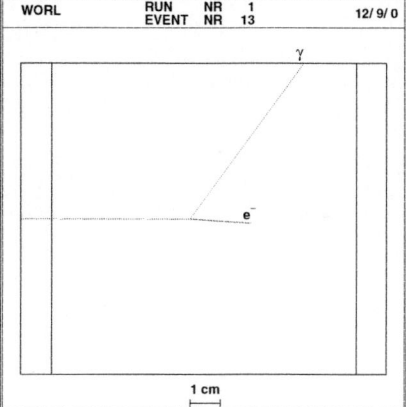

γ 10 MeV in 10 cm Aluminium: Compton scattering

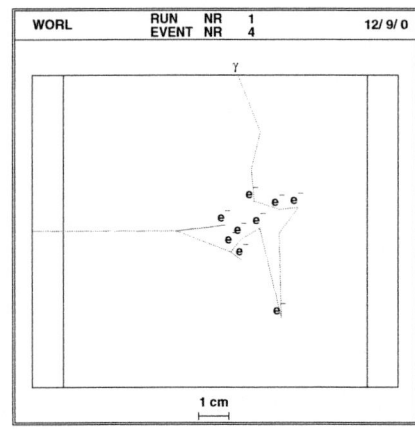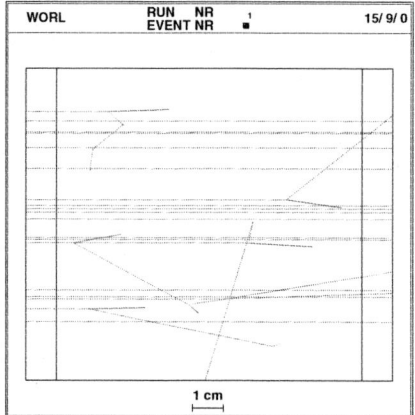

Compton scattering: γ 10 MeV in Aluminium. *Compton edge*:
energy spectrum of scattered photon (left) and emitted e^- (right)

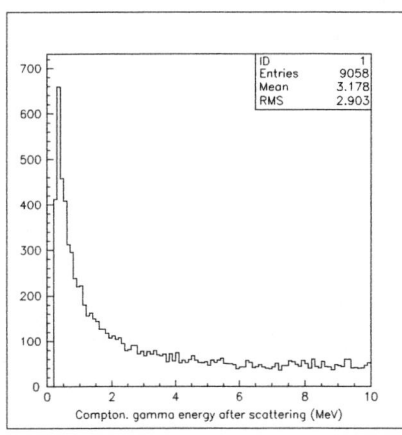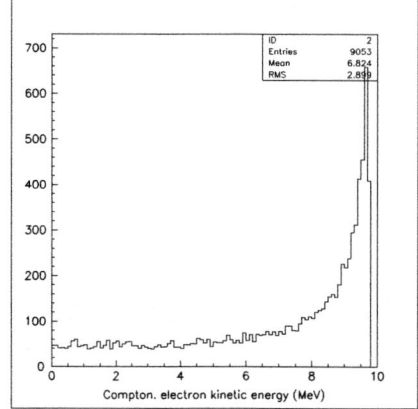

2.5 Transfer, Scattering, Attenuation

Only a fraction of the energy of the incident photon is transferred to the recoil electron, which is generally stopped into the material.
The mean kinetic energy of the electron is:

$$\langle T(k) \rangle = \frac{1}{\sigma(k)} \int_{k'_{min}}^{k} T \frac{d\sigma}{dk'} \, dk'$$

Then the energy transfer coefficient of Compton scattering is defined as

$$\mu_{tr} = n_{at} \; \sigma(k) \; \frac{\langle T(k) \rangle}{k} \qquad (n_{at} \text{ is the nb of atoms per volume})$$

$\dfrac{\mu_{tr}}{\rho}$ is the mass energy transfer coefficient.

Similar, from the mean energy of the scattered photon one defines the energy scattered coefficient of Compton scattering

$$\langle k' \rangle = \frac{1}{\sigma(k)} \int_{k'_{min}}^{k} k' \; \frac{d\sigma}{dk'} \; dk' \qquad \longrightarrow \mu_{sca} = n_{at} \; \sigma(k) \; \frac{\langle k' \rangle}{k}$$

The attenuation coefficient of Compton scattering is

$$\mu_{att} \stackrel{def}{=} n_{at} \sigma(k) = \mu_{tr} + \mu_{sca}$$

and similar relations for the mass coefficients.

2.6 Rayleigh Scattering

Rayleigh scattering is the scattering of photons by an atom as a whole: all the electrons of the atom participate in a coherent manner. It is an elastic collision: no energy transfer from photon to atom (no ionisation nor excitation). At x-rays and γ-rays energy region, Rayleigh scattering is small compared to the photo electric effect, and can be generally neglected.

References

[Klein29]   O. Klein and Y. Nishina Z. Phys. 52, 853 (1929)
[Cullen97]   D. Cullen et al., Evaluated photon library 97, UCRL-50400, 6, Rev.5 (1997)
              J.H. Hubbell et al., Rad. Phys. Chem. 50, 1 (1997)
[Salvat96]   F. Salvat et al., Penelope, Informes Técnicos Ciemat 799, Madrid (1996)
[GEANT3]   GEANT3 writeup, *Cern Program Library* (1993)

3 Photoelectric Absorption

A bound electron can absorb completely the energy of a photon:

$$\gamma + atom \rightarrow atom^+ + e^-$$

The electron is ejected with kinetic energy $T = E_\gamma - B_s$.
(E_γ: energy of the incident photon, B_s: binding energy of the corresponding sub-shell). The nucleus absorbs the recoil momentum. The cross section per shell can be parametrized [Biggs87]:

$$\sigma_s = r_e^2\, \alpha^4\, Z^5\, f\left(\frac{1}{E_\gamma^{a(E_\gamma)}}\right)$$

with f: nonsimple function of $1/E_\gamma$, and $1 \le a(E_\gamma) \le 4$.

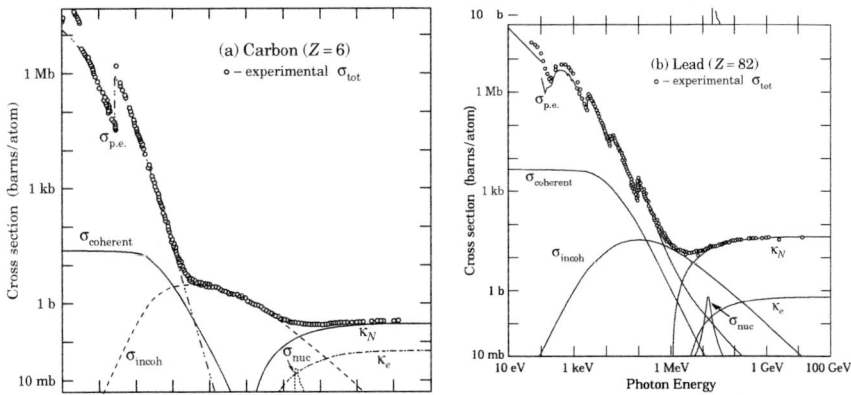

The total cross section has discontinuities at $E_\gamma = B_s$ (absorption edge). There are several parametrizations and tables of the cross sections. See [Cullen97,Biggs87]. If $E_\gamma > B_K$ the absorption occurs mainly on the K-shell (80% of the cases) [Gavri59]. The electron is emitted forward in the direction of the incident photon at high E_γ, and perpendicular to the photon at low E_γ [Sauter31,Gavri61]. Following the photoabsorption in the K-shell, characteristic X-rays or Auger electrons are emitted [Perkin91].

3.1 Attenuation

$$\sigma_{tot} = \sigma_{pair} + \sigma_{comp} + \sigma_{phot} + \sigma_{rayl} \qquad \longrightarrow \mu = n_{at}\,\sigma_{tot}$$

A beam of monoenergetic photons is attenuated in intensity (not in energy) according: $I(x) = I(0)\exp(-\mu x) = I(0)\exp(-x/\lambda)$

20 photons, 5 MeV, entering 10 cm of Al. 4 exit unaltered

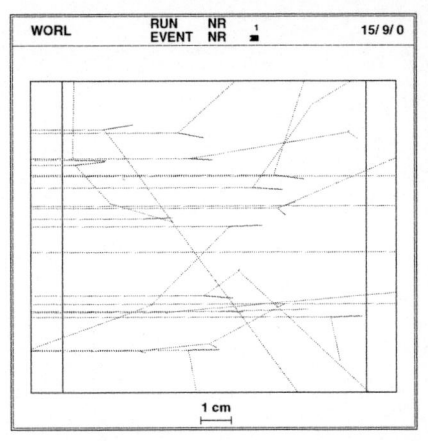 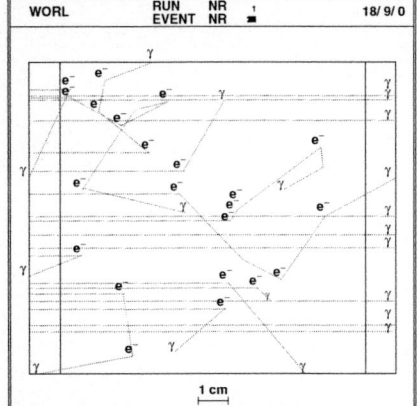

Macroscopic cross sections for photon in water ($\longrightarrow$ mean free path)

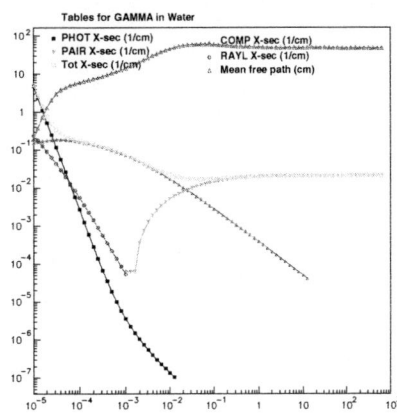 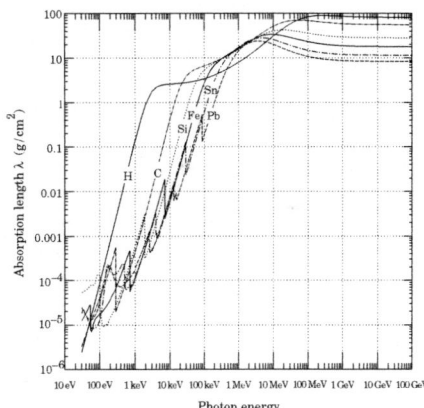

References

[Cullen97] D. Cullen et al., Evaluated photon library 97, UCRL-50400, 6, Rev. 5 (1997)
            J.H. Hubbell et al., Rad. Phys. Chem. 50, 1 (1997)
[Biggs87]   F. Biggs and R. Lighthill, Sandia Laboratory SAND 87-0070 (1987)
[Grichi94] V.M. Grichine, A.P. Kostin, S.K. Kotelnikov et al., Bulletin of the Lebedev Institute
            no. 2–3, 34 (1994)

[Sauter31]   F. Sauter, Ann. Physik. 11, 454 (1931)
[Gavri59]   M. Gavrila, Phys.Rev. 113, 514 (1959)
[Gavri61]   M. Gavrila, Phys. Rev. 124, 1132 (1961)
[Perkin91]   S.T. Perkin et al., UCRL-50400, 30 (1991)
[GEANT3]   GEANT3 writeup, *Cern Program Library* (1993)
[Hub00]   http://physics.nist.gov
[PDG00]   D.E. Groom et al., Particle Data Group. Review of Particle Properties. Eur. Phys. J. C15, 1 (2000) http://pdg.lbl.gov/

4 (e^+, e^-) Pair Creation by Photon

This is the transformation of a photon into an (e^+, e^-) pair in the Coulomb field of atoms (for momentum conservation). To create the pair, the photon must have at least an energy of $2mc^2(1 + m/M_{rec})$. Theoretically, (e^+, e^-) pair production is related to bremsstrahlung by crossing symmetry:

- Incoming $e^- \leftrightarrow$ outgoing e^+
- Outgoing $\gamma \leftrightarrow$ incoming γ

For $E_\gamma \geq$ few tens MeV, (e^+, e^-) pair creation is the dominant process for the photon, in all materials.

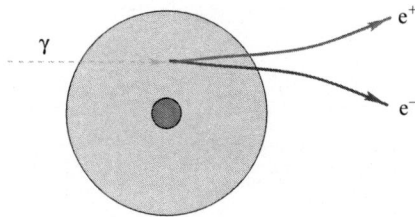

4.1 Differential Cross Section

The differential cross section is given by the Bethe-Heitler formula [Heitl57], corrected and extended for various effects:

- The screening of the field of the nucleus
- The pair creation in the field of atomic electrons
- The correction to the Born approximation
- The LPM suppression mechanism, etc.

See Seltzer and Berger for a synthesis of the theories [Sel85].

4.2 Corrected Bethe-Heitler Cross Section

Let E_γ the energy of the photon, E the total energy carried by one particle of the pair (e^+, e^-), and $\varepsilon = E/E_\gamma$. The kinematical limits of ε are:

$$\frac{m_e c^2}{E_\gamma} = \varepsilon_0 \leq \varepsilon \leq 1 - \varepsilon_0$$

The corrected Bethe-Heitler formula is written as in [Egs4]:

$$\frac{d\sigma(Z,\varepsilon)}{d\varepsilon} = \alpha r_e^2 Z[Z+\xi(Z)]\left\{[\varepsilon^2+(1-\varepsilon)^2]\left[\Phi_1(\delta(\varepsilon))-\frac{F(Z)}{2}\right]\right.$$
$$\left.+\frac{2}{3}\varepsilon(1-\varepsilon)\left[\Phi_2(\delta(\varepsilon))-\frac{F(Z)}{2}\right]\right\} \quad (2)$$

where α is the fine-structure constant and r_e the classical electron radius.

4.3 High Energies Regime: $E_\gamma \gg m_e c^2/(\alpha Z^{1/3})$

Above few GeV the energy spectrum formula becomes simple:

$$\left.\frac{d\sigma}{d\varepsilon}\right]_{Tsai} \approx 4\alpha\, r_e^2 \times \left\{\left[1-\frac{4}{3}\varepsilon(1-\varepsilon)\right]\left(Z^2\left[L_{rad}-f(Z)\right]+ZL'_{rad}\right)\right\} \quad (3)$$

where

E_γ	energy of the incident photon
E	total energy of the created e^+ (or e^-); $\quad \varepsilon = E/E_\gamma$
$L_{rad}(Z)$	$\ln(184.15/Z^{1/3})$ (for $z \geq 5$)
$L'_{rad}(Z)$	$\ln(1194/Z^{2/3})$ (for $z \geq 5$)
$f(Z)$	Coulomb correction function

4.4 Energy Spectrum

Limits: $E_{min} = mc^2$: no infrared divergence. $E_{max} = E_\gamma - mc^2$. The partition of the photon energy between e^+ and e^- is flat at low energy ($E_\gamma \leq 50\,MeV$) and increasingly asymmetric with energy.

For $E_\gamma > TeV$ the LPM effect reinforces the asymmetry.

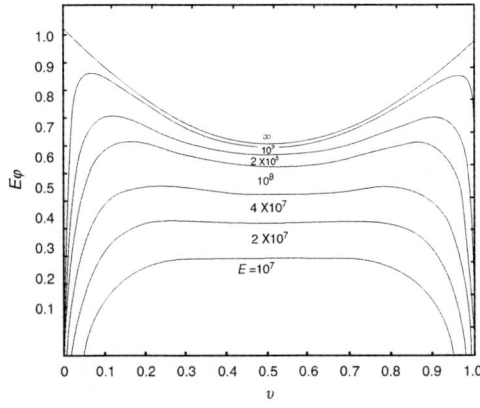

In the high energies regime, one can use the complete screened expression 9 of $d\sigma/d\varepsilon$ to compute the total cross section.

$$\sigma(E_\gamma) = \int_{\varepsilon_0 \approx 0}^{\varepsilon_{max} \approx 1} \frac{d\sigma}{d\varepsilon} d\varepsilon$$

which gives:

$$\sigma_{pair}(E_\gamma) \approx \frac{7}{9} \frac{1}{n_{at} X_0}$$

n_{at} is the number of atoms per volume. The total cross section is approximately constant above few GeV, for at least 4 decades.

Number of interactions per centimeter in Aluminium

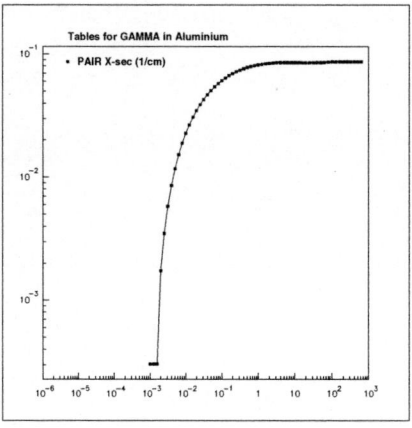

γ 200 MeV in 10 cm Aluminium (field 5 T)

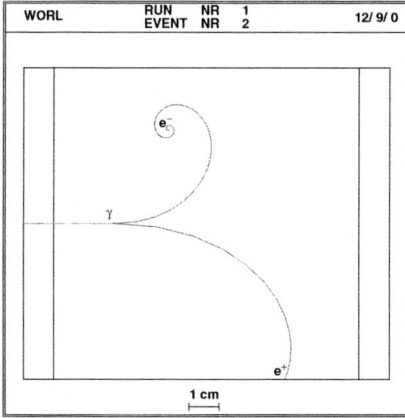

References

[Heitl57]   W. Heitler, The Quantum Theory of Radiation, Oxford University Press, Oxford
               (1957)
[Tsai74]   Y.S. Tsai, Rev. Mod. Phys. 46, 815 (1974)
               Y.S. Tsai, Rev. Mod. Phys. 49, 421 (1977)
[Egs4]   R. Ford and W. Nelson, SLAC-210, UC-32 (1978)
[Sel85]   S.M. Seltzer and M.J. Berger, MIN 80, 12 (1985)
[Antho96]   P. Anthony et al., Phys. Rev. Lett. 76, 3550 (1996)
[PDG00]   D.E. Groom et al., Particle Data Group. Review of Particle Properties. Eur. Phys.
               J. C15, 1 (2000) http://pdg.lbl.gov/
[Hubb80]   J.H. Hubbell, H.A. Gimm, and I. Overbo, J. Phys. Chem. Ref. Data 9, 1023 (1980)
[Geant3]   L. Urban in GEANT3 writeup, section PHYS-211. *Cern Program Library* (1993)

5 Multiple Coulomb Scattering

5.1 Single Coulomb Scattering

Single Coulomb deflection of a charged particle by a fixed nuclear target:

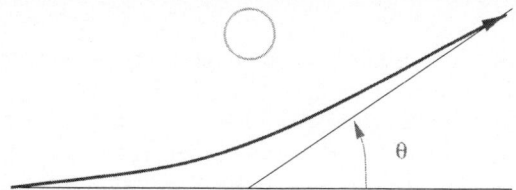

The cross section is given by the Rutherford formula:

$$\frac{d\sigma}{d\Omega} = \frac{r_e^2 z_p^2 Z^2}{4} \left(\frac{mc}{\beta p}\right) \frac{1}{\sin^4 \theta/2} \tag{4}$$

5.2 Multiple Coulomb Scattering

Charged particles traversing a finite thickness of matter suffer repeated elastic Coulomb scattering. The cumulative effect of these small angle scatterings is a net deflection from the original particle direction.

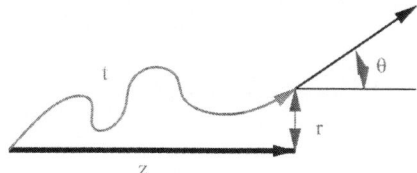

- Longitudinal displacement z (or geometrical path length)
- Lateral displacement r, Φ
- True (or corrected) path length t
- Angular deflection θ, ϕ

The practical solutions of the particle transport can be classified:

- Detailed (microscopic) simulation: exact, but time consuming if the energy is not small. Used only for low energy particles.
- Condensed simulation: simulates the global effects of the collisions during a macroscopic step, but uses approximations.
  EGS, Geant3 (both use Moliere theory), Geant4.
- Mixed algorithms: "hard collisions" are simulated one by one + global effects of the "soft collisions": Penelope.

5.3 Angular Distribution

If the number of individual collisions is large enough (>20) the multiple Coulomb scattering angular distribution is Gaussian at small angles and like Rutherford scattering at large angles.

The Molière theory [Molie48,Bethe53] reproduces rather well this distribution, but it is an approximation.

The Molière theory is accurate for not too low energy and for small angle scattering, but even for this case its accuracy is not too good for very low Z and high Z materials. (see e.g. [Ferna93,Gotts93])

The Molière theory does not give information about the spatial displacement of the particle, it gives only the scattering angle distribution.

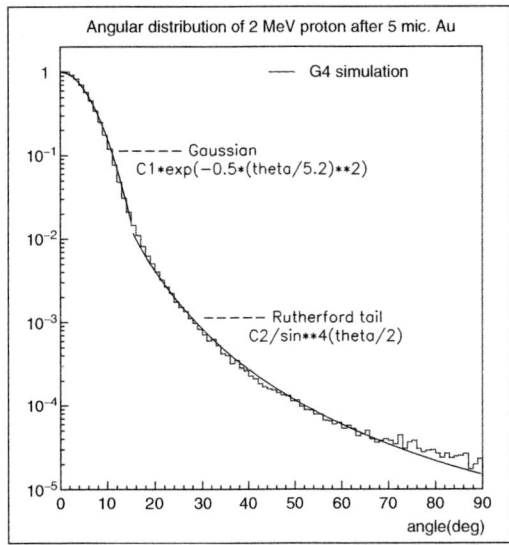

5.4 Gaussian Approximation

The central part of the spatial angular distribution is approximately

$$P(\theta)\,d\Omega = \frac{1}{2\pi\theta_0^2}\,\exp\left[-\frac{\theta^2}{2\theta_0^2}\right]\,d\Omega$$

with

$$\theta_0 = \frac{13.6\,\text{MeV}}{\beta\,pc}z\sqrt{\frac{l}{X_0}}\left[1 + 0.038\ln\left(\frac{l}{X_0}\right)\right] \tag{5}$$

where l/X_0 is the thickness of the medium measured in radiation lengths X_0 ([Highl75,Lynch91]).

This formula of θ_0 is from a fit to Molière distribution. It is accurate to $\leq 11\%$ for $10^{-3} < l/X_0 < 10^2$

This formula is used very often, but it is worth to note that this is an approximation of the Molière result for the small angle region with an error which can be as big as $\approx 10\%$.

5.5 Transport of Charged Particles

Let $p(r,d,t)$ denote the probability density of finding the particle at the point $r = (x,y,z)$ moving in the direction of the unit vector $\vec{d}$ after having travelled a path length t.

The transport is governed by the transport equation:

$$\frac{\partial p(r,d,t)}{\partial t} + \vec{d}.\overrightarrow{\nabla p(r,d,t)} = n_{at} \int [p(r,d',t) - p(r,d,t)]\frac{d\sigma(\chi)}{d\Omega}d\Omega \qquad (6)$$

which can be solved exactly for special cases only. but this equation can be used to derive different moments of p.

5.6 MSC Algorithm

Tasks of MSC algorithm are essentially the same for many condensed codes:

1. Selection of step length $\longleftarrow$ physics processes + geometry. MSC performs the $t \rightarrow z$ transformation
2. Transport to the initial direction (not MSC business)
3. Re-evaluate true step length: $z \rightarrow t$. Sample scattering angle (θ,ϕ)
4. Compute lateral displacement (with correlation). Relocate particle

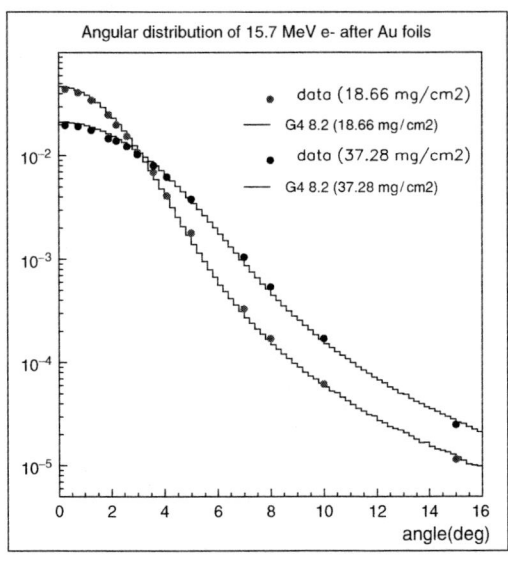

5.7 Backscattering of Low Energy Electrons

The incident beam is 10 electrons of 600 keV entering in 50 μm of Tungsten. 4 electrons are transmitted, 2 are backscattered.

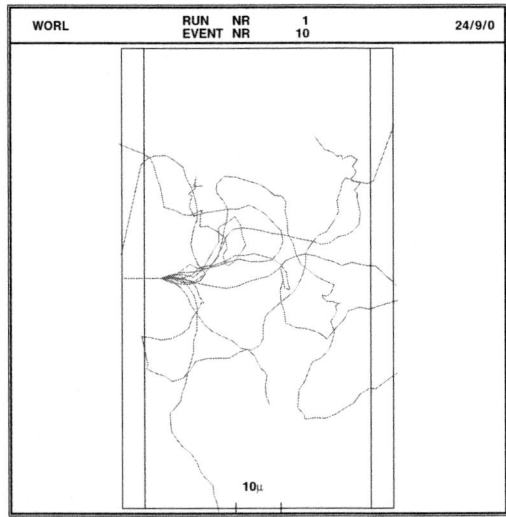

References

[Molie48]	G. Molière and Z. Naturforsch, 3a, 78 (1948)
[Bethe53]	H.A. Bethe, Phys. Rev. 89, 1256 (1953)
[Highl75]	V.I. Highland, NIM 129, 497 (1975)
[Lynch91]	G.R. Lynch and O.I. Dahl, NIM B58, 6 (1991)
[Gouds40]	S. Goudsmit and J.L. Saunderson, Phys. Rev., 57, 24 (1940)
[Lewis50]	H.W. Lewis, Phys. Rev. 78, 526 (1950)
[Kawra98]	I. Kawrakow and A.F. Bielajew, NIM B142, 253 (1998)
[Ferna93]	J.M. Fernandez-Varea et al., NIM B73, 447 (1993)
[Lilje87]	D. Liljequist and M. Ismail, J. Appl. Phys., 62, 342 (1987)
[Lilje90]	D. Liljequist et al., J. Appl. Phys., 68, 3061 (1990)
[Mayol97]	R. Mayol and F. Salvat, At. Data and Nucl. Data Tables, 55 (1997)
[Gotts93]	B. Gottschalk et al., NIM B74, 467 (1993)
[Seltz74]	S.M. Seltzer and M.J. Berger, NIM 119, 157 (1974)
[Hunge70]	H.J. Hunger and L. Kuchler, Phys. Stat. Sol. (a) 56, K45 (1970)
[Reste70]	D.H. Rester and J.H. Derrickson, J. Appl. Phys. 42, 714 (1970)
[Miche01]	C. Michelet et al., NIM, B181, 157 (2001)
[Shen79]	G. Shen et al., *Phys. Rev. D* 20, 1584 (1979)
[Attw06]	D. Attwood et al., *NIM B 251, 41 (2006)*

6 Ionization

The basic mechanism is an inelastic collision of the moving charged particle with the atomic electrons of the material, ejecting off an electron from the atom:

$$p + atom \rightarrow p + atom^+ + e^-$$

In each individual collision, the energy transferred to the electron is small. But the total number of collisions is large, and we can well define the average energy loss per (macroscopic) unit path length.

6.1 Mean Energy Loss and Energetic δ-Rays

$$\frac{d\sigma(Z,E,T)}{dT}$$

is the differential cross-section per atom for the ejection of an electron with kinetic energy T by an incident charged particle of total energy E moving in a material of density ρ.

One may wish to take into account separately the high-energy knock-on electrons produced above a given threshold T_{cut} (miss detection, explicit simulation ...).

$T_{cut} \gg I$ (mean excitation energy in the material).

$T_{cut} > 1$ keV in GEANT4

Below this threshold, the soft knock-on electrons are counted only as continuous energy lost by the incident particle.

Above it, they are explicitly generated. Those electrons must be excluded from the mean continuous energy loss count.

The mean rate of the energy lost by the incident particle due to the soft δ-rays is:

$$\frac{dE_{soft}(E,T_{cut})}{dx} = n_{at} \cdot \int_0^{T_{cut}} \frac{d\sigma(Z,E,T)}{dT} T \, dT \tag{7}$$

n_{at}: nb of atoms per volume in the matter. The total cross-section per atom for the ejection of an electron of energy $T > T_{cut}$ is:

$$\sigma(Z,E,T_{cut}) = \int_{T_{cut}}^{T_{max}} \frac{d\sigma(Z,E,T)}{dT} \, dT \tag{8}$$

where T_{max} is the maximum energy transferable to the free electron.

6.2 Mean Rate of Energy Loss by Heavy Particles

The integration of (7) leads to the well known Bethe-Bloch truncated energy loss formula [PDG]:

$$\frac{dE}{dx}\Bigg]_{T<T_{cut}} = 2\pi r_e^2 mc^2 n_{el} \frac{(z_p)^2}{\beta^2}$$

$$\times \left[\ln\left(\frac{2mc^2\beta^2\gamma^2 T_{up}}{I^2} \right) - \beta^2 \left(1 + \frac{T_{up}}{T_{max}} \right) - \delta - \frac{2C_e}{Z} \right]$$

where

r_e classical electron radius: $e^2/(4\pi\varepsilon_0 mc^2)$

mc^2 energy-mass of electron

n_{el} electrons density in the material

z_p charge of the incident particle

T_{up} $\min(T_{cut}, T_{max})$

I mean excitation energy in the material

δ density effect function

C_e shell correction function

$$n_{el} = Z n_{at} = Z \frac{\mathcal{N}_{av}\rho}{A} \qquad T_{max} = \frac{2mc^2(\gamma^2-1)}{1 + 2\gamma m/M + (m/M)^2}$$

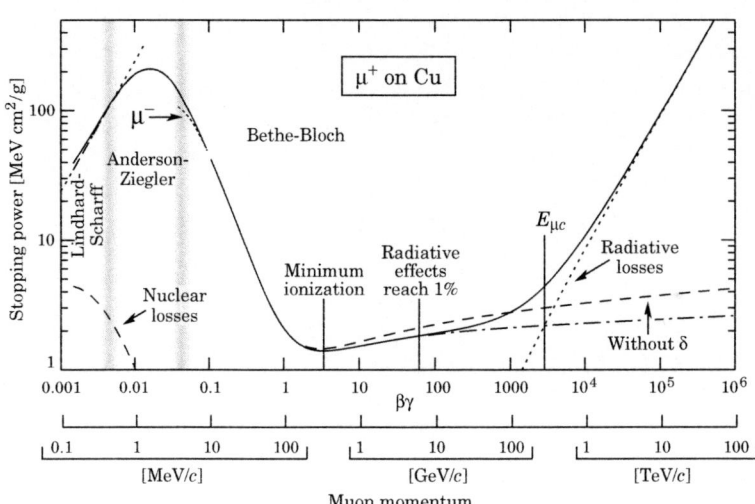

6.3 Mean Excitation Energy

There exists a variety of phenomenological approximations for I, the simplest being
$I = 10 eV \times Z$

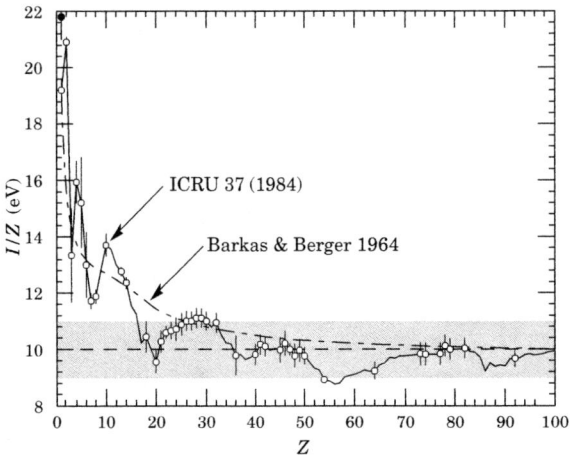

6.4 The Density Effect

δ is a correction term which takes into account of the reduction in energy loss due to the so-called density effect. This becomes important at high energy because media have a tendency to become polarised as the incident particle velocity increases. As a consequence, the atoms in a medium can no longer be considered as isolated. To correct for this effect the formulation of Sternheimer [Ster71] is generally used.

6.5 The Shell Correction

$2C_e/Z$ is a so-called *shell correction term* which accounts for the fact that, under certain conditions, the probability of collision with the electrons of the inner atomic shells (K, L, etc.) is negligible. The semi-empirical formula used in GEANT4, applicable to all materials, is due to Barkas [Bark62]:

$$C_e(I, \beta\gamma) = \frac{a(I)}{(\beta\gamma)^2} + \frac{b(I)}{(\beta\gamma)^4} + \frac{c(I)}{(\beta\gamma)^6}$$

6.6 Low Energies

The mean energy loss can be described by the Bethe-Bloch formula only if the projectile velocity is larger than that of orbital electrons. In the low-energy region where this is not verified, a different kind of parameterisation must be used.
For instance:

- Andersen and Ziegler [Ziegl77] for $0.01 < \beta < 0.05$
- Lindhard [Lind63] for $\beta < 0.01$

See ICRU Report 49 [ICRU93] for a detailed discussion of low-energy corrections.

6.7 Fluctuations in Energy Loss

$\langle \Delta E \rangle = (dE/dx).\Delta x$ gives only the average energy loss by ionization. There are fluctuations. Depending of the amount of matter in Δx the distribution of ΔE can be strongly asymmetric ($\rightarrow$ the Landau tail).

The large fluctuations are due to a small number of collisions with large energy transfers.

The figure shows the energy loss distribution of 3 GeV electrons in 5 mm of an Ar/CH4 gas mixture [Affh98]:

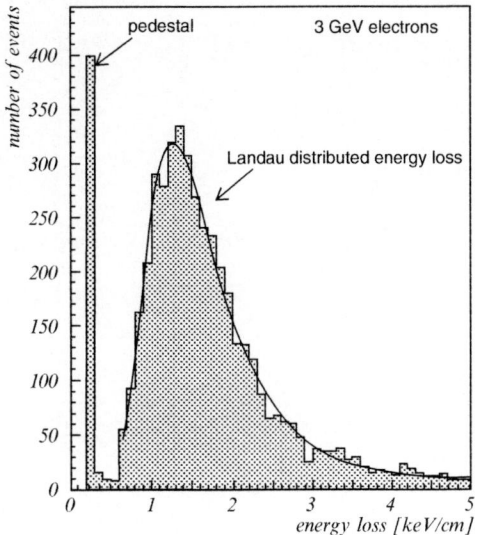

6.8 Energy-Range Relation

Mean total pathlength of a charged particle of kinetic energy E:

$$R(E) = \int_{\varepsilon=0}^{\varepsilon=E} \left[\frac{d\varepsilon}{dx} \right]^{-1} d\varepsilon$$

$$\left[\frac{dE}{dx} \right]_{tot} = \left[\frac{dE}{dx} \right]_{ioni} + \left[\frac{dE}{dx} \right]_{brem}$$

Ionization and Bremsstrahlung cannot be independent.

6.9 Compute the Mean Energy Loss of Charged Particles

The computation of the mean energy loss on a given step is done from the Range
and inverse Range tables.

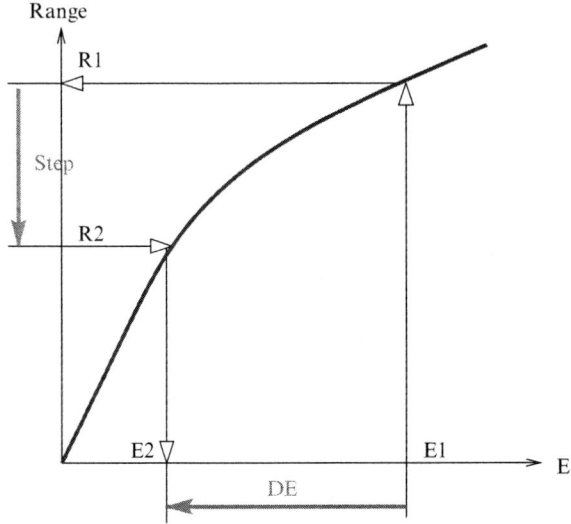

This is more accurate than $\Delta E = (dE/dx) * \text{stepLength}$.
Fluctuations on ΔE lead to fluctuations on the actual range (straggling).

Penetration of e^- (16 MeV) and proton (105 MeV) in 10 cm of water

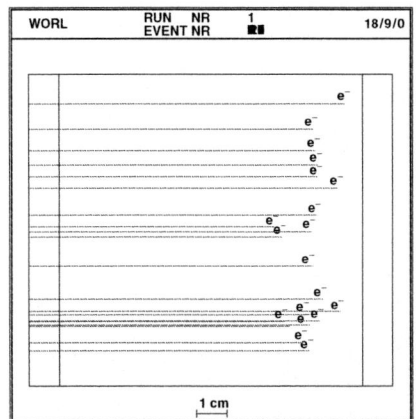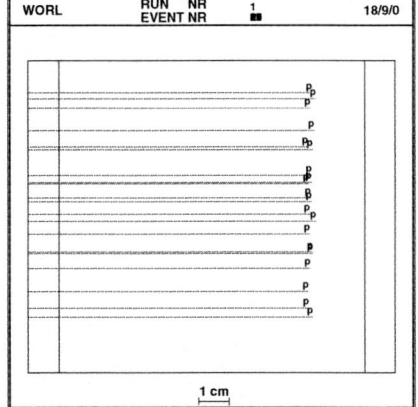

6.10 Bragg Curve

More energy per unit length are deposit towards the end of trajectory rather at its beginning.

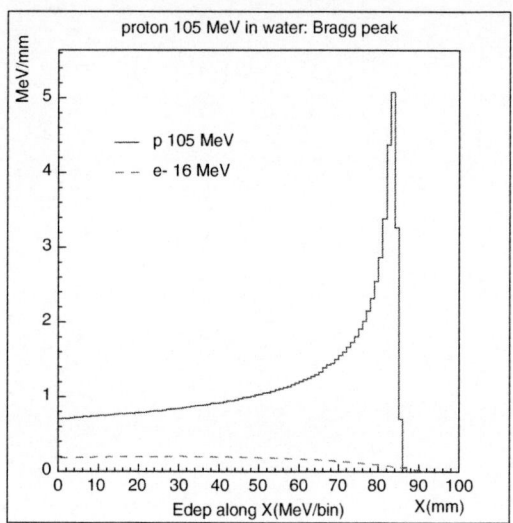

6.11 Energetic δ Rays

The differential cross-section per atom for producing an electron of kinetic energy T, with $I \ll T_{cut} \leq T \leq T_{max}$, can be written:

$$\frac{d\sigma}{dT} = 2\pi r_e^2 mc^2 Z \frac{z_p^2}{\beta^2} \frac{1}{T^2} \left[1 - \beta^2 \frac{T}{T_{max}} + \frac{T^2}{2E^2} \right]$$

(the last term for spin $1/2$ only). The integration of (8) gives:

$$\sigma(Z, E, T_{cut}) = \frac{2\pi r_e^2 Z z_p^2}{\beta^2} \left[\left(\frac{1}{T_{cut}} - \frac{1}{T_{max}} \right) - \frac{\beta^2}{T_{max}} \ln \frac{T_{max}}{T_{cut}} + \frac{T_{max} - T_{cut}}{2E^2} \right]$$

(the last term for spin $1/2$ only).

6.12 Delta Rays

200 MeV electrons, protons, alphas in 1 cm of Aluminium

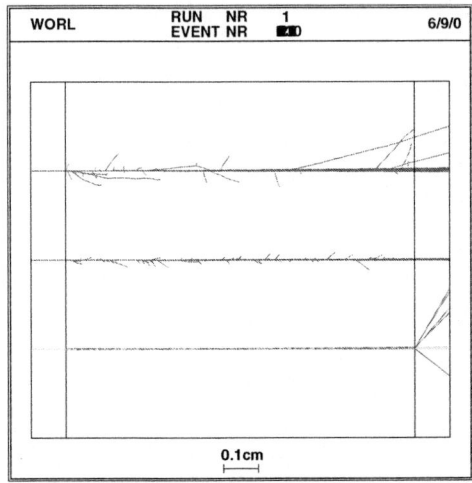

Muon: number of δ-rays per centimeter in Aluminium

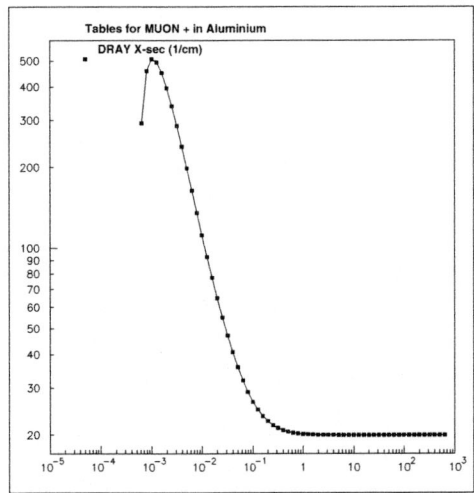

6.13 Incident Electrons and Positrons

For incident $e^{-/+}$ the Bethe Bloch formula must be modified because of the mass and identity of particles (for e^{-}). One use the Moller or Bhabha cross sections [Mess70] and the Berger-Seltzer dE/dx formula [ICRU84,Selt84].

6.14 de/dx Due to Ionization (Berger-Seltzer Formula)

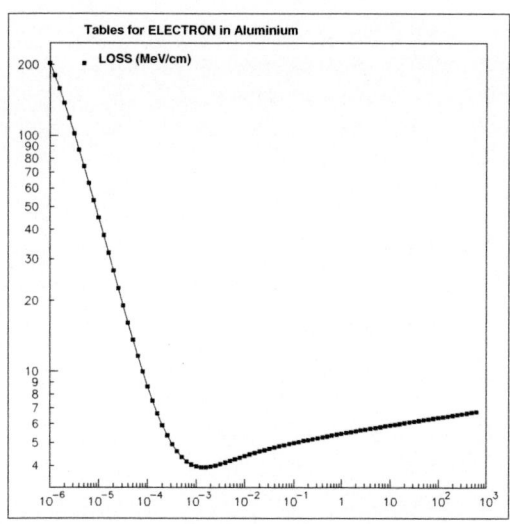

References

[Bark62]	W.H. Barkas, Technical Report 10292, UCRL, August 1962
[Lind63]	J. Lindhad et al., Mat.-Fys. Medd 33, No 14 (1963)
[Mess70]	H. Messel and D.F. Crawford, Pergamon Press, Oxford, 1970
[Ster71]	R.M. Sternheimer et al., Phys. Rev. B3 3681 (1971)
[Ziegl77]	H.H. Andersen and J.F. Ziegler, The Stopping and Ranges of Ions in Matter. Pergamon Press, Oxford 1977
[Selt84]	S.M. Seltzer and M.J. Berger, Int. J. Applied Rad. 35, 665 (1984)
[ICRU84]	ICRU Report No. 37 (1984)
[ICRU93]	ICRU Report No. 49 (1993)
[Affh98]	K. Affholderbach et al., NIM A410, 166 (1998)
[PDG]	D.E. Groom et al., Particle Data Group. Review of Particle Properties. Eur. Phys. J. C15, 1 (2000) http://pdg.lbl.gov/
[Bichs88]	H. Bichsel, Rev. Mod. Phys. 60 663 (1988)
[Urban95]	K. Lassila-Perini and L. Urbán, Nucl. Inst. Meth. A362 416 (1995)
[GEANT3]	GEANT3 manual *Cern Program Library* Long Writeup W5013 (1994)

7 Bremsstrahlung

A fast moving charged particle is decelerated in the Coulomb field of atoms. A fraction of its kinetic energy is emitted in form of real photons. The probability of this process is $\propto 1/M^2$ (M: mass of the particle). and $\propto Z^2$ (atomic number of the matter). Above a few tens MeV, bremsstrahlung is the dominant process for e- and e+ in most materials. It becomes important for muons (and pions) at few hundred GeV.

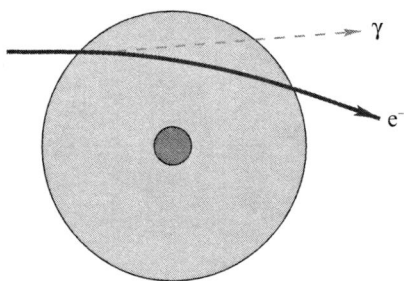

7.1 Differential Cross Section

The differential cross section is given by the Bethe-Heitler formula [Heitl57], corrected and extended for various effects:

- The screening of the field of the nucleus
- The contribution to the brems from the atomic electrons
- The correction to the Born approximation
- The polarisation of the matter (dielectric suppression)
- The so-called LPM suppression mechanism, etc.

See Seltzer and Berger for a synthesis of the theories [Sel85].

7.2 High Energies Regime: $E \gg mc^2/(\alpha Z^{1/3})$

Above few GeV the energy spectrum formula becomes simple:

$$\left.\frac{d\sigma}{dk}\right]_{Tsai} \approx 4\alpha\, r_e^2 \frac{1}{k} \times \left\{ \left(\frac{4}{3} - \frac{4}{3}y + y^2\right) \left(Z^2\left[L_{rad} - f(Z)\right] + ZL'_{rad}\right) \right\} \qquad (9)$$

where

$$
\begin{array}{ll}
k & \text{energy of the radiated photon ;} \quad y = k/E \\
\alpha & \text{fine structure constant} \\
r_e & \text{classical electron radius:} \quad e^2/(4\pi\varepsilon_0 mc^2) \\
L_{rad}(Z) & \ln(184.15/Z^{1/3}) \quad \text{\tiny(for } Z \geq 5) \\
L'_{rad}(Z) & \ln(1194/Z^{2/3}) \quad \text{\tiny(for } Z \geq 5) \\
f(Z) & \text{Coulomb correction function}
\end{array}
$$

7.3 Mean Rate of Energy Loss Due to Bremsstrahlung

$$-\frac{dE}{dx} = n_{at} \int_{k_{min}=0}^{k_{max}\approx E} k\,\frac{d\sigma}{dk}\,dk \tag{10}$$

n_{at} is the number of atoms per volume.
The integration immediately gives:

$$-\frac{dE}{dx} = \frac{E}{X_0} \tag{11}$$

with:

$$\frac{1}{X_0} \stackrel{def}{=} 4\alpha\, r_e^2\, n_{at} \left\{ Z^2 \left[L_{rad} - f(Z) \right] + Z L'_{rad} \right\}$$

7.4 Radiation Length

The radiation length has been calculated by Y. Tsai [Tsai74]

$$\frac{1}{X_0} = 4\alpha\, r_e^2\, n_{at} \left\{ Z^2 \left[L_{rad} - f(Z) \right] + Z L'_{rad} \right\}$$

where

$$
\begin{array}{ll}
\alpha & \text{fine structure constant} \\
r_e & \text{classical electron radius} \\
n_{at} & \text{number of atoms per volume:} \quad \mathcal{N}_{av}\rho/A \\
L_{rad}(Z) & \ln(184.15/Z^{1/3}) \quad \text{\tiny(for } Z \geq 5) \\
L'_{rad}(Z) & \ln(1194/Z^{2/3}) \quad \text{\tiny(for } Z \geq 5) \\
f(Z) & \text{Coulomb correction function}
\end{array}
$$

$$f(Z) = a^2[(1+a^2)^{-1} + 0.20206 - 0.0369a^2 + 0.0083a^4 - 0.002a^6 \cdots]$$

with $a = \alpha Z$ **main conclusion:** The relation 11 shows that the average energy loss per unit path length due to the bremsstrahlung increases linearly with the initial energy of the projectile. **equivalent:**

$$E(x) = E(0) \exp\left(-\frac{x}{X_0}\right)$$

This is the exponential attenuation of the energy of the projectile by radiation losses.

7.5 Critical Energy

The total mean rate of energy loss is the sum of ionization and bremsstrahlung. The *critical energy* E_c is the energy at which the two rates are equal. Above E_c the total energy loss rate is dominated by bremsstrahlung.

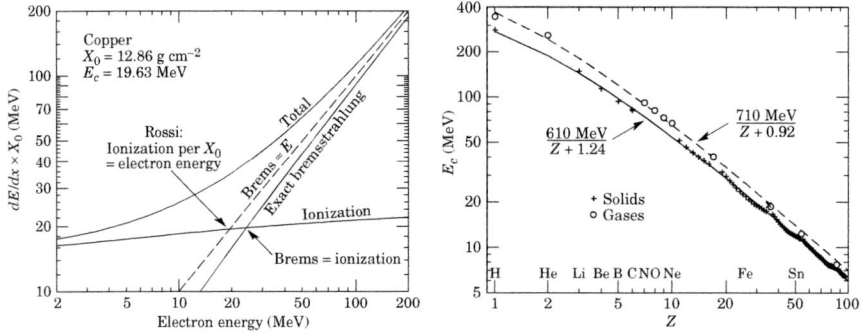

7.6 dE/dx for Electrons

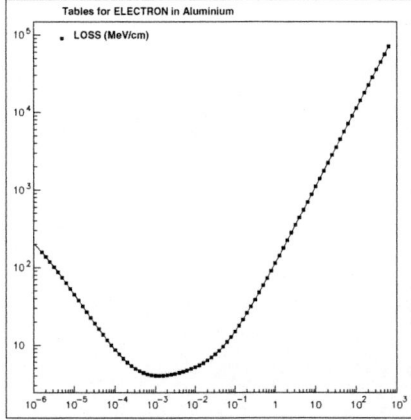

Fluctuations: Unlike the ionization loss which is quasicontinuous along the path length, almost all the energy can be emitted in one or two photons. Thus, the fluctuations on energy loss by bremsstrahlung are large.

7.7 Energetic Photons and Truncated Energy Loss Rate

One may wish to take into account separately the high-energy photons emitted above a given threshold k_{cut} (miss detection, explicit simulation...). Those photons must be excluded from the mean energy loss count.

$$-\frac{dE}{dx}\Bigg]_{k<k_{cut}} = n_{at} \int_{k_{min}=0}^{k_{cut}} k\,\frac{d\sigma}{dk}\,dk \tag{12}$$

n_{at} is the number of atoms per volume. Then, the truncated total cross-section for emitting 'hard' photons is:

$$\sigma(E, k_{cut} \leq k \leq k_{max}) = \int_{k_{cut}}^{k_{max}\approx E} \frac{d\sigma}{dk}\,dk \tag{13}$$

Number of interactions per millimeter in Lead (cut 10 keV)

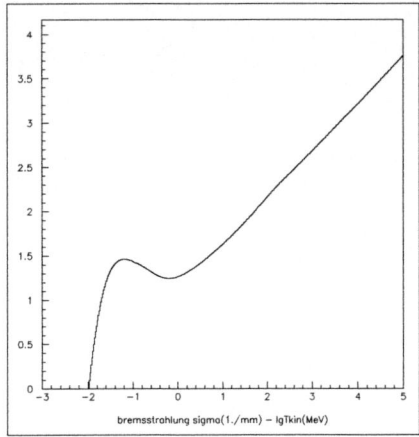

e^- 200 MeV in 10 cm Aluminium (cut: 1 MeV, 10 keV). Field 5 T

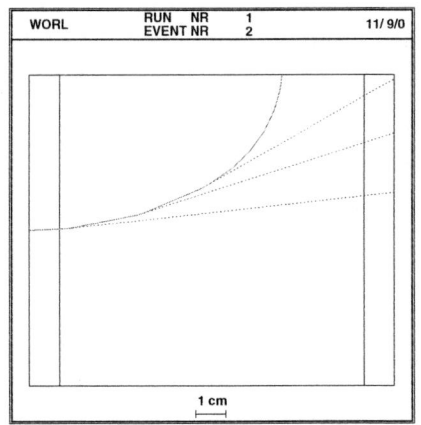 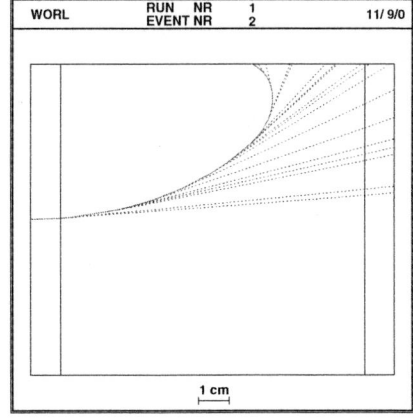

References

[Heitl57]   W. Heitler, The Quantum Theory of Radiation, Oxford University Press, Oxford (1957)
[Gal64]   V.M. Galitsky and I.I. Gurevich, Nuovo Cimento 32, 1820 (1964)
[Ter72]   M.L. Ter-Mikaelian, High Energy Electromagnetic Processes in Condensed Media, Wiley, New York (1972)
[Tsai74]   Y.S. Tsai, Rev. Mod. Phys. 46, 815 (1974)
             Y.S. Tsai, Rev. Mod. Phys. 49, 421 (1977)
[Geant3]   L. Urban in GEANT3 writeup, section PHYS-211. *Cern Program Library* (1993)
[Sel85]   S.M. Seltzer and M.J. Berger, MIN 80, 12 (1985)
[Antho96]   P. Anthony et al., Phys. Rev. Lett. 76, 3550 (1996)
[PDG00]   D.E. Groom et al., Particle Data Group. Review of Particle Properties. Eur. Phys. J. C15, 1 (2000). http://pdg.lbl.gov/
[Keln97]   S.R. Kelner, R.P. Kokoulin, and A.A. Petrukhin, Preprint MEPhI 024-95, Moscow, 1995; CERN SCAN-9510048
             S.R. Kelner, R.P. Kokoulin, and A.A. Petrukhin, Phys. Atom. Nucl., **60** 576 (1997)
             S.R. Kelner, R.P. Kokoulin, and A. Rybin. Geant4 Physics Reference Manual, Cern (2000)

8 (e^+, e^-) Annihilation into Two Photons

$$e^+ + e^- \rightarrow \gamma + \gamma$$

(need two γ for momentum conservation, if the e^- is assumed to be free). Theoretically, (e^+, e^-) annihilation is related to Compton scattering by crossing symmetry:

- Incoming $e^+ \leftrightarrow$ outgoing e^-
- Outgoing $\gamma \leftrightarrow$ incoming γ

8.1 Total Cross Section per Atom

The cross-section formula of Heitler is used [Heitl54]:

$$\sigma(Z, E) = \frac{Z\pi r_e^2}{\gamma + 1} \left[\frac{\gamma^2 + 4\gamma + 1}{\gamma^2 - 1} \ln\left(\gamma + \sqrt{\gamma^2 - 1}\right) - \frac{\gamma + 3}{\sqrt{\gamma^2 - 1}} \right]$$

E = total energy of the incident positron

$\gamma = E/mc^2$

r_e = classical electron radius

The cross section decreases with increasing E.
The nonrelativistic limit is:

$$\sigma_{nr}(Z, E) \sim \frac{Z\pi r_e^2}{\beta}$$

Number of interactions per centimeter in Aluminium

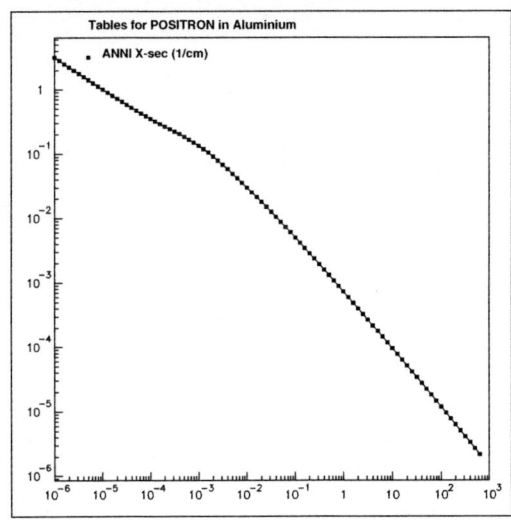

The annihilation in fly is not the dominant process. Most of the time the positron comes at rest and does a *positronium* with the electron. The positronium decays in two-photon (in 0.125 ns) or three-photon state (in 142 ns). The function AtRestDoIt treats this case. It generates two photons with energy $E_\gamma = mc^2$. The angular distribution is isotropic. The (e^+, e^-) can also annihilate in a single photon: the other photon is absorbed by the recoil nucleus. However this mechanism is suppressed by a factor α^4.

e^+ 30 MeV in 10 cm Aluminium. Annihilation in fly (left), at rest (right)

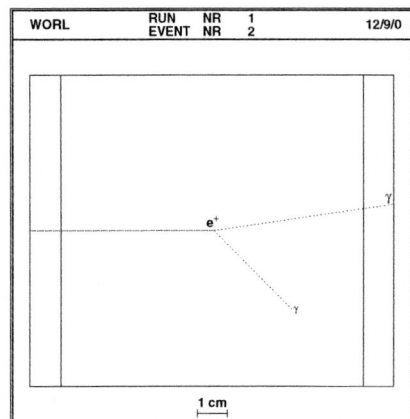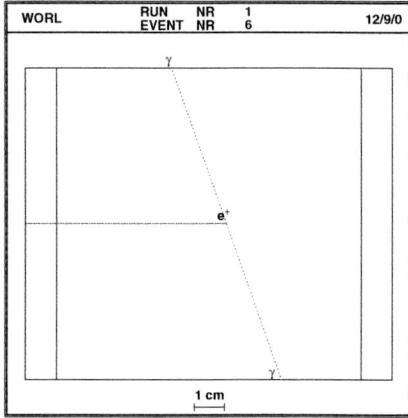

References

[Heitl54]   W. Heitler. The Quantum Theory of Radiation, Clarendon Press, Oxford (1954)

The Geant4 Simulation Toolkit and Applications
For the Geant4 Collaboration

John Apostolakis

Abstract Geant4 is a software toolkit for the simulation of the passage of particles through matter. It is used in a number of different application domains, including medical imaging and treatment, high energy physics experiments, and the assessment of radiation effects on satellites. Geant4 provides physics processes and models describing electromagnetic and hadronic interactions, and includes decay and optical processes. The energy range of processes spans from about 100 eV for electrons to 10 PeV for muons. Physics models are provided for electrons, positrons, gammas, hadrons, ions and optical photons. Often a choice of implementations is available for a physical process, providing different modelling approaches or a different level of approximation. Geant4 offers users a choice between prepared physics model configurations and the ability to create their own customised configuration, tailored for their requirements. Users can also choose which parts of the toolkit to utilise and assemble them, in a manner suitable for their particular application area. The Geant4 toolkit is open source, which enables its open distribution, and its incorporation into applications and frameworks. An overview of its capabilities is presented here spanning from its geometry and kernel capabilities, to its physics modelling and validation, and touching on the abilities it provides users to visualise setups and events and to configure applications interactively.

Keywords: Geant4 · simulation · radiation transport

1 Overview

The Geant4 toolkit[1] provides all the capabilities required to simulate the transport of radiation. It includes comprehensive physics modelling and is structured flexibly to enable users to adapt its use to their requirements. It offers a choice of physics

J. Apostolakis
CERN, CH-1211 Geneva 23, Switzerland
e-mail: john.apostolakis@cern.ch

Y. Lemoigne, A. Caner (eds.) *Molecular Imaging: Computer Reconstruction and Practice,*
and Experiments,
© Springer Science+Business Media B.V., 2008.

modelling options for most interactions. It has adaptability and functionality derived from a robust, structured kernel. This enables users to utilise it in and adapt it for diverse applications. Key parts of the toolkit and some recent developments undertaken by the Geant4 Collaboration[2] are described here, with emphasis on the capabilities of the Geant4 geometry and kernel, the modelling capabilities of its physics processes and their validation.

The Geant4 kernel provides the core upon which the toolkit is built. At the most basic level it includes a library of particles with their properties, a module for defining materials and volumes with which to create geometrical models of a detector or a setup, and a navigator for these geometry models. Building on these it defines an abstract interface for physics processes, and a tracking engine for combining their effects in flexible and powerful combinations. On top of these are the ability to stack tracks by priority, to manage as 'events' a set of primary particles and their resulting observables, and to create and to alter the configuration of a run, including the geometry setups and the generator of primary particles. In addition the kernel provides hooks with which to record the deposited energy, to count of particles, or otherwise record their passage as hits or scored observables. A framework for fast simulation for the parameterization of particle showers, and a concrete parameterisation for electromagnetic showers are included. A number of event biasing options and the ability to interface to external frameworks are also provided.

Physics processes included in Geant4 cover the range of physical interactions. They are organized as electromagnetic (EM), hadronic, decay and optical processes. Particles tracked include leptons, mesons, hadrons, ions and photons. Models for electrons, positrons and gammas are provided starting typically from 100 eV and extending up to 10 TeV. Physical processes of muons range up to 10 PeV, and for hadrons up to about 10 TeV. Neutrons are treated down to thermal energies for a variety of materials. A number of processes are provided for ions, with smaller energy coverage. Photons of optical wavelengths are treated separately from gammas. Different implementations of physics process are offered in several energy ranges for many physical processes. These typically use complementary modelling approaches and provide different trade-offs between accuracy and computing requirements.

To guide amongst the available choices, a number of configurations of physics models are provided tailored for specific application domains. Each configuration of physics processes (known as a 'physics list') provides coverage of all relevant physical processes, and ensures that each interaction is modeled by one process. The choices made are guided by the requirements for accuracy in important physics observables, with trade-offs varying levels of CPU resources. For example, models which accurately produce a particular type or energy range of particle, e.g. neutrons below 20 MeV, must be coupled with models which adequately describe the interaction of these particles. The expert user too can create a new physics lists or revise an existing one, to address specific requirements or check the dependence of physics results on a particular modelling choice.

The physics performance of Geant4 physics models has been documented by comparisons with established experimental data. Numerous comparisons undertaken of individual physics models and processes against reference data and thin-target experiments. There is also a growing body of independent validation efforts applying Geant4 in a variety of application domains and in physics experiments. A number of these results are noted in subsequent sections. Responding to this feedback, and to address known open issues, improvements in existing physics model are undertaken, and where possible new models are added.

In addition to the kernel and physics, Geant4 provides interfaces to enable users to interact with their application, and to save their results. A user can utilise one or more of the visualization drivers, which utilise OpenGL, Open Inventor, VRML and the high-quality DAWN system. Text-based and Graphical User Interfaces (GUIs) are included. For advanced applications, in particular those requiring storage for geometries or simulation results, a flexible framework for persistency is available. The user is free to utilise any package for creating histograms and saving the values of sets of observable within a Geant4 application. Examples are provided for using tools that follow the AIDA abstract interface for data analysis.

In order to meet the computing requirements of shielding and other applications, a number of established efficiency-enhancing techniques are offered in the toolkit: variance reduction (event biasing) techniques and options for fast simulation (shower parameterization).

The functionality of Geant4 and the set of physics models continue to be enhanced and extended.

The Geant4 toolkit is available with an open source license. New versions, with fixes, improvements and new features or physics models are released by the Geant4 collaboration, typically twice a year. Subject to the license terms, applicable since release 8.1 (June 2006), users can redistribute revised or customised versions, so long as they clearly identify that it is a changed version and they document and provide the changes under the same license. This enables incorporation into other open source applications and frameworks for specific areas. An additional clause allows the redistribution under other conditions, if a written license is made for each user. This was provided to enable some commercial applications which would not be possible otherwise.

The following sections cover the 'kernel', the electromagnetic physics processes, hadronic processes, the validation and the finally the visualisation and user-interaction capabilities.

2 The Geant4 Kernel

Any tool for radiation transport should provide the capabilities required to model the geometry and material composition of a setup, to describe a beam and to simulate the beam's interaction with the materials. The tool itself must undertake the key simulation tasks, tracking particles, navigating in the setup's geometry, propagating particles in the external electromagnetic field, modelling the physical interactions. In addition it can offer visualization options, and the ability to store information about key interactions and their energy deposition. Users are required only to describe the geometry and material composition of their setup, defining its geometry and EM field, specifying the radiation source and the details of sensitive regions.

The Geant4 toolkit, and in particular its kernel,[3] provides these key simulation capabilities, including geometry and field, particles, materials, tracking, and run and event management. It allows the simulation user to attach a set of physics models to a particle type, to create a geometrical model describing a particular setup, to configure a particle beam or source, and to manage most aspects of simulation behaviour. It provides hooks with which the user can choose the output required from measuring instruments or in radiation-sensitive parts. It also enables the use of visualization, which can show both the setup and event interactions using various graphics visualization systems. (Interactions, though, are undertaken by the physics modules.)

This section provides an overview of the kernel's modules, concentrating on geometry, field, and run management.

2.1 Geometry

The geometry modeller of Geant4 enables a user to describe the volumes in a setup or detector, and provides the ability to navigate in the resulting geometry model. A variety of shapes is available (Fig. 1). These include simple complete shapes, boxes, full tubes (cylinders), spheres and cones or their shells, and others which use the technique of Constructed Solid Geometry (CSG), and solids created by describing

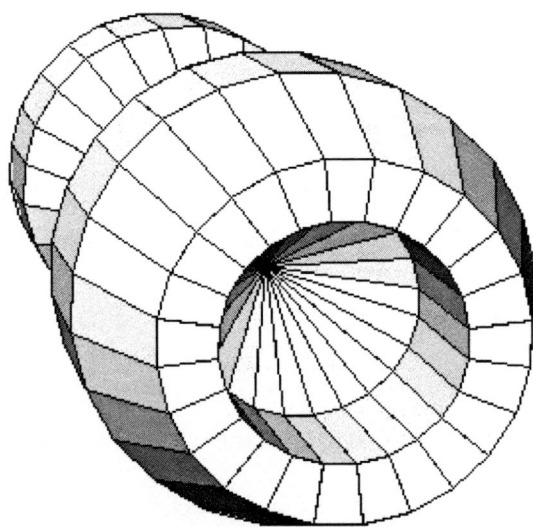

Fig. 1 An example of a Geant4 solid. This 'specific' solid, called a polycone, has cylindrical and conical sections which share a common axis. Inner and outer surfaces are possible (Courtesy O. Link, CERN)

their bounding surfaces, the Boundary Represented Solids (BREPS). In addition a set of 'specific' solids, found useful in different application areas are including, spanning connected cones, hyperboloids, ellipsoids and solids obtained by twisting several shapes. The user can also compose a customized solid by utilizing Boolean operations on simpler solids.

Adding material to a solid, and optionally an external electromagnetic field, creates a 'logical' volume. To create a volume in a specific location, a copy of a logical volume is placed with a translation and rotation inside a parent logical volume, to creating a 'physical volume'. Many such sub-volumes can be embedded inside a parent (logical) volume.

Each physical volume has enclosed inside the 'child' volumes which are contained in its logical volume. In this way two or more physical volumes which share the same logical volume will create apparent copies of a sub-structure in different parts of a setup. As a result the user can create a tree of volumes, or volume hierarchy, in which any set of structures can be repeated. Such multiple level hierarchies can used to describe a detector with repetitive elements, and many other types of complex setup.

The task of tracing straight-line paths (for model particle tracks) through a volume hierarchy, and of identifying the nearest boundary and computing its distance, is done by the Navigator. At the start of a simulation the geometry is scanned, and auxiliary information is automatically computed and stored in a set of 'voxels' for each volume in order to reduce computations required for navigation during tracking.

Advanced capabilities enable users to create complex setups more easily and to change them during run time. These include:

- Modifying parameters of the voxelisation, to reduce its memory cost at a small cost in tracking time
- Making a mirror image of a volume, including all its daughter volumes
- Modifying the setup geometry to create a new configuration in the same computer job
- Creating a different coexisting geometrical views of a single setup ('ghost geometries'), e.g. for fast parameterization, readout, biasing or scoring

These are abilities of the geometry module. The last two are coordinated with the simulation state, which is maintained by the Run module.

In a geometry setup special types of 'physical' volumes can be used, in which one volume in memory can represent several copies. Each represents a copy that is displaced, resized and rotated. This is used to save memory in case of a large number of repetitions. Different options are available for this. Replicas slice their parent volume completely along an axis; 'division' volumes extend its capabilities, enabling sub-dividing with offset and other features. Parameterized volumes require the user to define functions that computes the position and rotations of each volume copy; optional methods can also vary the solid type and size.

A geometry description mark-up language, GDML, [4] was created to describe in XML the volumes and material composition of a setup. A module for interfacing with Geant4 reads and writes geometry models from/into XML text files.

A major challenge for users is to create a consistent description of the geometry of a setup. It requires effort to avoid placing a volume that protrudes from its parent volume or overlaps with 'sibling' volumes which are also contained in the same parent logical volume. Geant4 provides tools to identify such overlaps, and thus enable users to correct the geometry model of their setup. These tools include the DAVID [5] tool, which intersects the graphical representations of volumes, a verification [11] sub-module inside the geometry and optional checks during construction. All have adjustable intersection tolerances.

Geant4 does not check for malformed geometries during tracking, in order to gain performance and to simplify the implementation. A new option now enables a number of checks during tracking which can identify some problems in the user's geometry model.

2.2 Field

Charged particles in Geant4 are affected by external electromagnetic fields. The intersections of a track's curved trajectory with potential geometry boundaries are calculated by approximating the trajectory by linear segments using a user specified precision.

A 'global' field for a full setup and the option for 'local' fields are available. Magnetic fields, varying with position and time or constant, electric fields and combined EM fields can be described and simulated. Field maps or user methods can be used to describe these fields.

The user can adjust a number of algorithm parameters to obtain the desired accuracy and/or computing performance. These parameters control the allowed integration error, the maximum distance between the exact curved and approximate segmented trajectory and the accuracy required for the intersection with a volume boundary. Their values may be set globally and refined for any volume.

2.3 Processes, Tracks, Events and Runs

The architecture of Geant4 is open and well documented. Geant4 physics processes are implemented using an abstract interface. This enables an advanced user to create a process for a particular application domain's physics or other needs. The process can then be registered to be applied for one or more Geant4 particle types using the process management.

The same tracking code is utilized in Geant4 for all particle types. The list of processes is different for each particle type (e.g. electrons and protons), and a particular order must be used to poll the processes for their cross-sections or step size, and to enable them to occur.

The tracking for all particles in Geant4 is done by one class G4SteppingManager. It utilises the same logic for all particle types. Each particle type (electron, gamma, positron, proton, pion+, ion, ..) has its own process manager which contains the list of processes representing physical interaction or alternative actions, such as the encountering of geometrical boundaries (transportation). Each particle's process manager maintains three lists of the processes, one each for actions that take place along the step (e.g. the continuous part of energy loss or the emission of Cerenkov light), at the endpoint of a step (any discrete process, e.g. hadronic interaction, hard Bremmstrahlung gamma) or in time (if the particle is at rest). The ordering of processes is important, and is carefully specified to enable the processes' effects to be combined correctly.

In addition to physics processes, other 'general' processes are used to update the simulation state or apply a different way of simulation. Examples include event biasing for variance reduction and the parameterization of showers inside calorimeters for fast simulation. The transportation process communicates with the geometry navigator and the field propagation module to identify the step to the next boundary.

Geant4 classes enable a user to create and revise a Geant4 run configuration, which brings together a particle source, the setup's geometry and a set of physics processes. In a single computing job, several run can be made with the same or different configuration. Run attributes can be changed, except for physics configurations.

The choice of the configuration of physics models (the 'physics' list) provides the ability to make choices between accuracy and computing performance.

3 Electromagnetic Physics

The electromagnetic (EM) physics modules in Geant4 aim to provide for the simulation of the electromagnetic interactions of charged particles and photons from the eV energies to TeV energies. EM processes implement the interactions of electrons and positrons, photons, muons, charged hadrons and ions. Typically alternative models are provided in the same physics interaction, providing different degrees of precision and simulation speed. This enables users to choose models adapted to the requirements of a particular application.

The alternatives exploit the features of the Object Oriented technology to enable transparency of physics results. Physics processes obey the same abstract interface, and, whenever available, use evaluated databases, distributed by a variety of international sources.

Two packages have been created to address different sets of requirements. The first, the *standard* electromagnetic package[6] aims to model all relevant EM processes which do not require producing secondary particles with energy below 1 KeV, or depend on modelling of atomic excitations. If a particle is produced, it is tracked particles down to zero energy. In addition, a number of implementation choices are made, including modelling cross-sections using parameterizations, to enable high computing performance. The *low energy* electromagnetic package[7] addresses especially the requirements of precise simulation, extending Geant4 capabilities first down to 250 eV (and recently below to 100 eV and below), and modelling in detail atomic relaxation effects.

Geant4 Electromagnetic Physics is exercised in a wide variety of simulation domains; a selection of applications. Early surveys of the different areas of applicability were undertaken by Daly et al.[8] for space, and by Chauvie et al.[9] for medical domains.

3.1 'Standard' Electromagnetic Physics

The *standard* electromagnetic package was aimed initially to provide equivalent (and subsequently improved) modelling of interactions compared to the earlier, established GEANT, version 3, while maintaining efficient use of computing resources. Models and algorithms have been created, refined and are being improved in order to provide better stability for key simulation results (in typical use cases) with respect to variation of model parameters and the production thresholds of secondary particles. The requirements for accuracy and computing performance of HEP applications have been the primary driving force in many of the standard processes, but feedback from validation for applications in other domains has also been encouraged and greatly valued.

Standard electromagnetic physics provides implementations of electron, positron, photon, charged hadron interactions, ions and muons. Photon (gamma) processes include Compton scattering, gamma conversion, photoelectric effect and muon pair production. Electron and positron processes span Bremsstrahlung, ionisation and delta ray production, positron annihilation, synchrotron radiation and transition radiation.

The energy loss process undertakes the continuous energy loss of particles due to ionisation and Bremsstrahlung and electron-positron pair production. A production threshold is used by electromagnetic processes, which have infrared divergences or can create large numbers of low energy secondaries, in particular by Ionization and Bremsstrahlung. Separate thresholds (or production cuts) are available for photons and gammas. A choice has been made to utilise production thresholds expressed in length,[1] which represents the range of the particle in the current material. This choice enables easier enforcement of uniform ranges between electrons and photons, and reflects the need to obtain the best accuracy improvement for available computing resources. It serves to concentrate its use on radiation that is more likely to 'leak' from each volume. The module also utilizes geometrical information, for example in determining the displacement due to multiple scattering.

EM processes now allow the use of different particle production thresholds in different geometrical regions. This 'cuts per region' functionality is useful in large setups, in which precisions requirements can differ significantly in different regions.

Starting from a particle cut, given as a length representing a stopping range, the module computes the cut values in kinetic energy for each material. In case one or more regions with alternative cut values have been created, this is done only for the pairs of materials and cut values which exist in the setup.

For charged hadrons tables for the stopping power and range are created for the proton and antiproton. These quantities are computed for other hadrons and utilising scaling laws and those tables. Separate tables are utilised for pions. Corrections have been incorporated for the molecular structure of materials[31] and due to the nuclear stopping power.[30]

Alternative models are available for simulating the ionisation and energy loss of hadrons. An implementation of the Photo-Absorption Interaction (PAI)[14] model is targeted to detailed modelling of ionisation in gases. Models are available to describe transition radiation in a variety of radiators.[15]

An optional algorithm[12] is available to optimise the generation of delta rays during steps which are near boundaries. Benchmarks of a new implementation of this algorithm show improved the stability of the simulation of calorimeter setups, when varying the production thresholds.[13]

The Geant4 multiple scattering process and model[16] are used for all charged particles. This model computes the mean path length correction, simulates the angular scattering and proposes a lateral displacement for a step. Its performance is compared with experimental data.[1,16]

The package's design has evolved from its first implementation[1] and is based on an approach that uses EM models[2] to encapsulate physics cross-sections and interactions. It utilises a small set of common process classes to implement book-keeping aspects. This simplifies maintenance and eases the creation of refinements, extensions and the creation of alternative models. The package's applications are particularly sensitive to computing performance, and to address efforts continue on performance assessment and optimization. Recent physics refinements include a revision of the multiple scattering model, improvements to the tail of the angular distribution for multiple scattering for light material, and improved calculation of the radiative corrections for the Bethe-Bloch model for muon energies above 1 GeV. The multiple scattering process for ions has been separated, to ensure accuracy and reproducibility.

In Geant4 optical photons are treated separately from higher energy gammas. This implementation allows their processes to utilise the wave like property of electromagnetic radiation, implementing refraction and reflection at medium boundaries, bulk absorption and Rayleigh scattering. Cerenkov photon emission and scintillation processes generate optical photons. A special process implements wavelength shifting. The optical properties of a material medium, which are required for these processes, are stored in tables connected to the material.

Geant4 is the first tool to span the generation of optical photons from the propagation of charged particles through to the detection of the ensuing optical photons on photo sensitive areas, in a single event loop. It is an effective and comprehensive tool capable of realistically modelling the optics of scintillation and Cerenkov detectors and their associated light guides.

3.2 Low-Energy Extensions

The *low energy* electromagnetic package[6,7,9] includes a set of physics processes is implemented to extend the validity range of electromagnetic interactions down to lower energy than the standard Geant4 electromagnetic processes. Initially the extensions covered processes[19] for electrons, photons, positive and negative charged hadrons and positive ions and extended the energy range coverage down to 250 eV for elements with atomic number up to 99.

It included photoelectric effect, Compton scattering, Rayleigh effect, Bremsstrahlung and ionization, and photon conversion. Fluorescence emission from excited atoms is generated, and the Auger effect is modelled. The implementation is based on the exploitation of evaluated data libraries (EPDL97,[24] EEDL[25] and EADL[26]). Utilising their evaluated data the cross-sections are determined and the final state are sampled.

Processes for positrons and alternative processes for electrons, implement physics models originally developed for the Penelope[27] Monte Carlo code. This implementation corresponds to the 2001 release of Penelope. Other developments provided new, alternative, high precision models for the angular distribution of Bremsstrahlung photons from incident electron energies below 500 keV 20, and the implementation of X-ray fluorescence emission induced by protons (PIXE).

A simulation based on Geant4 Low Energy processes for photons and electrons is compared with experimental data in Fig. 2, with evidence of shell effects.

A Low Energy process exists for the ionisation by hadrons[22] and ions.[23] It adopts different models depending on the energy range and the particle charge. In the high energy ($E > 2\,\text{MeV}$) domain the Bethe-Bloch formula and in the low energy one ($E < 1\,\text{keV}$ for protons) the free electron gas model are applied respectively. In the intermediate energy range parameterised models based on experimental data from the ICRU[30] review are implemented; corrections due to the molecular structure of materials[31] and to the effect of the nuclear stopping power[30] are taken into account. The Barkas effect is also described by means of a specialised model.[29]

4 Hadronic Physics

Geant4 includes cross sections and physics models for hadronic interactions from thermal energies (for neutrons) to hundreds of GeV. For many regimes, a choice of physics models is available, enabling a user to choose between more precision and better CPU performance.

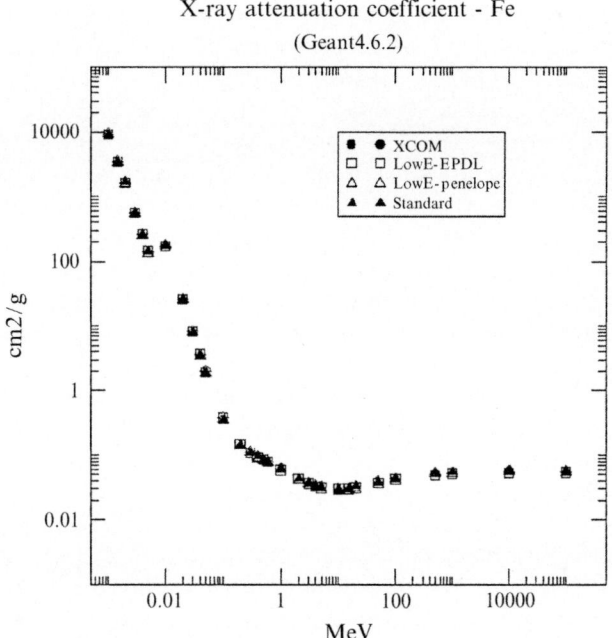

Fig. 2 Validation of Geant4 electromagnetic physics: comparison of various Geant4 physics models against the NIST data of the photon attenuation coefficient in Fe Courtesy

Direct modelling of low-energy interactions and the de-excitation of nuclei created in higher energy collisions is handled by an evaporation and pre-compound module developed within Geant4.[66] Within this framework several types of nuclear de-excitation channels, including fission, Fermi break-up and multi-fragmentation, are available.

Two cascade models are provided. One following the Bertini approach.[45] An alternative cascade code is the Binary cascade,[46] native to Geant4. These are discussed in the next section.

At high energies two theory-driven interactions models are used. For energies above about 15 GeV the Quark-Gluon String (QGS)[677] model is usable, while for energies above 5–10 GeV it is possible to use an alternative Fritiof-like String model (FTF). An overview of these higher energy models is provided elsewhere.[68] After the escape of high-energy particles, the nucleus can be treated by a choice of approaches: using the pre-compound model directly (the basis of the QGSP physics list), or by reabsorbing the slowest particles and utilizing the Chiral Invariant Phase space (CHIPS) model.

For photo-nuclear and electro-nuclear interactions processes are provided that use the CHIPS model.[69] Stopping hadrons are treated utilizing a CHIPS-based capture process.[47]

Covering all long-lived particles at all energies are the Low Energy Parameterized (LEP) and High Energy Parameterized (HEP) models which have their origins in the GHEISHA hadronic package[70] which was used with Geant3. The GHEISHA Fortran code was cast into C++, re-engineered and split into the current high- and low-energy parts. A number of fixes and improvements have been undertaken. These processes utilize simplified descriptions of interaction mechanisms, with key quantities parameterized for speed. They cover all long-lived hadrons, and were designed to describe hadronic showers reasonably well. They are also intended to conserve energy and momentum on average but not event by event.

The LEP models are especially important because they cover the energy region above the cascade models, typically for energy above between 2 and 15 GeV. For less frequently-occuring particles (beyond protons, neutrons and pions), the existing implementations of the theory-driven models usually do not offer an alternative. In addition until the onset of the string models (5–20 GeV) the LEP models are utilised in reference physics model configurations (physics lists), if case the chosen theoretical model(s) do not cover all energies. The LEP (and HEP) models also provide faster alternatives to the theory-driven models for calorimeters and other applications for which some precision can be traded for speed.

Several elastic scattering models are available which have been optimized for various energy ranges. A universally applicable elastic model, derived from GHEISHA, has been available since early Geant4 releases and was utilised in most physics lists configuration through Geant4 version 8.3. Coherent elastic scattering of hadrons from nuclei has recently been added. It covers energies above 400 MeV and is valid for protons, neutrons, pions and kaons. Parameterisations of cross-sections and angular distributions for proton and neutron elastic scattering have created in the CHIPS module, and released for light elements in Geant4 version 8.1 and for the remainder in 8.2.

At the lowest energies, the data-driven photo-evaporation module and an alternative set of neutron processes rely on data libraries. A custom database is utilised for each of these. For neutrons the process cross-sections and final states are stored in G4NDL. This data library was created by a selection of data from evaluated neutron data libraries, including ENDF/B-VI,[36] JENDL3.2,[39] and FENDL2.2.[37]

In addition there are data-driven models for neutron and proton induced isotope production.[1] These models run parasitically to Geant4 transport models, and can used to replace final state production for use in isotope production studies.

The evaluated data libraries that are the basis of the Geant4 neutron transport and activation library G4NDL version 3.11 are Brond-2.1,[32] CENDL2.2,[33] EFF-3,[34] ENDF/B-VI.0,[36] ENDF/B-VI.1, ENDF/B-VI.5, ENDF/B-VI.R8, ENDF/B-VI Thermal Scattering Data, ENDF/B-VII, FENDL/E2.0,[37] JEF2.2,[38] JENDL-FF,[39] JENDL-3.1, JENDL-3.2, JENDL3.3, JENDL High Energy File 2004 and MENDL-2.[40]

The approach is limited by the available data to neutron kinetic energies up to 20 MeV, with extensions to 150 MeV for some isotopes. The codes for neutron interactions are generic sampling codes, based on the ENDF/B-VI data format, and can be extended to use additional data.

Below is a brief description of three of these models, the Bertini-like cascade, the Binary cascades, and the CHIPS model.

4.1 Cascade Energy-Range Models

4.1.1 Bertini-Like Cascade

For hadronic interactions in the cascade energy range Geant4 includes a Bertini-style cascade model[45] which handles incident nucleons, pions, kaons and hyperons up to 10 GeV. It follows the Bertini approach[42] for an intra-nuclear cascade, as implemented in INUCL by N. Stepanov.[43] The nucleus is modelled using shells of constant density. Experimental cross sections and angular distributions are used, enabling the model to be extended to any particle for which there are sufficient measurements.

The projectile and any secondary particles created are transported along straight line paths through the nuclear medium. The interaction length is sampled using the density of the medium and the total cross section of a free hadron-nucleon interaction. The nuclear medium is approximated by a set of to three or fewer concentric shells. Pauli blocking is used to restrict collisions. Also a particle can be reflected or transmitted at boundaries between nuclear shells.

As cascade collisions occur, an excited residual nucleus is built up. This cascade has its own exciton routine which follows the approach of Griffin.[44] For light, highly-excited nuclei Fermi breakup may occur, and a fission channel is available. A custom nuclear evaporation, for neutrons and alphas, is followed by gamma emission at the lowest energies, below 0.1 MeV.

4.1.2 Binary Cascade

The Geant4 Binary cascade model,[46] uses a three dimensional model of the nuclear with discrete individual nucleons. For particle-particle collisions within the nucleus, free cross-sections are used. Nucleon momentum is taken into account for the evaluation of cross sections and collision probabilities. The approach is valid for incident protons and neutrons with kinetic energy $E_{kin} < 3\,GeV$, pions with $E_{kin} < 1.5\,GeV$, and light ions with $E_{kin} < 3\,GeV/A$ per nucleon. In some cases it provides an adequate descript for kinetic energies up to 10 GeV.

The incident particle and secondaries propagate through inside the nucleus along curved paths. These are calculated using numerical integration of the motion in the model nuclear potential.

Elastic nucleon-nucleon scattering are modelled. Inelastic collisions proceed by resonance formation and decay. Meson-nucleon inelastic scattering, other than absorption, is modelled as s-channel resonance excitation. Resonances may also propagate and interact with other nucleons before they decay.

When the cascade phase is finished, the residual nucleus is passed to the native Geant4 pre-compound model for de-excitation.

4.1.3 Chiral Invariant Phase Space Model

The 'CHIPS' model has its origins as an event generator incorporated into Geant4 at first for the treating the anti-baryon-nucleon annihilation,[47] capture of negatively charged hadrons at rest,[48] gamma- and lepto-nuclear reactions.[49,69] It is also used in some Geant4 models to handle the nuclear fragmentation part of nuclear de-excitation.

CHIPS is based the quasmon, a bunch of massless free partons, which form hadrons. A critical temperature T_c relates the quasmon mass m_q to the average number of a bunch's partons n. Hadronization proceeds via quark fusion and quark exchange. In the model u, d, and s quarks are massless and related by chiral symmetry.

Revised models for stopping muons, pions and other hadrons are utilised in most physics lists since Geant4 release 9.0.

5 Validation

A large number of physics observables have been compared between reference data and corresponding simulation data for the experimental validation of Geant4. These validation studies, undertaken by the Geant4 Collaboration, typically cover key features of the Geant4 physics models. Simulated distributions are compared to established references and data in literature. Additional contributions of user groups primarily compare results for full simulation with experimental data from test beams or on-going experiments. These provide valuable feedback on the accuracy of Geant4, which is complementary.

The full set of electromagnetic models provided by Geant4 for electrons, photons, protons and α particles have been compared[50] to the National Institute of Standards and Technology (NIST) database,[50–60] which represents a well known, authoritative reference in the field. This database is adopted in the definition of medical physics protocols. This systematic test involved quantitative comparisons between simulation and reference data, using statistical goodness-of-fit methods.[51] It confirmed good agreement of all Geant4 electromagnetic models with the NIST reference and highlighted details peculiar to each model, which cause their behaviour to differ as a function of the incident particle energy. Figure 2 shows the comparison of the photon attenuation coefficient obtained with various Geant4 simulation models against the corresponding NIST reference data. The p-value resulting from the χ^2 test represents the probability that the test statistics has a value at least as extreme as that observed, assuming that the null hypothesis (equivalence between Geant4 simulation and NIST data distributions) is true; p-values greater than the confidence level of 0.05 led to the acceptance of the null hypothesis.

Other comprehensive validation studies concerning Geant4 electromagnetic physics addressed transmission and backscattering distributions[50] and other specific applications of the standard package.[55]

Many comparisons have been undertaken in order to validate the use of Geant4 for applications in imaging and radiotherapy.[57] Recent improvements of the multiple scattering process,[16] available since Geant4 releases 8.0, have addressed key issues, and provide results which are more accurate stable,[58] including when varying production thresholds. Experimental verification has been undertaken of the Geant4 simulation of the does effects of tranverse magnetic fields.[59]

Two complementary approaches are adopted for the validation of simulation results in the domain of hadronic physics. The first utilises thin target setups with simple geometry, which tests single interactions in isolation. Each physics model is tested in this manner against established experimental data. The second relies on calorimeter test beam setups,[60] in which the observables are convolutions of many complex processes and interactions. Figure 3 shows an example of the comparison of the e/π ratio between a Geant4 simulation and preliminary experimental data coming from three different calorimeter test beams (ATLAS HEC Liquid Argon Calorimeter, ATLAS Tile Calorimeter.)

6 Steering and Visualising Applications

Geant4 provides the capability for the user to create an interactive application, able to steer key characteristics of the incident particles or to revise the setup, and to visualise the geometry and simulated events.

The toolkit provides a built-in text interaction mechanism,[1] usable across computing platforms including Linux, Unix and Windows. The toolkit provides a set of commands, to steer key elements. These include the option to provide options for simple logging of particle steps, and the ability to choose the energy, type, position and direction of incident particles. Example applications extend this to steer custom elements of a setup, for example to choose the material or length of a volume. Users can also extend this set with new user commands.

7 Geant4 Applications

A significant number of diverse applications have been created utilising Geant4 for high-energy and nuclear physics experiments, for space engineering and space science, and for medical imaging, research and therapy.

7.1 A Sample of Geant4 Applications

A majority of applications of Geant4 are specific to a single setup or detector, and are created to measure particular physical observables.

In contrast a number of flexible applications, include 'engineering tools' that utiliseGeant4 and simulate many potential setup of a particular domain. These depend on the open architecture of Geant4, and its extensibility, embedding part of the toolkit inside a tailored user interaction shell. This shell provides commands which enable a user to specify a problem's parameters and obtain estimate for observables in volume or location which he/she determines. Tools such as this have been created for a variety of domains: one of the first was MULASSIS for the optimization of shielding of spacecraft and for effects analysis. It is integrated in the widely used SPENVIS package.[62] MULASSIS includes an additional Java user interface, to enable its use by the space community over the World Wide Web.

A number of tools built on top of Geant4 specialise in simulating beamlines for accelerator beam delivery, including BDSIM[63] created for the future Linear Collider. Another tool, G4beamline,[64] enables the user to create beamline elements via text user commands, and to "virtual" detectors which sample the tracks.

For tomographic emission the GATE[65] application enables the simulation of a variety of PET and SPECT detectors. GATE provides customised extensions of Geant4 to create geometries, undertake customised counting and scoring in coincidence. Users drive the GATE application using only text commands.

It is beyond the scope of this article to survey the application domain of the large number of specific application which have been created utilising Geant4.

7.2 Creating a Geant4 Application

The challenge of creating a Geant4 application for a particular domain and task depends on the setup and task's complexity and its similarity to available examples and tools. There are several way to obtain a Geant4 simulation which undertakes a particular task. The simplest utilises a ready-made application or tool which is adapted to the application domain, and provides the necessary capabilities for creating the setup or detector and measuring the observables of interest. Such ready-made applications have been created in a few areas, as described above.

A second way involves starting from an existing application. For this purpose a variety of examples are provided with the Geant4 toolkit, including simple novice applications and examples which focus on particular elements of the toolkit. Starting from one of these, e.g. an electromagnetic example, it is possible in many cases to simulate simple setups with one or two materials with little or no coding, and to determine the energy deposition in a few volumes. Starting from such an application it is typically possible to create an application of low to medium complexity by extending it.

These applications involve some coding in C++. For example scoring the energy deposition in one or a few volumes only in coincidence (or anti-coincidence) with a signal in another volume typically will require coding.

The most advanced way, corresponds to setups or applications of significant complexity, or those which have outgrown the prototype stage. In this users can create a custom application, utilising or extending the main existing Geant4 capabilities, or even choosing to utilise only a particular set. For example some applications may choose to replace the G4RunManager with a custom object, in order to interface with an existing application or framework. Depending on where the complexity occurs, a custom application could involve the creation (or adaptation) of a hierarchy of classes for creating the geometry, classes for customising the physics process configuration, and/or classes to undertake scoring taking into account specific requirements.

Advanced applications can also utilise unique or specialised capabilities of the Geant4 toolkit. One of these is provided by the Geant4 geometry, also provided by Geant version 3 for its simpler geometry setups, [10] allows changes of the geometrical setup within the same computing process. It is supported by the kernel's run management, which treats them as different run configuration and triggers the geometry model to re-optimise the geometrical setup. This has been utilised[20] to enable the modelling of organ motion.

8 Outlook

The Geant4 toolkit's structure and openness, the wealth of physics models and customised configuration (physics lists), its open architecture and use of open source which result in the transparency of physics results, its simulation capability together have enabled a large diversity of application in domains from high energy physics experiments, medical physics and space applications. Its open source model for distribution and revision have enabled advanced users to adapt and extend it, and build customisation applications for a number of application domains. Many cases extensions have been contributed, adapted and included in the Geant4 toolkit, e.g. the GFLASH-derived shower parameterisation models. The toolkit is maintained, improved and further developed by the Geant4 collaboration.Future improvements will include performance improved for complex, voxel geometries, and the refinement of existing and of alternative physics models.

References

1. S. Agostinelli, J. Allison, K. Amako, J. Apostolakis, H. Araujo, P. Arce et al., Geant4: a simulation toolkit, Nucl. Instrum. Meth. A 506, 250–303 (2003).
2. J. Allison, K. Amako et al., Geant4 developments and applications, IEEE Trans. Nucl. Sci. 53(1), 270–278 (2006).
3. J. Apostolakis, G. Cosmo, and M. Asai, The Geant4 Kernel: Status and Recent Developments, in the Monte Carlo Method: Versatility Unbounded in a Dynamic Computing World, Chattanooga, Tennessee, April 17–21, 2005, on CD-ROM, American Nuclear Society, LaGrange Park, IL (2005).
4. R. Chytracek, The Geometry Description Markup Language, Proceedings of CHEP 2001, Beijing, R. Chytracek, J. Mccormick, W. Pokorski, G.Santin, Nuclear Science, IEEE Transactions on October 2006, Volume: 53, 2892–2896.
5. S. Tanaka, K. Hashimoto, and Y. Sawada, Proceedings of the Computing in High-energy and Nuclear Physics Conference, CHEP'98, Chicago, IL (1998).

6. H. Burkhardt et al., Geant4 Standard Electromagnetic Physics, IEEE NSS-33-179 Conference Record (2004).

7. S. Chauvie, G. Depaola, V. Ivanchenko, F. Longo, P. Nieminen, and M.G. Pia, Geant4 low energy electromagnetic physics, in Proceedings of the Computing in High Energy and Nuclear Physics, Beijing, China, pp. 337–340 (2001).

8. E. Daly et al., Space Applications of the Geant4 Simulation Toolkit, Proceedings of the MC2000 Conference, Lisbon (2000).

9. S. Chauvie et al., Medical Applications of the Geant4 Simulation Toolkit, Proceedings of the MC2000 Conference, Lisbon (2000).

10. R. Brun, F. Bruyant, A.C. McPherson, M. Maire, and P. Zanarini, 'Geant 3 (Users Guide)', CERN Data Handling Division, DD/EE/84–1 (1987).

11. V.N. Ivanchenko, M. Maire, and L. Urban, Geant4 standard electromagnetic package for HEP applications, in Conf. Rec. 2004 IEEE Nuclear Science Symposium, N33–165 (2004).

12. J. Apostolakis et al., CERN-OPEN-99-299 (1999).

13. O. Kadri, V.N. Ivanchenko, F. Gharbi, and A. Trabelsi, Geant4 simulation of electron energy deposition in extended media, in Nucl. Instr. Meth. B 258 (2), 381–387 (2007).

14. J. Apostolakis, S. Giani, V. Grichine et al., Nucl. Instr. Meth. A453, 597 (2000).

15. V.M. Grichine and S.S. Sadilov (Lebedev Inst. & CERN), GEANT4 X-ray transition radiation package, Nucl. Instrum. Meth. A563 (2), 299–302 (2006).

16. L. Urban, A multiple scattering model in Geant4, CERN-OPEN-2006-077 (December 2006).

17. A.G. Bogdanov, H. Burkhardt, V.N. Ivanchenko, S.R. Kelner, R.P. Kokoulin, M. Maire, A.M. Rybin, and L. Urban, Geant4 simulation of production and interaction of muons, in IEEE Trans. Nucl. Sci. 53 (2), 513–519 (2006).

18. J. Apostolakis, V.M. Grichine, V.N. Ivanchenko, M. Maire, and L. Urban, The recent upgrades in the Standard electromagnetic physics package, in Proceedings of the CHEP'06 Conference, Mumbai, India, February 2006.

19. J. Apostolakis et al., CERN-OPEN-99-034 and INFN/AE-99/18 (1999).

20. H. Paganetti, H. Jiang, J.A. Adams, G.T.Y. Chen, and E. Rietzel, Monte Carlo simulations with time-dependent geometries to investigate effects of organ motion with high temporal resolution, Int. J. Radiat. Oncol., Biol., Phys. 60, 942–950 (2004).

21. P. Rodrigues, R. Moura, C. Ortigao, L. Peralta, M. G. Pia, A. Trindade, et al., Geant4 applications and developments for medical physics experiments, IEEE Trans. Nucl. Sci. 51(4), 1412–1419 (August 2004).

22. K. Amako, S. Guatelli, V. Ivanchenko, et al., Validation of Geant4 Electromagnetic Physics against Protocol Data, IEEE Nuclear Science Symposium 2004 Conference Record, Rome, Italy (2004).

23. H. Araujo et al., Geant4 low energy electromagnetic physics, in The Monte Carlo method: versatility unbounded in a dynamic computing world, Proceedings of the Monte Carlo 2005 MC2005 Conference, Chattanooga, TN, April 17–21, 2005, on CD-ROM, American Nuclear Society, LaGrange Park, IL (2005).

24. D. Cullen et al., EPDL97: the Evaluated Photon Data Library, 97 version, UCRL–50400, Vol. 6, Rev. 5 (1997).

25. S.T. Perkins et al., Tables and Graphs of Electron-Interaction Cross Sections from 10 eV to 100 GeV Derived from the LLNL Evaluated Electron Data Library (EEDL), UCRL-50400 Vol. 31 (1997).

26. S.T. Perkins et al., Tables and Graphs of Atomic Subshell and Relaxation Data Derived from the LLNL Evaluated Atomic Data Library (EADL), Z=1-100, UCRL-50400 Vol. 30 (1997).

27. J. Baró, J. Sempau, J. M. Fernández-Varea, and F. Salvat, Penelope: an algorithm for Monte Carlo simulation of the penetration and energy loss of electrons and positrons in matter, Nucl. Instr. Meth. B, 100(1), 31–46 (1995). J. Sempau, E. Acosta, J. Baro, J.M. Fernandez-Varea, and F. Salvat, Nucl. Instr. Meth. B 132, 377–390 (1997); J. Sempau, J.M. Fernandez-Varea, E. Acosta, and F. Salvat, Nucl. Instr. Meth. B 207, 107–123 (2003).

28. S. Guatelli, A. Mantero, B. Mascialino, P. Nieminen, M.G. Pia, and S. Saliceti, Geant4 Atomic Relaxation, in Conf. Rec. 2004 IEEE Nuclear Science Symposium, N44–4 (2004).

29. S. Chauvie, P. Nieminen, and M.G. Pia, Geant4 model for the stopping power of low energy negatively charged hadrons, IEEE Nuclear Science Symposium 2006 Conference Record, San Diego, CA (2006).

30. ICRU A. Allisy et al., ICRU Report 49 (1993).

31. J.F. Ziegler and J.M. Manoyan, Nucl. Instr. Meth. B 35, 215 (1988).

32. Brond-2.2: A.I Blokhin et al., Current status of Russian Nuclear Data Libraries, Nuclear Data for Science and Technology, Vol. 2, p. 695, edited by J.K. Dickens, American Nuclear Society, LaGrange, IL (1994).

33. CENDL-2: Chinese Nuclear Data Center, CENDL-2, The Chinese Evaluated Nuclear Data Library for Neutron Reaction Data, Report IAEA-NDS-61, Rev. 3, International Atomic Energy Agency, Vienna, Austria (1996).

34. H.D. Lemmel (IAEA), EFF-2.4, The European Fusion File 1994, including revisions up to May 1995, Summary Documentation, IAEA-NDS-170 (June 1995).

35. JEF-2: C. Nordborg, M. Salvatores, Status of the JEF Evaluated Data Library, Nuclear Data for Science and Technology, edited by J.K. Dickens American Nuclear Society, LaGrange, IL (1994).

36. ENDF/B-VI: Cross Section Evaluation Working Group, ENDF/B-VI Summary Document, Report BNL-NCS-17541 (ENDF-201), edited by P.F. Rose, National Nuclear Data Center, Brookhaven National Laboratory, Upton, NY (1991).

37. FENDL/E2.0, The processed cross-section libraries for neutron-photon transport calculations, version 1 of February 1998. Summary documentation H. Wienke and M. Herman, report IAEA-NDS-176 Rev. 0 (International Atomic Energy Agency, April 1998). Data received on tape (or: retrieved on-line) from the IAEA Nuclear Data Section.

38. JEF-2.2: C. Nordborg, M. Salvatores, Status of the JEF Evaluated Data Library, Nuclear Data for Science and Technology, edited by J.K. Dickens, American Nuclear Society, LaGrange, IL (1994).

39. JENDL-3: T. Nakagawa, et al., Japanese Evaluated Nuclear Data Library, Version 3, Revision 2, J. Nucl. Sci. Technol. 32, 1259 (1995).

40. Yu. N. Shubin, V.P. Lunev, A.Yu. Konobeyev, and A.I. Ditjuk, Cross section data library MENDL-2 to study activation as transmutation of materials irradiated by nucleons of intermediate energies, report INDC(CCP)-385, International Atomic Energy Agency (May 1995).

41. M.R. Bhat, Evaluated Nuclear Data File (ENSDF), Nuclear Data for Science and Technology, p. 817, edited by S.M. Qaim, Springer, Berlin, Germany (1992).

42. M.P. Guthrie, R.G. Alsmiller, and H.W. Bertini, Nucl. Instr. Meth. 66, 29 (1968); H.W. Bertini and P. Guthrie, Nucl. Phys. A169 (1971).

43. N.V. Stepanov, ITEP Preprint ITEP-55, Moscow (1988).

44. J.J. Griffin, Phys. Rev. Lett. 17, 478 (1966).

45. A. Heikkinen, A., N. Stepanov, and J.P. Wellisch, Bertini Intra-Nuclear Cascade implementation in Geant4, in Proceedings of CHEP03 Conference, La Jolla, CA nucl-th/0306008 (March 2003).

46. G. Folger, V.N. Ivanchenko, and J.P. Wellisch, Eur. Phys. J. A21, 407 (2004).

47. M.V. Kossov, Manual for the CHIPS event generator, KEK internal report 2000-17, Feb 2001 H/R; P.V. Degtyarenko, M.V. Kossov, and H.P. Wellisch, Eur. Phys. J. A8 (2), 217 (2000).

48. P.V. Degtyarenko, M.V. Kossov, and H.P. Wellisch, Eur. Phys. J. A9 (2), 211 (2001).

49. P.V. Degtyarenko, M.V. Kossov, and H.P. Wellisch, Eur. Phys. J. A9 (2), 221 (2001).

50. K. Amako, S. Guatelli, V. Ivanchenko, B. Mascialino, K. Murakami, L. Pandola, et al., Comparison of Geant4 electromagnetic physics models against the NIST reference data, IEEE Trans. Nucl. Sci. 52(4), 910–918 (2005).

51. M.J. Berger, J.H. Hubbell, S.M. Seltzer, J.S. Coursey, and D.S. Zucker, XCOM: photon cross section database (version 1.2), National Institute of Standards and Technology, Gaithersburg, MD (1999). URL: http://physics.nist.gov/PhysRefData/Xcom/Text/ XCOM.html.

52. M.J. Berger, J.S. Coursey, and D.S. Zucker, ESTAR, PSTAR, and ASTAR: computer programs for calculating stopping-power and range tables for electrons, protons and helium ions (version 1.2.2), National Institute of Standards and Technology, Gaithersburg, MD (2000). url: http://physics.nist.gov/PhysRefData/Star/Text/contents.html.

53. G.A.P. Cirrone, S. Donadio, S. Guatelli, A. Mantero, B. Mascialino, S. Parlati, et al., A Goodness-of-Fit Statistical Toolkit, IEEE Trans. Nucl. Sci. 51 (5), 2056–2063 (Oct 2004).

54. G.A.P. Cirrone, G. Cuttone, S. Donadio, V. Grichine, S. Guatelli, P. Gumplinger, et al., Precision validation of Geant4 electromagnetic physics, in Conf. Rec. 2003 IEEE Nuclear Science Symposium, N23–2 (2003).

55. V.N. Ivanchenko, H. Burkhardt, M. Maire, V.M. Grichine, R.P Kokoulin, L. Urban, et al., "Overview and new developments on Geant4 electromagnetic physics", in Proceedings of the Computing in High Energy and Nuclear Physics, Interlaken, Switzerland, 320 (2004).

56. J.-F. Carrier, L. Archambault, and L. Beaulieu, Validation of GEANT4, an object-oriented Monte Carlo toolkit, for simulations in medical physics, Med. Phys. 31, 484–492 (2004).

57. E. Poon and F. Verhaegen, Accuracy of the photon and electron physics in GEANT4 for radiotherapy applications, Med. Phys. 32 (6), 1696–1711 (2005); E. Poon, J. Seuntjens, and F. Verhaegen Consistency test of the electron transport algorithm in the GEANT4 Monte Carlo code, Phys. Med. Biol. 50, 681–694 (2005).

58. S. Elles, V. Ivanchenko, M. Maire, and L. Urban, Geant4 and Fano cavity : where are we? LAPP-TECh-2004-04, and to appear in Proceedings of Third McGill International Workshop on Monte carlo techniques in Radiotherapy Delivery and Verification, Montreal, Canada (2007), Journal of Physics: Conference Series, IOP Publishing Lim.

59. A.J.E. Raajmakers et al., Experimental verification of magnetic field does effects for the MRI-accelerator, Phys. Med. Biol. 52, 4283–4291.

60. C. Alexa, J. Apostolakis, S. Banerjee, S. Constantinescu, A. De Roeck, S. Dita, A. Dotti, D. Elvira, F. Gianotti, A. Kiryunin, A. Lupi, C. Roda, D. Salihagic, P. Schacht, P. Strizenec, and H.-P. Wellisch, G4 Hadronic Physics Validation with LHC test-beam data: First Conclusions CERN-LCGAPP-2004-10 (July 2004); A. Ribon, Physics validation of the simulation packages in a LHC-wide effort, in Proceedings of the Computing in High Energy and Nuclear Physics, Interlaken, Switzerland, 493 (2004).

61. G. Santinab, P. Nieminena, H. Evansab, E. Dalya, F. Leic, P.R. Truscottc, C.S. Dyerc, B. Quaghebeurd, and D. Heynderickx, New Geant4 based simulation tools for space radiation shielding and effects analysis, Nucl. Phys. B 125, 69–74 (2003).

62. D. Heynderickx, B. Quaghebeur, E. Speelman, and E. Daly, ESA's Space Environment Information System (SPENVIS) - A WWW interface to models of the space environment and its effects, 38th Aerospace Sciences Meeting and Exhibit, Reno, NV, AIAA 2000-0371 (2000). SPENVIS web-site is http://www.spenvis.oma.be/spenvis/.

63. I. Agapov, J. Carter, G. A. Blair, and O. Dadoun, BDSIM - Beamline Simulation Toolkit Based on GEANT4, Proceedings of EPAC 2006, Edinburgh, Scotland, WEPCH124 (2006).

64. K. Beard, S.A. Bogacz, Y. Derbenev, K. Yonehara, R.P. Johnson, K. Paul, and T.J. Roberts, International Workshop on Beam Cooling and Related Topics - COOL05. AIP Conference Proceedings, Vol. 821, pp. 453–457 (2006); T.J. Roberts, D.M. Kaplan, THPAN103, Proceedings of IEEE Particle Accelerator Conference PAC07, Albuquerque, NM.

65. S. Jan et al., GATE: a simulation toolkit for PET and SPECT, Phys. Med. Biol. 49, 4543 (2004).

66. V. Lara and J. P. Wellisch, CHEP 2000: Computing in High Energy and Nuclear Physics, pp. 52–55 (2000).

67. N.S. Amelin and L.V. Bravina, Yad. Fiz. 51, 211 (1990) [Sov. J. Nucl. Phys. 51, 133 (1990)]; N.S. Amelin, K.K. Gudima, and V.D. Toneev, Yad. Fiz. 51, 512 (1990) [Sov. J. Nucl. Phys. 51, 327 (1990)]; G. Folger and J.-P. Wellisch, String Parton Models in Geant4, Proceedings of CHEP03, La Jolla, CA, nucl-th/0306007 (Mar 2003).

68. Low and High Energy Modeling in Geant4, T. Koi, D.H. Wright, G. Folger, V. Ivanchenko, M. Kossov, N. Starkov, A. Heikkinen, P. Truscott, F. Lei, and H.P. Wellisch, in Hadronic Shower SimulationWorkshop, 6–8 Sept. 2006, Batavia, IL, edited by M. Albrow and R. Raja, AIP Conference Proceedings 896.

69. P.V. Degtyarenko, M.V. Kossov, and H.P. Wellisch, Chiral invariant phase space event generator, III Photonuclear reactions below Δ(3.3) excitation, Eur. Phys. J. A 9 (2001).

70. H. Fesefeldt, GHEISHA: The Simulation of Hadronic Showers, RWTH/PITHA 85/02 (1985).

An Overview of Some Mathematical Methods for Medical Imaging

Michel Defrise and Christine De Mol

Abstract This chapter based on a series of lecture notes proposes an overview of mathematical concepts commonly used in medical physics and in medical imaging. One goal is to summarize basic and, in principle, well-known tools such as the continuous and discrete Fourier transforms, Shannon's sampling theorem, or the singular value decomposition of a matrix or operator. Tomographic reconstruction is covered in another chapter. Due to differences in vocabulary, notations and context, it is often difficult to grasp the links between inverse problems as different as image reconstruction in emission tomography,[18] photoacoustic tomography,[25] image registration,[12] or simply data fitting in the analysis of time sequences. The second goal of this chapter is therefore to give a synthetic and coherent view of the structure of the inverse problems encountered in this field, by rigorously defining concepts such as ill-posedness, stability, and regularization. This will allow a better understanding of the rationale behind data processing methods described in the literature.

1 Image Representations

The analysis of an imaging problem requires defining the class of mathematical objects to which the various quantities belong (measured data, restored or reconstructed images, etc.). Selecting these classes of objects in a way that reflects our prior knowledge of the problem is an essential prerequisite, which enables to unambiguously answer questions concerning the existence or the uniqueness of the

M. Defrise
Department of Nuclear Medicine, Vrije Universiteit Brussel, B-1090 Brussels, Belgium,
e-mail: Miche.Defrise@vub.ac.be

C. De Mol
Département de Mathématique and ECARES, Université Libre de Bruxelles, B-1050, Bruxelles, Belgium

Y. Lemoigne, A. Caner (eds.) *Molecular Imaging: Computer Reconstruction and Practice, and Experiments,*
© Springer Science+Business Media B.V., 2008.

solution to the problem, its stability in the presence of noise or its robustness for incorrect modeling of the data acquisition.

Object classes will be denoted by H, with – when needed – a subscript to distinguish the data class H_d from the solution class H_o. A class H is defined as a collection of objects equipped with a measure of distance between these objects. This distance serves to quantify "physical" concepts such as measurement error, reconstruction accuracy, or convergence speed for iterative algorithms. These concepts are poorly represented by a single scalar distance, and more sophisticated mathematical techniques are needed to objectively quantify the "quality" of an image in the context of a specific application.[13,20] Scalar distances are nevertheless useful for the mathematical analysis of the problem.

A convenient and sufficiently general structure to define H is that of a *Hilbert space*. A Hilbert space is a vector space of elements $f \in H$, equipped with a *scalar product* $H \times H \rightarrow C$, which to every pair of elements $f_1, f_2 \in H$, associates a complex number denoted (f_1, f_2). This scalar product defines a *norm* $H \rightarrow \mathbb{R}$ which to each element $f \in H$ associates the non-negative real number $||f|| = \sqrt{(f,f)}$. The distance between two elements f_1 and f_2 is then the norm of their difference, $||f_1 - f_2||$. The vector space H equipped with this norm is a Hilbert space if it is *complete*, i.e. if any Cauchy sequence of elements of H converges in norm to an element of H. Otherwise, it is a pre-Hilbert space.

Imaging problems can be described in a *discrete* or a *continuous* framework according to whether the Hilbert space H is finite or infinite. Although numerical implementation, as well as image storage and display, require a discrete set-up, the continuous framework is useful because

- In most cases the physical nature of the image is continuous.
- Stability or convergence results obtained in a continuous framework are often the key for guaranteeing the stability or convergence of discretized versions of the problem, independently of the level of discretization.
- Many results, such as inversion formulae, are more easily derived in a continuous framework.

Many problems involve more than one type of objects, e.g. data and solution, or sampled image and interpolated image, etc. For such problems, mixed frameworks are also possible, where the data are digital, hence inherently discrete, whereas the solution is more appropriately described in a continuous set-up, with an infinite dimensional Hilbert space of functions. We will discuss this point further in the section (5) on inverse problems.

1.1 Continuous Functions Spaces

In a continuous description, an object is represented by a function $f(\mathbf{x})$ of a variable $\mathbf{x}$, which takes values in some subset $\Omega \subset \mathbb{R}^d$, where d is the dimension of the space in which the object is living, e.g. $d = 2$ or $d = 3$ for 2-D or 3-D images. Note that Ω

is the region of space that contains the physical object under study, and therefore Ω is bounded. However, the mathematical analysis is sometimes simplified by working in an infinite domain such as $\Omega = \mathbb{R}^d$. Restriction to the bounded domain of interest is then done at the time of numerical implementation.

A common Hilbert space used in imaging is $L^2(\Omega)$, the vector space of (Lebesgue integrable) functions $f(\mathbf{x}), \mathbf{x} \in \Omega$ that are *square-integrable*, i.e. such that the integral

$$\int_\Omega |f(\mathbf{x})|^2 \, d\mathbf{x} \tag{1}$$

exists and is finite. This space is equipped with the scalar product

$$(f_1, f_2) = \int_\Omega f_1(\mathbf{x}) \bar{f}_2(\mathbf{x}) \, d\mathbf{x}, \quad f_1, f_2 \in L^2(\Omega) \tag{2}$$

Let us recall the *Schwarz inequality* $|(f_1, f_2)| \leq ||f_1|| . ||f_2||$. The L^2-norm appears naturally when quantifying the distance between two images by the mean-square error. The Hilbert space $L^2(\mathbb{R}^d)$ is a natural framework to define several integral transforms, such as the Fourier transform (see section 2).

Functions in $L^2(\Omega)$ are not defined pointwise and are not necessarily differentiable. When it is known a priori that the images to be represented are in some sense smooth, the class of admissible images may be restricted to spaces of more regular functions, as for example the *Schwartz space* $\mathcal{S}(\mathbb{R}^d)$ of infinitely differentiable functions on $\mathbb{R}^d$, which decay at infinity faster than any polynomial. The Schwartz space when equipped with the L^2-norm is not complete for that norm. It is however a dense subspace of $L^2(\mathbb{R}^d)$: for any element $f \in L^2(\mathbb{R}^d)$ it is possible to find an element $f' \in \mathcal{S}(\mathbb{R}^d)$ that approximates f to an arbitrary precision $\varepsilon > 0$ in the sense that $||f - f'|| \leq \varepsilon$.

Other subspaces of $L^2(\mathbb{R}^d)$ are also used; let us mention the Sobolev spaces $\mathcal{H}_\alpha$ [22] which provide a natural framework for the stability analysis of many inverse imaging problems such as the inversion of the Radon transform in tomography.

1.2 Discrete Spaces

Signal or image data collected by digital detectors are discrete and are represented by vectors in finite-dimensional vector spaces $\mathbb{R}^N$, where N is the number of discrete measurement values collected in one imaging experiment.

The situation is more complicated for reconstructed or restored images, which are discretized for the sake of numerical processing or display. Normally, the physical imaging problem is first modeled in a continuous framework, and the discrete image then represents some kind of sampling, or approximation, of a continuous (in the sense defined in section 1.1) underlying function. Translating the model into the discrete setup requires to unambiguously characterize the link between the discrete image space $\mathbb{R}^N$ and the underlying function space, e.g. $L^2(\mathbb{R}^d)$ or $\mathcal{S}(\mathbb{R}^d)$.

Implicitly or explicitly, the discrete image space $\mathbb{R}^N$ is often identified with some finite dimensional subspace $V_N = \mathrm{span}\{\psi_1, \psi_2, \cdots, \psi_N\} \subset H$ spanned by N linearly independent functions $\psi_j \in H$. The discrete data $(f_1, f_2, \cdots, f_N) \in \mathbb{R}^N$ is then associated with the function

$$f_a(\mathbf{x}) = \sum_{j=1}^{N} f_j\, \psi_j(\mathbf{x}) \tag{3}$$

where the subscript a indicates that $f_a \in V_N$ is only an approximation (in some sense) of the physical underlying function $f \in H$. Substituting f_a for f into the continuous equations modeling the problem then defines a discretized model, which can be further processed numerically.

Given a function $f \in H$, the discrete coefficients f_j have to be determined in order to guarantee that f_a is a reasonable model for f. Two standard criteria for this are

- $\|f - f_a\|$ is minimum: f_a is the *best approximation* of f in V_N, in the sense of the norm of the space H.
- $f(\mathbf{x}) = f_a(\mathbf{x})$ for a set of points $\mathbf{x}_1, \mathbf{x}_2, \cdots, \mathbf{x}_N$: f_a *interpolates* f. If the ψ_j satisfy the interpolating relation $\psi_j(\mathbf{x}_k) = \delta_{j,k}$, the coefficients are then simply the samples of f: $f_j = f(\mathbf{x}_j)$.

More sophisticated criteria for defining the coefficients f_j may involve additional constraints on f_a, e.g. a smoothness constraint, or a positivity constraint.

The basis functions ψ_j can sometimes be selected in function of the specific problem, for instance to diagonalize the imaging operator (see section 5.5). In most cases, however, one uses standard families of basis functions chosen for their attractive properties for approximation and interpolation.

1. *Interpolation and approximation with B-splines.* The B-spline of order $n = 0, 1, 2\ldots$ is a piecewise polynomial of degree n defined recursively by [9]

$$B_0(x) = \begin{cases} 1 & |x| < 1/2 \\ 0 & |x| \geq 1/2 \end{cases} \tag{4}$$

$$B_n(x) = \int_{-\infty}^{\infty} B_0(y) B_{n-1}(x-y)\, dy \equiv (B_0 * B_{n-1})(x) \tag{5}$$

where $*$ denotes the convolution product. The main properties of the B-splines are

- $B_n(x)$ has $n-1$ continuous derivatives and has its spectrum well concentrated.
- $B_n(x) = 0$ when $x \notin [-(n+1)/2,\ (n+1)/2]$.
- $B_n(x) \geq 0$.
- $\sum_{k \in \mathbb{Z}} B_n(x-k) = 1$ where for each x the sum over all integers k has at most $n+1$ non-zero terms.

These properties make of the splines excellent basis functions to represent signals and images. In the simplest case, one uses a set of B-splines of fixed order n, centered at uniformly spaced points $x = k\Delta$:

$$f_a(x) = \sum_{k} f_k\, B_n(x/\Delta - k) \tag{6}$$

The coefficients can be determined to interpolate or to approximate the underlying function f.[9,26] The B-splines satisfy the interpolating relation $B_n(j) = \delta_{j,0}$ only for $n = 0$ (*nearest-neighbour interpolation*) and $n = 1$ (*linear interpolation*). For any $n > 1$, interpolation requires solving a set of linear equations, but efficient recursive algorithms are available for this purpose.[28]

2. *Multi-dimensional splines.* Multi-dimensional splines are built as tensor products of the B-splines, e.g.

$$B_{d,n}(\mathbf{x} = (x_1, x_2, \cdots, x_d)) = B_n(x_1) B_n(x_2) \cdots B_n(x_d) \qquad (7)$$

For $n = 0$, this basis function is the indicator function of a $d-$dimensional *voxel*, and corresponds to the most widely used representation of images. As in the 1D case, the image coefficients for nearest-neighbour interpolation ($n = 0$) and for linear interpolation $n = 1$ are simply the image samples: a set of function samples $f(\mathbf{k}\Delta), \mathbf{k} \in I \subset Z^d$ is interpolated by

$$f_a(\mathbf{x}) = \sum_{\mathbf{k} \in I} f(\mathbf{k}\Delta) B_{d,n}(\mathbf{x}/\Delta - \mathbf{k}) \qquad n = 0 \text{ or } 1 \qquad (8)$$

where $\mathbf{k} = (k_1, k_2, \cdots, k_d)$ is a $d-$dimensional integer index, and I is some finite set of such indices. Using higher order interpolation, e.g. cubic splines $n = 3$, considerably improves the accuracy of image transformations such as rotations or dilations.

3. *Kaiser-Bessel blobs.* When the object has no a priori privileged direction, a natural representation should make use of basis functions which are invariant for rotation, contrary to the tensor product B-splines. The Kaiser-Bessel functions have been used as basis functions for iterative image reconstruction in emission and transmission tomography.[17,34] They share several interesting properties of the B-splines: non-negativity, good concentration in frequency space and bounded support, and, in addition, they are radially symmetrical. The generalized Kaiser-Bessel function of order m depends on two other parameters, the radius a of its support and a shape parameter α:

$$b_{m,a,\alpha}(r) = \begin{cases} (1/I_m(\alpha)) \left[1 - (r/a)^2\right]^{m/2} I_m(\alpha\sqrt{1 - (r/a)^2}) & 0 \le r \le a \\ 0 & r > a \end{cases} \qquad (9)$$

where $r = |\mathbf{x}| = \sqrt{x_1^2 + x_2^2 + \cdots + x_d^2}$ and I_m is the modified Bessel function of order m. Analytic expressions are known for the Fourier transform and for the Radon transform of the $b_{m,a,\alpha}$. A $d-$dimensional image is then represented as a linear combination of Kaiser-Bessel functions centered at the knots of a grid. This grid can be either the standard cubic grid (the same as the grid used with voxels), or the body-centered grid, which allows to reduce the number of basis functions required to represent an image (see section 4.4).

4. *Finite elements* are well suited for the numerical solution of partial differential equations; they are applied for example in elastography (e.g.,[10]) and in magnetic resonance electrical impedance tomography (e.g.,[16]). Finite elements have also

been applied to maximum-likelihood or bayesian reconstruction from 2D gated cardiac SPECT data.[5]

5. *Wavelets.* Orthogonal wavelet bases can be constructed via a Multiresolution Analysis, i.e. a sequence of closed subspaces V_j such that $\cdots \subset V_2 \subset V_1 \subset V_0 \subset V_{-1} \subset V_{-2} \subset \cdots \subset L^2(R)$ and having scaling properties, so that if $\phi_{0k}(x) = \phi(x-k)$, $k \in Z$, form an orthonormal basis in V_0, their scaled versions $\phi_{jk}(x) = 2^{-j/2}\phi(2^{-j}x-k)$ form an orthonormal basis in V_j. The function ϕ is called the scaling function. The orthogonal complement W_j of V_j in V_{j-1} is then spanned by the following functions generated by the "mother" wavelet ψ (which can be constructed from ϕ by some well-defined recipe): $\psi_{jk}(x) = 2^{-j/2}\psi(2^{-j}x-k)$. Taking the collection $\{\psi_{jk}\}$ for all $j,k \in Z$, one gets an orthonormal basis of $L^2(R)$. The translation parameter k represents a discrete "time" whereas the dilation parameter j is called the "scale". In practice, assuming that the function f belongs to the subspace V_{-1}, one uses the following expansion

$$f = \sum_{k=-\infty}^{+\infty} c_{0k}(f)\,\phi_{0k} + \sum_{j=0}^{+\infty}\sum_{k=-\infty}^{+\infty} d_{jk}(f)\,\psi_{jk} \,. \tag{10}$$

where $c_{jk}(f) = (f,\phi_{jk})$ and $d_{jk}(f) = (f,\psi_{jk})$. Starting from the coefficients $c_{jk}(f)$ at some scale, say $j = 0$, the coefficients at coarser scales can be easily computed by means of repeated applications of some discrete convolution operators H and G, given respectively by $c_{j+1,k}(f) = \sum_l h_{l-2k}\,c_{jl}(f)$ and $d_{j+1,k}(f) = \sum_l g_{l-2k}\,c_{jl}(f)$, followed at each step by a decimation or downsampling (i.e. keeping only the even samples). The filter coefficients h_k and g_k are determined from the scaling function and mother wavelet. If these filters are of finite length (which corresponds to the case of compactly supported wavelets), the number of multiplications required to perform this discrete wavelet decomposition is essentially proportional to the number N of samples used at the initial scale. Hence, the complexity of this *cascade algorithm* or Fast Wavelet Transform (FWT) is $O(N)$. The reconstruction can be performed in a similar fashion, with the same complexity. Many extensions of this scheme are available such as bi-orthogonal or nonorthogonal ("frame") expansions. Wavelet bases in higher dimension can be build e.g. by tensor products. Basic references on the subject are.[7,19]

2 The Fourier Transform

2.1 The Continuous Fourier Transform

The Fourier transform (FT) plays a central role in any problem that is, exactly or approximately, invariant for translation. See[6,27] for more details.

For any function $f(\mathbf{x}) \in L^1(\mathrm{IR}^d)$, i.e. any function such that

$$\int_{\mathrm{IR}^d} |f(\mathbf{x})| \, d\mathbf{x} < \infty \tag{11}$$

the Fourier transform $\mathcal{F} : f \to \mathcal{F}f$ is defined by

$$(\mathcal{F}f)(\mathbf{v}) = \int_{\mathrm{IR}^d} f(\mathbf{x}) \, e^{-2\pi i \mathbf{x} \cdot \mathbf{v}} \, d\mathbf{x} \qquad \mathbf{v} \in \mathrm{IR}^d \tag{12}$$

where $\mathbf{x} \cdot \mathbf{v} = x_1 v_1 + x_2 v_2 + \cdots x_d v_d$ is the euclidean scalar product in IR^d. The vector $\mathbf{v}$ is the *frequency* and $\Omega = 2\pi\mathbf{v}$ is the *angular frequency*. The Fourier transform is also called the *spectrum* of f, and it will also be denoted by $\hat{f} = \mathcal{F}f$. The conjugate transform is

$$(\mathcal{F}^*f)(\mathbf{v}) = \int_{\mathrm{IR}^d} f(\mathbf{x}) \, e^{2\pi i \mathbf{x} \cdot \mathbf{v}} \, d\mathbf{x} \tag{13}$$

The main properties of the Fourier transform are:

- *Inverse Fourier transform.* When $\mathcal{F}f \in L^1(\mathrm{IR}^d)$ one has, at least for f continuous, that

$$f(\mathbf{x}) = \int_{\mathrm{IR}^d} (\mathcal{F}f)(\mathbf{v}) \, e^{2\pi i \mathbf{x} \cdot \mathbf{v}} \, d\mathbf{v} \tag{14}$$

that is, $f = \mathcal{F}^{-1}\mathcal{F}f$ with $\mathcal{F}^{-1} = \mathcal{F}^*$.
- $\mathcal{F}f(\mathbf{v})$ is bounded and $\lim_{|\mathbf{v}| \to \infty} \mathcal{F}f(\mathbf{v}) = 0$ (Riemann-Lebesgue lemma), but is not necessarily integrable.
- For any $f, h \in L^1(\mathrm{IR}^d)$,

$$\int_{\mathrm{IR}^d} f(\mathbf{x}) \, (\mathcal{F}h)(\mathbf{x}) \, d\mathbf{x} = \int_{\mathrm{IR}^d} (\mathcal{F}f)(\mathbf{x}) \, h(\mathbf{x}) \, d\mathbf{x} \tag{15}$$

- *Differentiation.* Under appropriate assumptions of smoothness and decay at infinity one has,

$$\frac{\partial^{k_1}}{\partial v_1^{k_1}} \frac{\partial^{k_2}}{\partial v_2^{k_2}} \cdots \frac{\partial^{k_d}}{\partial v_d^{k_d}} (\mathcal{F}f)(\mathbf{v}) = \mathcal{F}\left(\left(-2\pi i x_1^{k_1} \right) \left(-2\pi i x_2^{k_2} \right) \cdots \left(-2\pi i x_d^{k_d} \right) f(\mathbf{x}) \right)(\mathbf{v})$$

and

$$\mathcal{F}\left(\frac{\partial^{k_1}}{\partial x_1^{k_1}} \frac{\partial^{k_2}}{\partial x_2^{k_2}} \cdots \frac{\partial^{k_d}}{\partial x_d^{k_d}} f(\mathbf{x}) \right)(\mathbf{v}) = (2\pi i v_1)^{k_1} (2\pi i v_2)^{k_2} \cdots (2\pi i v_d)^{k_d} (\mathcal{F}f)(\mathbf{v})$$

- *Translation* by a vector $\mathbf{b} \in \mathrm{IR}^d$:

$$(\mathcal{F}f(\mathbf{x} - \mathbf{b}))(\mathbf{v}) = e^{-2\pi i \mathbf{b} \cdot \mathbf{v}} \, (\mathcal{F}f(\mathbf{x}))(\mathbf{v}) \tag{16}$$

$$(\mathcal{F}e^{2\pi i \mathbf{b} \cdot \mathbf{x}} f(\mathbf{x}))(\mathbf{v}) = (\mathcal{F}f(\mathbf{x}))(\mathbf{v} - \mathbf{b}) \tag{17}$$

- *Dilation* by a scalar $a \in \mathbb{R}, a \neq 0$:

$$(\mathcal{F}f(a\mathbf{x}))(\mathbf{v}) = \frac{1}{|a|}(\mathcal{F}f(\mathbf{x}))\left(\frac{\mathbf{v}}{a}\right) \tag{18}$$

- *Symmetry*. The Fourier transform of a real function $f(\mathbf{x}) = \overline{f(\mathbf{x})}$ satisfies:

$$\overline{(\mathcal{F}f)(\mathbf{v})} = (\mathcal{F}f)(-\mathbf{v}) \tag{19}$$

The Schwartz space $\mathcal{S}(\mathbb{R}^d)$ of infinitely differentiable functions with rapid decay is stable under $\mathcal{F}$, with the inversion $\mathcal{F}^{-1} = \mathcal{F}^*$. The Fourier transform is an *isometry* of $\mathcal{S}(\mathbb{R}^d)$, i.e. it satisfies the Parseval-Plancherel equality

$$\int_{\mathbb{R}^d} |f(\mathbf{x})|^2 \, d\mathbf{x} = \int_{\mathbb{R}^d} |\hat{f}(\mathbf{v})|^2 \, d\mathbf{v} \tag{20}$$

and also preserves the scalar product

$$\int_{\mathbb{R}^d} f(\mathbf{x}) \, h^*(\mathbf{x}) \, d\mathbf{x} = \int_{\mathbb{R}^d} \hat{f}(\mathbf{v}) \, \hat{h}^*(\mathbf{v}) \, d\mathbf{v} \tag{21}$$

The Fourier transform is extended to the space $L^2(\mathbb{R}^2)$ of square-integrable functions as the limit

$$(\mathcal{F}f)(\mathbf{v}) = \lim_{n \to \infty} \int_{|\mathbf{x}|<n} f(\mathbf{x}) \, e^{-2\pi i \mathbf{x} \cdot \mathbf{v}} \, d\mathbf{x} \tag{22}$$

The limit in L^2 is not defined pointwise but as a mean-square convergence. With this definition, $\mathcal{F}$ defines an isometry of $L^2(\mathbb{R}^d)$ and satisfies the Parseval-Plancherel equality.

In distribution theory, the Fourier transform can be extended to be a bijective and bicontinuous mapping in the space $\mathcal{S}'$ of tempered distributions.[27] This allows to give a sound mathematical interpretation e.g. to the following rules which are often used in practice

$$\mathcal{F}\delta = 1 \qquad \mathcal{F}1 = \delta \tag{23}$$

where δ is the Dirac distribution which maps any function $f \in \mathcal{S}(\mathbb{R}^d)$ onto its value $f(0)$ at the origin. This relation can also be written, with the usual abuse of notation (handling a distribution as a function), as

$$\delta(\mathbf{x}) = \int_{\mathbb{R}^d} e^{2\pi i \mathbf{x} \cdot \mathbf{v}} \, d\mathbf{v} \tag{24}$$

Another important relation involves the *Dirac comb*, defined as

$$C_\Delta(\mathbf{x}) = \sum_{\mathbf{k} \in \mathbb{Z}^d} \delta(\mathbf{x} - \mathbf{k}\Delta) \tag{25}$$

The Fourier transform of this distribution plays a key role in sampling theory. It is given by the Poisson formula:

$$(\mathcal{F}C_\Delta)(\mathbf{v}) = \sum_{\mathbf{k} \in \mathbb{Z}^d} e^{-2\pi i \Delta \mathbf{k} \cdot \mathbf{v}} = \frac{1}{\Delta^d} \sum_{\mathbf{k}' \in \mathbb{Z}^d} \delta\left(\mathbf{v} - \frac{\mathbf{k}'}{\Delta}\right) = \frac{1}{\Delta^d} C_{1/\Delta}(\mathbf{v}) \tag{26}$$

These d−dimensional relations are built simply as the product of d one-dimensional relations, one for each coordinate x_j. Generalization to the case where the sampling step is different along each axis is straightforward.

2.2 The Fourier Series

A continuous function $f(x) \in L^1([-\Delta/2, +\Delta/2])$ on a finite interval of length Δ, with $f(-\Delta/2) = f(\Delta/2)$, can be represented by its Fourier series

$$f(x) = \sum_{k=-\infty}^{\infty} f_k e^{2\pi i k x/\Delta} \qquad |x| \le \Delta/2 \qquad (27)$$

where the equality holds almost everywhere and the *Fourier coefficients* are

$$f_k = \frac{1}{\Delta} \int_{-\Delta/2}^{\Delta/2} f(x) e^{-2\pi i k x/\Delta} dx \qquad k = 0, \pm 1, \pm 2, \dots \qquad (28)$$

When extended to $x \in \mathbb{R}$, the series (27) defines a periodic function with period Δ.

2.3 The Discrete Fourier Transform

Consider a vector $f = (f_0, f_1, \cdots, f_{N-1}) \in C^N$. The discrete Fourier transform (DFT) of f is the vector $\hat{f} = (\hat{f}_0, \hat{f}_1, \cdots, \hat{f}_{N-1}) \in C^N$ defined by

$$\hat{f}_j = \sum_{k=0}^{N-1} f_k e^{-2\pi i k j/N} \qquad j = 0, \cdots, N-1 \qquad (29)$$

The main properties of the DFT are:

- The inverse relation

$$f_k = \frac{1}{N} \sum_{j=0}^{N-1} \hat{f}_j e^{2\pi i k j/N} \qquad k = 0, \cdots, N-1 \qquad (30)$$

  (A different normalization of the DFT is sometimes used, with a factor $1/\sqrt{N}$ in both (29) and (30)).
- When extended to all integers $j \in Z$, (29) defines a periodic sequence of period N, i.e. $\hat{f}_j = \hat{f}_{j\%N}$, where $\%$ denotes the modulo operation.
- Circular translation property:

$$\widehat{T_m f}_j = e^{-2\pi i j m/N} \hat{f}_j \qquad j = 0, \cdots, N-1 \qquad (31)$$

where T_m is a circular translation by an integer m: $(T_m f)_k = f_{(k-m)\%N}$.

- The DFT of a real input vector has the symmetry $\hat{f}_j^* = \hat{f}_{N-j}$.

Numerically the DFT is calculated using the Fast Fourier Transform (FFT) algorithm,[26] which reduces the number of operation from $O(N^2)$ (for a straightforward implementation of (29)) to $O(N \log N)$. The FFT exists for any N but is most efficient when N is a power of two. The numerical efficiency is also improved when the input vector is known to be real (Fast Hartley Transform). When using FFT routines from numerical libraries, one should be careful because different routines may use different ways of indexing the components of the DFT, so that the index in the output vector generated by a given routine might not coincide with the index j in the definition (29) of the DFT.[26]

A major application of the DFT is to provide a discrete approximation of the Fourier transform. Consider a 1-D function $f(x) \in \mathcal{S}(\mathbb{R})$ which is zero outside an interval $x \in [-\Delta/2, (N-1/2)\Delta]$, and a vector of uniformly spaced samples $f^s = (f_0^s = f(0), f_1^s = f(\Delta), \cdots, f_k^s = f(k\Delta), \cdots, f_{N-1}^s = f((N-1)\Delta))$. Using the trapezoidal quadrature rule, the DFT of f^s yields a set of approximate samples of the continuous Fourier transform of f for the frequencies $v = j/(N\Delta), j = 0, \pm 1, \pm 2, \cdots, |j| \leq N/2$, as

$$\widehat{f^s}_j = \sum_{k=0}^{N-1} f_k^s e^{-2\pi i k j/N} \simeq \frac{1}{\Delta} \int_{-\Delta/2}^{(N-1/2)\Delta} f(x) e^{-2\pi i x j/(N\Delta)} dx = \frac{1}{\Delta}(\mathcal{F}f)\left(v = \frac{j}{N\Delta}\right) \tag{32}$$

Using Poisson's formula (26) one shows that the exact relation between the continuous and the discrete Fourier transforms is

$$\widehat{f^s}_j = \frac{1}{\Delta} \sum_{l \in \mathbb{Z}} (\mathcal{F}f)\left(v = \frac{j}{N\Delta} - \frac{l}{\Delta}\right) \qquad j = 0, \pm 1, \pm 2, \cdots, |j| \leq N/2 \tag{33}$$

Therefore the approximation (32) is accurate provided $(\mathcal{F}f)(v)$ is small for $|v| \geq 1/(2\Delta)$, because in that case the only significant contribution in the infinite sum (with $|j| \leq N/2$) is the term $l = 0$ (see also section 4).

3 Linear Systems

3.1 Impulse Response and Imaging Operators

For many purposes, imaging systems such as optical or electronic devices, lenses, telescopes, microscopes, medical scanners, etc ..., can be viewed as black boxes transforming an input signal into an output signal.[1,2] In optics the input signal is usually referred to as the *object*, and the output signal as the *image*. Terminology is somewhat different in medical imaging: the object is often called the image (since it will eventually be displayed on screen), whereas most of the time the output signal generated by the scanner is not an image. To avoid ambiguities, we will therefore describe the input signal as the "object" and the output signal as the "data":

$$\text{object} \rightarrow \text{imaging system} \rightarrow \text{data} \tag{34}$$

The object is represented mathematically by an element f of some Hilbert space H_o, and the data by an element g of another (possibly the same) Hilbert space H_d. The imaging system is modeled by an operator A mapping H_o onto H_d. In many applications with a continuous setting a natural choice is to take $H_o = L^2(\Omega_o)$ and $H_d = L^2(\Omega_d)$ where $\Omega_o \subset \mathbb{R}^{d_o}$ and $\Omega_d \subset \mathbb{R}^{d_d}$ are usually bounded subsets.

An imaging system is *linear* if the operator A is linear, i.e. if

$$A(c_1 f_1 + c_2 f_2) = c_1 A(f_1) + c_2 A(f_2) \qquad \text{for any } f_1, f_2 \in H_o, \ c_1, c_2 \in C \tag{35}$$

The operator is *bounded* if there is a real number $||A|| \geq 0$, called the norm of A, such that

$$||A|| = \sup_{f \in H_o} \frac{||Af||}{||f||} < \infty \tag{36}$$

The *impulse response* of a linear system is defined as the output data corresponding to an impulse object. If H_o and H_d are function spaces, the impulse response is the function $h(\mathbf{x}, \mathbf{x}_0)$ representing the data produced at point $\mathbf{x} \in \Omega_d$ by a point source $\delta(\mathbf{x} - \mathbf{x}_0)$ located at point $\mathbf{x}_0 \in \Omega_o$. A linear system is entirely defined by the set of its impulse responses for all locations $\mathbf{x}_0 \in \Omega_o$. The linear operator is then defined by

$$g(\mathbf{x}) = (Af)(\mathbf{x}) = \int_{\Omega_o} h(\mathbf{x}, \mathbf{x}_0) f(\mathbf{x}_0) \, d\mathbf{x}_0 \tag{37}$$

which is written more concisely as $g = Af$. When describing idealized models of imaging systems, the impulse response $h(\mathbf{x}, \mathbf{x}_0)$, for a fixed source position $\mathbf{x}_0$, is usually not an element of H_d, but rather a distribution.

3.2 Translation-Invariant Systems and Convolution Operators

Consider a linear imaging system such that the object and data space coincide, e.g. $H_o = H_d = L^2(\mathbb{R}^d)$. This system is invariant for translation (*shift-invariant*) if the impulse response $h(\mathbf{x}, \mathbf{x}_0)$ only depends on the vector linking the two points, i.e.

$$h(\mathbf{x}, \mathbf{x}_0) = h(\mathbf{x} - \mathbf{x}_0) \tag{38}$$

The operator is then equal to the convolution with the impulse response,

$$g(\mathbf{x}) = (Af)(\mathbf{x}) = (h * f)(\mathbf{x}) = \int_{\mathbb{R}^d} h(\mathbf{x} - \mathbf{x}_0) f(\mathbf{x}_0) \, d\mathbf{x}_0 \tag{39}$$

Two remarks are in order:

- An imaging system with a finite field-of-view (i.e. $\Omega_o \neq \mathbb{R}^d$ in (39)) is never shift-invariant. Shift-invariance holds however to a good approximation for many systems provided the field-of-view is large enough to contain the support of the object and of the data function.
- The definition can be extended to cases where the object and data spaces do not coincide. With the Radon or x-ray transform for instance, a translation of the object by a vector $\mathbf{a} \in \mathbb{R}^{d_o}$ results in a translation of the data (plane or line integrals) by a related vector $\mathbf{b} \in \mathbb{R}^{d_d}$.

The Fourier transform diagonalizes any convolution operator: indeed, by the *convolution theorem*, convolution reduces to a multiplication in the Fourier domain, between the FT (spectrum) of the object and the FT of the PSF. Equation (39) is therefore equivalent to

$$\hat{g}(\mathbf{v}) = \hat{h}(\mathbf{v})\,\hat{f}(\mathbf{v}) \tag{40}$$

The function $\hat{h}(\mathbf{v})$ is called the *transfer function*, or *modulation transfer function*, and the linear system A is then called a *filter*. A *band-limited* filter is such that $\hat{h}(v) = 0$ for $|v| \geq v_{max}$, with v_{max} the *cut-off frequency* of the system. Many imaging system are well modelled by a band-limited filter, the cut-off frequency being determined by the limited resolution of the detectors (see section 4).

3.3 Examples

- The ideal 1D low-pass filter is the convolution by the impulse response

$$h(x) = \frac{\sin 2\pi v_{max}x}{\pi x} \;\leftrightarrow\; \hat{h}(v) = B_0(v/2v_{max}) = \begin{cases} 1 & |v| < v_{max} \\ 0 & |v| \geq v_{max}. \end{cases} \tag{41}$$

- Conversely, the blurring effect of an idealized rectangular detector of width Δ is modelled by a filter with impulse response

$$h(x) = B_0(x/\Delta) = \begin{cases} 1 & |x| < \Delta/2 \\ 0 & |x| \geq \Delta/2 \end{cases} \;\leftrightarrow\; \hat{h}(v) = \frac{\sin \pi \Delta v}{\pi v}. \tag{42}$$

- The ideal circular 2D low-pass filter is the convolution by the impulse response

$$h(\mathbf{x}) = \pi v_{max}^2 \frac{J_1(2\pi|\mathbf{x}|v_{max})}{2\pi|\mathbf{x}|v_{max}} \;\leftrightarrow\; \hat{h}(\mathbf{v}) = \begin{cases} 1 & |\mathbf{v}| < v_{max} \\ 0 & |\mathbf{v}| \geq v_{max} \end{cases} \tag{43}$$

where J_1 is a Bessel function, $|\mathbf{x}| = \sqrt{x_1^2 + x_2^2}$ and $|\mathbf{v}| = \sqrt{v_1^2 + v_2^2}$.

- The *Hilbert transform* $\mathcal{H}$ plays an important role in the inversion of the Radon transform. It is defined as the convolution with the impulse response $h(x) = 1/(\pi x)$,

$$(\mathcal{H}f)(x) = \lim_{\varepsilon \to 0} \left(\int_{-\infty}^{x-\varepsilon} + \int_{x+\varepsilon}^{\infty} \right) \frac{1}{\pi(x-x_0)} f(x_0)\, dx_0 \qquad -\infty < x < \infty \quad (44)$$

where the integral is a Cauchy principal value to handle the singularity of $x = x_0$. The corresponding transfer function is $\hat{h}(v) = -i\,sgn(v)$. The inverse transform is $-\mathcal{H}$, so that $-\mathcal{H}\mathcal{H}f = f$. See [24,35] for recent applications of the Hilbert transform in tomography with limited data.

3.4 Discrete Operators, Circular Convolutions, and Zero-padding

The previous definitions can be extended to linear operators on finite-dimensional spaces. A linear system $A : \mathbb{R}^{N_o} \to \mathbb{R}^{N_d}$ is represented by a $N_d \times N_o$ matrix A. The j-th column of this matrix is the (discrete) impulse response of the system for an impulse object located at j.

The system is invariant for translation if the matrix elements $A_{j,k}$ only depend on the difference $j - k$. The output vector Af is then the discrete convolution of f with a shift-invariant impulse response. Two different discrete convolutions must be defined:

- The *non-circular convolution* corresponds to a matrix $A_{j,k} = h_{j-k}$ with discrete impulse response h_j, $j \in Z$.
- The *circular convolution* with $N_d = N_o = N$ corresponds to a matrix $A_{j,k} = h_{(j-k)\%N}$ with discrete impulse response $h_j, j = 0, \cdots, N-1$.

The circular convolution

$$g_j = (Af)_j = \sum_{k=0}^{N-1} h_{(j-k)\%N}\, f_k \quad j = 0, \cdots, N-1 \tag{45}$$

is equivalent with a non-circular discrete convolution with a periodized impulse response $\tilde{h}_j = h_{j\%N}$. The main interest of the circular convolution is that it is diagonalized by the DFT:

$$\hat{g}_k = \sum_{j=0}^{N-1} g_j\, e^{-2\pi i jk/N} = \hat{h}_k.\hat{f}_k = \sum_{j=0}^{N-1} h_j\, e^{-2\pi i jk/N}.\sum_{j'=0}^{N-1} f_{j'}\, e^{-2\pi i j'k/N} \tag{46}$$

Extension to multiple dimensions is straightforward.

In most applications the discrete convolution is used as an approximation to a continuous convolution with impulse response $h(x)$, using typically uniformly spaced samples of the input and output signal functions. Because the continuous convolution is not circular, the desired discrete operator is

$$g_j = (Af)_j = \sum_{k=0}^{N-1} h_{j-k}\, f_k \quad j = 0, \cdots, N-1 \tag{47}$$

with typically $h_j = h(j\Delta), j \in Z$, rather than the circular convolution (45). To exploit the numerically efficient diagonalization by means of the DFT, (46), we need to cast the non-circular convolution (47) in the form of a circular convolution. This is possible provided

1. The filter has a finite impulse response (FIR), i.e. there is some $L < N$ such that $h_j = 0$ unless $0 \leq j \leq L$.
2. The input discrete signal has a finite support and is appropriately *padded* with zeros, i.e. $f_j = 0, j = N - L, \cdots, N - 1$.

If these two conditions are satisfied, one easily checks that

$$\sum_{k=0}^{N-1} h_{j-k} f_k = \sum_{k=0}^{N-1} h_{(j-k)\%N} f_k \quad j = 0, \cdots, N-1 \tag{48}$$

hence the discrete convolution can be implemented using the DFT. Failure to satisfy the two above conditions often results in very severe artefacts when implementing discrete convolutions by means of the FFT.

4 Sampling

We consider in this section the problem of recovering a function $f(\mathbf{x})$ from its samples taken at points $\mathbf{x} = \mathbf{x}_j, j = 1, \cdots, N$. The major questions to be addressed are

- Does the set of samples determine f in a unique way?
- Can one determine minimum sets of samples that uniquely determine f?
- Is there an efficient *interpolation* method to recover f from its samples?
- How robust is the interpolation with respect to noise on the sample values and to uncertainties ("jitter") on the sample positions $\mathbf{x}_j$?

The answer to these questions depends on the given set of sample positions and on the class of functions H to which f belongs. Clearly, an infinite number of functions f can be found with specified values at a discrete set of sample points, and selecting – and clearly stating- to which class of functions f belongs is essential. The literature on sampling considers various classes of functions (e.g. the class of solutions to certain types of differential equations), different sets of sample locations (equidistant or not, random or not, located on both sides or only on one side of the point where f must be interpolated, interlaced sampling grids, etc.), or different type of samples (e.g. the function f or both f and its derivative at the sample points, etc.). In addition, when the sample values are noisy, the "best" estimate of f might not be one that exactly coincides with the samples, but rather one that only approximately coincides with the samples. See [29] for an overview.

Here we only summarize the most classical theorems on sampling, and restrict the discussion to:

1. The class of band-limited functions

$$L^2_{v_{max}}(\mathbb{R}^d) = \left\{ f \in L^2(\mathbb{R}^d) | \hat{f}(\mathbf{v}) = 0 \text{ when } |\mathbf{v}| > v_{max} \right\} \tag{49}$$

2. Uniformly spaced samples (in 1D) or samples located on a regular grid (in d−D)
3. A set of samples which covers the whole region where $f(\mathbf{x}) \neq 0$

These conditions are never strictly satisfied because a function with bounded support cannot be band-limited (Paley-Wiener theorem), and hence the first and third conditions are incompatible if the set of sample points is to be kept finite. Often, however, the measured samples are those of a function that is "seen" through a detecting device with an essentially band-limited impulse response, e.g. (for a shift-invariant system), $f = f_0 * h$ where f_0 denotes the modelled physical quantity, and h the impulse response of the detector, with $\hat{h}(\mathbf{v}) \simeq 0$, and therefore also $\hat{f}(\mathbf{v}) \simeq 0$, when $|\mathbf{v}| > v_{max}$.

4.1 The Sampling Theorem of Whittaker-Shannon-Kotelnikov

A function $f \in L^2_{v_{max}}(\mathbb{R})$ is uniquely determined by its samples $f(k\Delta), k \in Z$, where $\Delta = 1/(2v_{max})$ is the Nyquist distance. The function can be recovered using the Whittaker-Shannon-Kotelnikov interpolation formula

$$\hat{f}(v) = 0 \text{ for } |v| > v_{max} = \frac{1}{2\Delta} \Rightarrow f(x) = \sum_{k=-\infty}^{\infty} f(k\Delta) \operatorname{sinc}((x/\Delta) - k) \quad (50)$$

where the interpolating function $\operatorname{sinc}(x) = \sin(\pi x)/(\pi x)$ is called the cardinal sine, and has the orthonormality property

$$\int_{-\infty}^{\infty} \operatorname{sinc}(x - k) \operatorname{sinc}(x - j) \, dx = \delta_{j,k} \qquad j, k \in Z \quad (51)$$

Due to the slow decay of the sinc function at infinity, an impractically large number of terms must be kept in the series (50) to recover f with a given accuracy at any point x. This is why the Shannon series is rarely used in its original form.

4.2 Undersampling and Aliasing

If the Shannon interpolation (50) is applied to the samples of a function f that is not band-limited with cut-off frequency $v_{max} = 1/2\Delta$, the interpolated function $f_I(x)$ is not equal to f. Using the Poisson formula (26), the error can be characterized in terms of the Fourier transforms $\hat{f}_I$ and $\hat{f}$ as

$$\hat{f}_I(v) = \sum_{k=-\infty}^{\infty} f(k\Delta) \operatorname{sinc}((x/\Delta) - k) = \begin{cases} \hat{f}(v) + \sum_{k=-\infty, k\neq 0}^{\infty} \hat{f}(v - \frac{k}{\Delta}) & |v| < v_{max} \\ 0 & |v| \geq v_{max} \end{cases}$$

$$(52)$$

This equation shows that the interpolation error does not only affect the high frequencies $|v| \geq v_{max}$, but also the low frequency components, which are corrupted by the repetitions of the spectrum of f at shifted frequencies $v - k/\Delta$, $k = \pm 1, \pm 2, \cdots$. This phenomenon is called *aliasing* and appears whenever a signal is undersampled.

4.3 Oversampling

One speaks of *oversampling* when a band-limited function $f \in L^2_{v_{max}}(\mathbb{R})$ is sampled with a sampling step smaller than the Nyquist distance, i.e. when $\Delta < 1/(2v_{max})$. The interpolation function is no longer unique in this case, and $f \in L^2_{v_{max}}(\mathbb{R})$ can be recovered exactly as

$$v_{max} < \frac{1}{2\Delta} \text{ and } \hat{f}(v) = 0 \text{ for } |v| > v_{max} \Rightarrow f(x) = \sum_{k=-\infty}^{\infty} f(k\Delta)\, w(x - k\Delta) \quad (53)$$

where $w(x)$ is any function with a Fourier transform satisfying

$$\hat{w}(v) = \begin{cases} 1 & |v| \leq v_{max} \\ 0 & |v| \geq 1/\Delta - v_{max} \end{cases} \quad (54)$$

and taking arbitrary values in the interval $v_{max} < v < 1/\Delta - v_{max}$. If the values in this interval are such that $\hat{w}(v)$ has $n-1$ continuous derivatives, the interpolating function $w(x) = (\mathcal{F}^{-1}\hat{w})(x)$ decays as $|x|^{-(n+1)}$ at large $|x|$. As a result, the number of terms required to recover $f(x)$ with a prescribed accuracy can be much smaller than with the Shannon interpolation. An example of a family of interpolation functions satisfying this requirement is

$$w(x) = \text{sinc}\left(\frac{x}{\Delta}\right) \text{sinc}^n\left((\frac{1}{\Delta} - 2v_{max})\frac{x}{n}\right) \quad n = 1, 2, 3, \cdots \quad (55)$$

which decays as $|x|^{-(n+1)}$. This function reduces to sinc when $v_{max} = 1/(2\Delta)$ and this is the only case where $w(x)$ satisfies the orthonormality relation (51).

4.4 Multidimensional Sampling

A multi-dimensional function $f \in L^2(\mathbb{R}^d)$ is said to be band-limited if its Fourier transform vanishes outside some bounded subset $K \subset \mathbb{R}^d$. Thus one defines the space of functions

$$L^2_K(\mathbb{R}^d) = \{f \in L^2(\mathbb{R}^d) | (\mathcal{F}f)(\mathbf{v}) = 0, \mathbf{v} \notin K\} \quad (56)$$

The simplest way to sample a multi-dimensional band-limited function $f(\mathbf{x}), \mathbf{x} \in \mathbb{R}^d$ consists in applying Shannon's interpolation to each component $x_i, i = 1, 2, \cdots, d$ separately. This can be done by enclosing the support K of $\hat{f}$ into an hyper-cube of sufficiently large side $2v_{max}$. Then f can be recovered from its samples on a cubic grid with sampling step $\Delta \leq 1/(2v_{max})$ as

$$f(\mathbf{x}) = \sum_{\mathbf{k} \in \mathbf{Z}^d} f(\mathbf{k}\Delta)\, \mathbf{w}(\mathbf{x} - \mathbf{k}\Delta) \tag{57}$$

where $\mathbf{k} = (k_1, \cdots, k_d)$ is a d-dimensional vector of integers, and

$$\mathbf{w}(\mathbf{x} - \mathbf{k}\Delta) = \prod_{j=1}^{d} w(x_j - k_j \Delta) \tag{58}$$

where w is an interpolation function satisfying the conditions in previous section, e.g. (55). Generalization for the case where K is enclosed within an hyper-parallelipiped (i.e. v_{max} depends on the axis j) is straightforward.

If the support K of the spectrum of a band-limited function is not a paralleliped, the previous approach with samples on a cubic grid is not optimal in the sense that the number of samples required, per unit volume (in $\mathbb{R}^d$) is larger than needed to allow exact recovery of f. More efficient interpolation formulae can be derived using the following theorem due to Petersen and Middleton (see e.g.[22,23]).

- Let W be a $d \times d$ non-singular matrix, and assume that the $d-$th dimensional signal $f(\mathbf{x})$ is sampled on a grid $\mathbf{x_j} = W\mathbf{j}$, with $\mathbf{j} = (j_1, \cdots, j_d) \in Z^d$. If the sets

$$K_L = K + (W^{-1})^t \mathbf{l} \qquad\qquad \mathbf{l} \in Z^d \tag{59}$$

are disjoint, exact interpolation is possible as

$$f(\mathbf{x}) = \sum_{\mathbf{j} \in Z^d} f(W\mathbf{j})\, w_K(\mathbf{x} - W\mathbf{j}) \det W \tag{60}$$

where w_K is the $d-$th dimensional inverse Fourier transform of the indicator function of the set K:

$$w_K(\mathbf{x}) = \int_K \exp(2\pi i x \cdot \boldsymbol{v})\, d\boldsymbol{v} \tag{61}$$

If the support of $\mathcal{F}f$ is a smaller domain K_0 strictly contained within K there is oversampling, and a technique similar to the one described above (section 4.3) can be used to design an interpolation function $w(\mathbf{x})$ with fast decay at infinity.

Example : In the absence of privileged direction, one expects a natural 2D band-limited image to have its spectrum restricted to a disc $K = \{\boldsymbol{v} \in \mathbb{R}^2 | \sqrt{v_1^2 + v_2^2} \leq v_{max}\}$ (Fig. 1). We take $v_{max} = 1/2$ for simplicity. The standard sampling on a regular square grid corresponds to the matrix

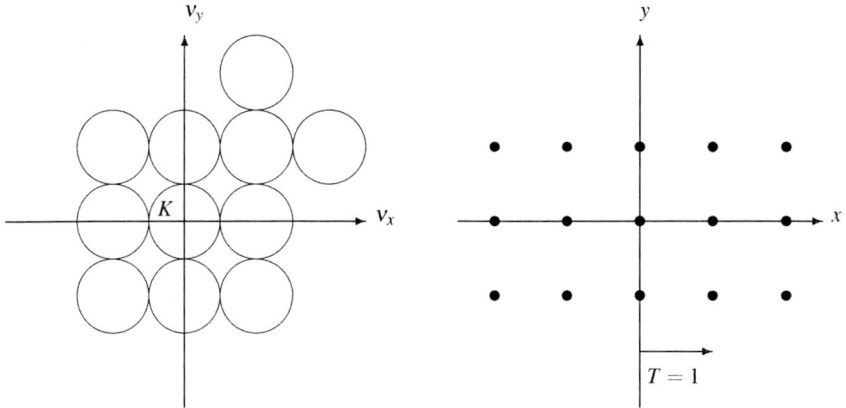

Fig. 1 Square sampling grid for a 2D image with a spectrum band-limited in the centered disk K of radius $v_{max} = 1/2$. Left: non-overlapping replica of the band K on a square grid. Right: corresponding square grid of samples in image space, with uniform spacing $T = 1$

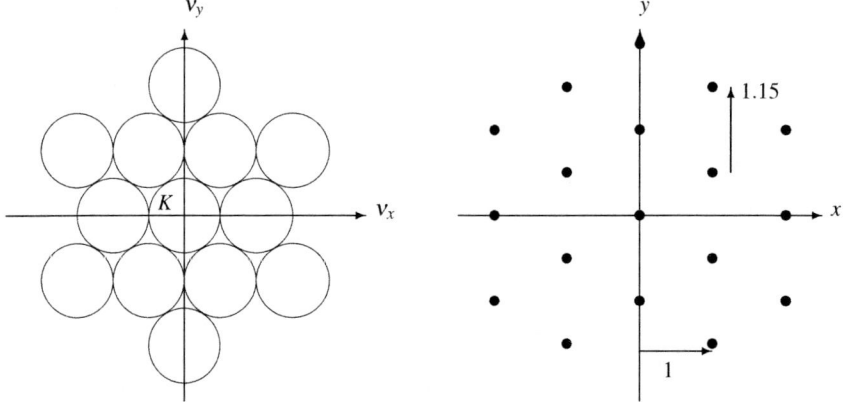

Fig. 2 Interlaced sampling grid for a 2D image with a spectrum band-limited in the centered disk K of radius $v_{max} = 1/2$. Left: non-overlapping replica of the band K on a hexagonal grid. Right: corresponding dual hexagonal grid of samples in image space

$$W = \begin{pmatrix} \Delta & 0 \\ 0 & \Delta \end{pmatrix} \tag{62}$$

To allow exact interpolation the sets $K + (W^{-1})^t \mathbf{l}, \, \mathbf{l} \in Z^2$ must not overlap; hence the sampling condition is $\Delta \le 1$.

Applying the theorem of Petersen and Middleton, one checks that *hexagonal sampling* with Fig. 2

$$(W^{-1})^t = \begin{pmatrix} 1 & \cos\frac{\pi}{3} \\ 0 & \sin\frac{\pi}{3} \end{pmatrix} \tag{63}$$

also ensures that the shifted copies of the disc K do not overlap. The corresponding sampling grid is defined by transposing and inverting this matrix:

$$W = \begin{pmatrix} 1 & 0 \\ -\cot\frac{\pi}{3} & 1/\sin\frac{\pi}{3} \end{pmatrix} \tag{64}$$

and the interpolating function is then the 2D ideal low-pass filter (43).

The number of sampling points per unit area is equal to the determinant of W^{-1}:

- $\det W^{-1} = 1$ for standard square sampling
- $\det W^{-1} = \sin\pi/3 < 1$ for the hexagonal sampling

Thus, the hexagonal sampling pattern reduces the number of samples by a factor $\sqrt{3}/2$, without loss of accuracy.

For a 3D band-limited function $f(x,y,z)$ with spherical band

$$K = \{v \in \mathbb{R}^3 | \sqrt{v_1^2 + v_2^2 + v_3^2} \leq v_{max} = 1/2\}, \tag{65}$$

the *body-centered cubic* (BCC) sampling grid defined by the matrix

$$W = \begin{pmatrix} \sqrt{2} & 0 & 1/\sqrt{2} \\ 0 & \sqrt{2} & 1/\sqrt{2} \\ 0 & 0 & 1/\sqrt{2} \end{pmatrix} \tag{66}$$

allows exact interpolation, with $\det W^{-1} = 1/\sqrt{2}$ samples per unit volume compared to $\det W^{-1} = 1$ for a cubic sampling grid. Optimal sampling schemes have also been proposed for the Radon transform.[22,23] We stress finally that the "optimal" sampling grids require using the appropriate interpolating function (61). Failure to do so, using e.g. linear interpolation, may result in significant errors. In addition, for some problems such as the Radon transform, the optimal grid is less robust to violations of the assumption that the function is strictly band-limited.

4.5 Function Approximation

Interpolation is not necessarily the best way to estimate a function from noisy samples, because it makes obviously little sense to fit exactly sampled values which are known with a limited accuracy. Approximation may then be preferred to interpolation. We conclude this section on sampling with a note on the problem of approximating a function $f \in L^2(\mathbb{R}^d)$ as a linear combination f_r of functions $\Psi(\mathbf{x} - \mathbf{k}\Delta), \mathbf{k} \in Z^d$ centered in the nodes of a cubic grid of side Δ:

$$f_r(\mathbf{x}) = \sum_{\mathbf{k} \in Z^d} f_{\mathbf{k}} \Psi(\mathbf{x} - \mathbf{k}\Delta) \tag{67}$$

The coefficients $f_{\mathbf{k}}$ which minimize the L^2 error

$$\int_{\mathbb{R}^d} |f(\mathbf{x}) - f_r(\mathbf{x})|^2 \tag{68}$$

are given by

$$f_{\mathbf{k}} = (\xi * f)(\mathbf{x} = \mathbf{k}\Delta) = \int_{\mathbb{R}^d} \xi(\mathbf{k}\Delta - \mathbf{x}) f(\mathbf{x}) d\mathbf{x} \tag{69}$$

where $*$ denotes the d-dimensional convolution, and the analysis function ξ associated to Ψ is defined by its d-dimensional Fourier transform,[29]

$$(\mathcal{F}\xi)(\mathbf{v}) = \Delta^d \frac{(\mathcal{F}\Psi)^*(\mathbf{v})}{\sum_{\mathbf{k}\in Z^d} |(\mathcal{F}\Psi)(\mathbf{v} - \mathbf{k}/\Delta)|^2} \tag{70}$$

The two sets of functions $\{\Psi(\mathbf{x} - \mathbf{k}\Delta), \mathbf{k} \in Z^d\}$ and $\{\xi(\mathbf{x} - \mathbf{k}\Delta), \mathbf{k} \in Z^d\}$ satisfy the *bi-orthonormality* relation

$$\int_{\mathbb{R}^d} \Psi(\mathbf{x} - \mathbf{k}\Delta) \xi(\mathbf{x} - \mathbf{k}'\Delta) d\mathbf{x} = \delta_{\mathbf{k},\mathbf{k}'} = \begin{cases} 1 & \mathbf{k} = \mathbf{k}' \\ 0 & \mathbf{k} \neq \mathbf{k}' \end{cases} \tag{71}$$

As a special case, one easily checks using (70) that the Shannon interpolating function $\Psi(x) = \text{sinc}(x)$ coincides with its analysis function, $\xi(x) = \Psi(x)$.

5 Linear Inverse Problems

5.1 Terminology

Many situations in medical imaging require estimating a physical quantity f from measurements g collected by some system such as a scanner. This system is modelled mathematically by some operator A mapping the *object space* H_o onto the *data space* H_d and the problem is then represented by the equation

$$g = A(f) \tag{72}$$

Recovering f from g is called an *inverse problem*. In general the operator A is not linear, because measuring devices involve non-linear effects such as the detector dead-time (but also, for example, the Lambert law in radiology or the random coincidences in positron emission tomography). Whenever possible, the data are preprocessed and/or some effects are neglected to obtain a simplified model where the operator A is linear. Even for highly non-linear problems (e.g. in magnetoencephalography, electroencephalography, or bio-medical optics[30]), iterative methods are often used which, at each iteration, apply some kind of linearization. The discussion below is restricted to *linear inverse problems*.

As discussed previously, the model of the inverse problem must also specify the spaces H_o and H_d to which the object f and the data g belong. Three types of inverse problems can be considered: [1,3,4]

1. *Continuous-Continuous (CC) problems*: the object and the data are both represented by functions of one or several real variables, and the object and data spaces are typically Hilbert spaces such as $L^2(\mathbb{R}^d)$.
2. *Continuous-Discrete (CD) problems*: the object is represented by a function but the data are represented by a vector $g \in \mathbb{R}^{N_d}$.
3. *Discrete-Discrete (DD) problems*: the object and the data are both represented by vectors with dimensions N_o and N_d, and the operator A is a $N_d \times N_o$ matrix.

In almost all applications the data are digital, while the physical quantity f to be estimated is inherently continuous. From this perspective, the CD description appears the most natural. For practical reasons however, the CC and DD descriptions are used more frequently:

- The CC description allows the system to be modelled by mathematical operators, such as the Radon transform for tomography, for which inversion formulae and other relevant properties can be obtained using functional analysis. Analytical results, however, are only available for operators which correspond to simplified models of the system, implying that many physical effects must be neglected or pre-corrected. In addition, numerical implementation requires interpolating the digital data to approach the data function g, which is assumed continuous in the CC approach. Despite these drawbacks, the CC approach provides valuable information over general properties of the inverse problem concerning for instance the uniqueness or stability of the solution.
- The DD model appears natural in the sense that the actual data are discrete, and that data processing must ultimately be digitized. In contrast with the CC description, the DD approach does not restrict modelling to well-known (or at least mathematically tractable) operators, and in principle any linear effect can be incorporated in the matrix A. Monte-Carlo simulation techniques can be used to calculate the element of the matrix A with an accurate modeling of the scanner (see [15] for a recent example in SPECT). With large dimensional problems however, the quest for numerical efficiency often forces to simplify the model, for instance to ensure that the matrix A is sparse or diagonally dominant. Another difficulty with the DD approach is the necessity to enforce a priori a discrete representation of an object, which, physically, is continuous.

5.2 Well-Posed and Ill-Posed Problems

An inverse problem modelled by the equation $g = Af, f \in H_o, g \in H_d$ is said to be *well-posed* if the three following conditions are satisfied

1. *Existence of a solution.* For any $g \in H_d$, there is a solution $f \in H_o$ such that $g = Af$.

2. *Uniqueness of the solution.* For any $f_1, f_2 \in H_o, Af_1 = Af_2 \Rightarrow f_1 = f_2$.
3. *Stability in the presence of noise.* A small perturbation of the data g should only cause a small perturbation of the solution.

The stability condition is translated mathematically as follows. Denotes by $g_0 = Af_0$ the ideal, noise-free data corresponding to some object f_0. Consider the solution $f = f_0 + \delta f$ which corresponds to data $g = g_0 + \delta g$ corrupted by measurement noise δg. *Stability* then means that there exists some constant C such that the reconstruction error is bounded by an inequality $||\delta f||_{H_o} \leq C ||\delta g||_{H_d}$. Recall that $||.||_H$ denotes the norm in the Hilbert space H.

The three conditions for well-posedness have been defined by Hadamard in 1917. When one of the three conditions does not hold, the inverse problem is *ill-posed*. A regrettable law of nature dictates that the majority of practical inverse problem are ill-posed. Examples include the image reconstruction problems in emission and transmission tomography, the non-linear problems in MEG and EEG, the super-resolution problems (PSF deblurring), and even the simple problem of estimating the derivative of a signal.

Example: Integral equations of the first type with integrable impulse response

A large class of imaging inverse problems can be represented by an integral operator (see section 3.1):

$$A : L^2(\mathbb{R}^{d_o}) \rightarrow L^2(\mathbb{R}^{d_d}) \; : \; f(\mathbf{y}) \rightarrow g(\mathbf{x}) = (Af)(\mathbf{x}) = \int_{\mathbb{R}^{d_o}} h(\mathbf{x}, \mathbf{y}) f(\mathbf{y}) \, d\mathbf{y} \quad (73)$$

with an impulse response $h(\mathbf{x}, \mathbf{y})$ that is integrable, i.e.

$$\int_{\mathbb{R}^{d_o}} |h(\mathbf{x}, \mathbf{y})| \, d\mathbf{y} < \infty \quad \text{for every } \mathbf{x} \quad (74)$$

The associated integral equation of the first kind, $g = Af$, is an ill-posed problem for most reasonable choices of the norms in the object and data spaces. This is easily seen because the lemma of Riemann-Lebesgue tells us that

$$\lim_{n \to \infty} \int_{\mathbb{R}} h(y) \sin(ny) \, dy = 0 \quad (75)$$

for any integrable function $h(y)$.

Example: Deconvolution problems

Deconvolution problems are also represented by (89), now with $d_o = d_d$ and a shift-invariant impulse response $h(\mathbf{x}, \mathbf{y}) = h(\mathbf{x} - \mathbf{y})$, here assumed to be integrable. The corresponding equation $g = Af$ is diagonalized by the Fourier transform (see (40)), and can formally be solved as $\hat{f}(\mathbf{v}) = \hat{g}(\mathbf{v})/\hat{h}(\mathbf{v})$. Ill-posedness is then apparent because the Fourier transform of any integrable impulse response $h(\mathbf{x})$ decays towards zero at large frequencies, and there is therefore no upper bound on the noise amplification caused by the division by $\hat{h}(\mathbf{v})$.

Example: Continuous-discrete inverse problem

A linear CD inverse problems is described by a bounded linear operator A mapping some object space H_o onto $\mathbb{R}^{N_d}$, where as above N_d is the total number of discrete data collected by the scanner. Typically, H_o is a Hilbert space of functions of d_o real variables, e.g. $H_o = L^2(\mathbb{R}^{d_o})$:

$$A : H_o \to \mathbb{R}^{N_d} \ : \ f(\mathbf{x}) \to g = Af = \left(g_1 = (Af)_1, \cdots, g_{N_d} = (Af)_{N_d}\right) \quad (76)$$

For each data $j = 1, \cdots, N_d$, the mapping $f \to (Af)_j$ is a bounded linear functional on H_o, and by the Riesz theorem, there is an element $\psi_j \in H_o$ such that $(Af)_j = (f, \psi_j)$, where $(.,.)$ is the scalar product in H_o. Thus, if $H_o = L^2(\mathbb{R}^{d_o})$, each discrete data is associated to some function $\psi_j(\mathbf{x})$ such that

$$(Af)_j = (f, \psi_j) = \int_{\mathbb{R}^{d_o}} f(\mathbf{x}) \, \psi_j(\mathbf{x}) \, d\mathbf{x} \quad (77)$$

Physically, the function $\psi_j(\mathbf{x})$ models the sensitivity profile of the "detector" number j. A simple example is that of a signal (or image) measured by an array of identical detectors centered on a grid $\mathbf{x}_j, j = 1, \cdots, N_d$ and characterized by a response function $h(\mathbf{x} - \mathbf{x}_j)$. This inverse problem amounts to recovering f from samples of $f * h$ on this grid.

The adjoint of the CD operator (76) is:

$$A^* : \mathbb{R}^{N_d} \to H_o \ : \ g = (g_1, \cdots, g_{N_d}) \to (A^* g)(\mathbf{x}) = \sum_{j=1}^{N_d} g_j \, \psi_j(\mathbf{x}) \quad (78)$$

The functions $\psi_j(\mathbf{x})$ are sometimes called *natural pixels* for some reconstruction algorithms which define approximate solutions of the form $A^* r$ for some $r \in \mathbb{R}^{N_d}$.

Since a function in an infinite-dimensional function space cannot be determined by a finite number of real numbers, the solution to the equation $g = Af$ is not unique and the CD inverse problem is necessarily ill-posed.

5.3 Existence: Pseudo-Solutions of an Inverse Problem

The first condition of Hadamard concerns the existence of the solution to the equation $g = Af$.

Existence of a solution for any data g is guaranteed if the operator A is surjective, that is, if the *range* of the operator,

$$\text{range}(A) = A(H_o) = \{g = Af \,|\, f \in H_o\} \quad (79)$$

coincides with the data space, i.e. $\text{range}(A) = H_d$. Note that the data space must be large enough to contain any data g that could be generated by the measurement, including data corrupted by noise. This is why the existence problem cannot in general be avoided by simply defining the data space as $H_d = \text{range}(A)$. The range of the operator can often be characterized explicitly by a set of conditions which must be satisfied by any "consistent" (noise-free) data, i.e. by any $g \in \text{range}(A)$.

These conditions are called *range conditions*, or *consistency conditions*, and they reflect the fact that the data are to some extent redundant. Consider as an example a DD inverse problem, where A is a $N_d \times N_o$ matrix. Assume that the rows of A are not linearly independent, but satisfy a set of R relations

$$\sum_{k=1}^{N_d} c_{r,k} A_{k,j} = 0 \quad j = 1, \cdots, N_o \quad r = 1, \cdots, R. \tag{80}$$

Then, the data consistency conditions are simply

$$\sum_{k=1}^{N_d} c_{r,k} g_k = 0 \quad r = 1, \cdots, R. \tag{81}$$

Another example, for a CC problem, are the consistency conditions of Helgason-Ludwig for the 2D Radon transform.[22]

In some applications where the operator $A = A_{p_1, p_2, \cdots, p_m}$ depends on a small number m of unknown parameters $p_1, p_2, \cdots, p_m$ (e.g. geometric calibration parameters), the consistency conditions can be exploited to estimate these parameters, provided the noise level is not too high. This is achieved by finding the values of the parameters $p_1, p_2, \cdots, p_m$ for which value of the measured data belong to range$(A_{p_1, p_2, \cdots, p_m})$. An application to attenuation correction in emission tomography can be found in.[32]

Measured data g corrupted by noise do not in general belong to range(A), and in that case the equation $g = Af$ does not admit any solution. The concept of an "exact" solution must be abandoned. An approximate solution is defined as follows. The data g are replaced by modified data $\tilde{g}$ that are consistent (i.e. $\tilde{g} \in$ range(A)) and that are in some sense "close" to g. The standard approach consists in projecting g onto range(A), i.e.

$$\tilde{g} = \arg \min_{g' \in \text{range}(A)} ||g' - g||^2 \tag{82}$$

A *pseudo-solution* of the inverse problem is then defined as a solution to the equation $\tilde{g} = A\tilde{f}$. Such a solution is also called a *least-squares* solution when working with L^2 spaces, because (82) can be rewritten as

$$\tilde{f} = \arg \min_{f' \in H_o} ||Af' - g||^2 = \arg \min_{f' \in H_o} \int_{\mathbb{R}^{d_o}} |(Af')(\mathbf{x}) - g(\mathbf{x})|^2 \, d\mathbf{x} \tag{83}$$

This corresponds to a finding a $\tilde{f}$ which best fits the measured data. The functional $||Af - g||$ is often called the *residual error*. In a stochastic framework the least-squares solution corresponds to a maximum-likelihood solution for a gaussian model of the noise (see section 6.4).

By computing the Euler-Lagrange equations for the minimization of the functional $||Af - g||^2$, one finds that the pseudo-solutions of $Af = g$ are the solutions to the *normal equation* associated to the problem,

$$A^* A f = A^* g \tag{84}$$

where A^* is the adjoint of the operator A.

5.4 Uniqueness: The Generalized Inverse

The second condition of Hadamard concerns the uniqueness of the solution to the equation $g = Af$.

Uniqueness of the solution (if it exists) is guaranteed if the operator A is injective, that is, if the *null-space* of the operator,

$$\mathcal{N}(A) = \{f \in H_o \,|\, Af = 0\} \tag{85}$$

is trivial, i.e. $\mathcal{N}(A) = \{0\}$. When the null-space is non-trivial, the measured data $g = Af$ do not provide sufficient information to determine in a unique way the solution f, or the pseudo-solution $\tilde{f}$. Indeed, if $\tilde{f}$ minimizes $||Af' - g||$ and if $f_N \in \mathcal{N}(A)$, then $||A\tilde{f} - g|| = ||A(\tilde{f} + f_N) - g||$ and therefore $\tilde{f} + f_N$ is also a pseudo-solution. An object $f_N \in \mathcal{N}(A)$ is sometimes said to be *transparent* because it is unseen by the measuring system modeled by the operator A.

Some remarks are in order.

- The *null-space* depends on the definition of the object space H_o, a definition which should reflect our prior knowledge of the problem. For example, when the problem of image reconstruction from projections is modeled using the 2D Radon transform with limited angular acceptance, uniqueness of the solution is guaranteed if the object is assumed to have a compact support, but not otherwise.
- When A is not injective, the data g may still convey sufficient information to determine in a unique way the object $f(\mathbf{x})$ in some restricted region of interest, or the value of some specific functionals of f. Recent developments in tomography provide interesting examples of such situations.[24,35]

When the operator is not injective, the pseudo-solution to the inverse problem is not unique. Indeed, the residual error $||Af - g||$ is minimized by any f in the set

$$S_g = \tilde{f} + \mathcal{N}(A) = \{f \in H_o \,|\, f = \tilde{f} + f_N \,, f_N \in \mathcal{N}(A)\} \tag{86}$$

where $\tilde{f}$ is any particular pseudo-solution. If no additional prior information is available, one selects in this set the element of minimum norm. This unique element is denoted by f^+ and is called the *generalized solution*, or *Moore-Penrose solution* of the inverse problem $g = Af$:

$$f^+ = \arg \min_{f' \in S_g} ||f'||_{H_o} \tag{87}$$

The rationale for this choice is that any pseudo-solution $\tilde{f}$ can be decomposed in a unique way as $\tilde{f} = f_v + f_N$ where $f_v \in (\mathcal{N}(A))^\perp$ and $f_N \in \mathcal{N}(A)$. If no additional prior information is available to partly or fully determine f_N, it appears logical to take $f_N = 0$, and therefore to select f_v as our solution. As $||\tilde{f}||^2 = ||f_v||^2 + ||f_N||^2$, taking $f_N = 0$ amounts to selecting the pseudo-solution of minimum norm, so $f_v = f^+$.

The operator that maps the data $g \in H_d$ onto the generalized solution is called the *generalized inverse* of A, and is denoted A^+:

$$f^+ = A^+ g \tag{88}$$

5.5 The Singular Value Decomposition of a Compact Operator

Consider two Hilbert spaces H_o and H_d and a bounded linear operator $A : H_o \rightarrow H_d$. The operator is *compact* if, for each bounded sequence $(f_1, f_2, \cdots, f_i, \cdots)$ of elements of H_o, it is possible to extract a subsequence $(f_{i_1}, f_{i_2}, \cdots, f_{i_j}, \cdots)$ such that the corresponding sequence of data $(A f_{i_1}, A f_{i_2}, \cdots, A f_{i_j}, \cdots)$ converges in the norm of H_d. Examples of compact operators include

- Integral operators with square-integrable impulse response $h(\mathbf{x}, \mathbf{y})$,

$$A : L^2(\mathbb{R}^{d_o}) \rightarrow L^2(\mathbb{R}^{d_d}) \; : \; f(\mathbf{y}) \rightarrow g(\mathbf{x}) = (Af)(\mathbf{x}) = \int_{\mathbb{R}^{d_o}} h(\mathbf{x}, \mathbf{y}) f(\mathbf{y}) \, d\mathbf{y} \tag{89}$$

with

$$\int_{\mathbb{R}^{d_o}} |h(\mathbf{x}, \mathbf{y})|^2 \, d\mathbf{y} < \infty \tag{90}$$

Note that this condition cannot be satisfied if $h(\mathbf{x}, \mathbf{y}) = h_c(\mathbf{x} - \mathbf{y})$, and therefore convolution operators are not compact. However, they usually become so if the domain is restricted to a bounded region $\Omega \subset \mathbb{R}^{d_0}$, with

$$h(\mathbf{x}, \mathbf{y}) = \begin{cases} h_c(\mathbf{x} - \mathbf{y}) & \mathbf{x} \in \Omega, \mathbf{y} \in \Omega \\ 0 & \text{otherwise} \end{cases} \tag{91}$$

In this case of course the Fourier transform no longer diagonalizes the operator.
- All operators with finite-dimensional ranges, hence all DD and CD inverse problems have compact A. All matrices, in particular, are compact operators.
- Limits of sequences A_k, $k = 1, 2, \ldots$ of compact operators A_k (e.g. belonging to one of the two above categories), since

$$\lim_{k \to \infty} \|Af - A_k f\|_{H_d} = 0 \quad \forall f \in H_o \text{ and } A_k \text{ compact} \Rightarrow A \text{ compact} \tag{92}$$

The Radon transform on a bounded spatial domain can be shown to be compact.

A compact operator admits a *singular value decomposition* (SVD), characterized by

- A finite set or a sequence of elements $u_k \in H_o$, $k = 1, 2, \cdots, R$, called the *singular vectors in object space*,
- A finite set or a sequence of elements $v_k \in H_d$, $k = 1, 2, \cdots, R$, called the *singular vectors in data space*,

- A finite set or a sequence of real *singular values*, strictly positive by convention, which are ordered by decreasing magnitudes $\sigma_1 \geq \sigma_2 \geq \cdots \geq \sigma_R > 0$

and such that the following properties hold:

1. The (u_k) form an orthonormal basis of $\mathcal{N}(A)^\perp$, with $(u_k, u_j) = \delta_{k,j}, j, k = 1, \cdots, R$,
2. The (v_k) form an orthonormal basis of range(A), with $(v_k, v_j) = \delta_{k,j}, j, k = 1, \cdots, R$,
3. $Au_k = \sigma_k v_k, \ k = 1, \cdots, R$
4. $A^* v_k = \sigma_k u_k, \ k = 1, \cdots, R$
5. The u_k are eigenelements of the operator A^*A, with eigenvalue σ_k^2
6. The v_k are eigenelements of the operator AA^*, with eigenvalue σ_k^2

For a CC problem, the rank R is either finite or infinite. If the rank is infinite, then $\lim_{k\to\infty} \sigma_k = 0$. The rank is always finite for DD and CD problems. For a DD problem, $R \leq \min\{N_o, N_d\}$. For a CD problem, $R \leq N_d$, and the σ_k^2 can be determined as the non-zero eigenvalues of the $N_d \times N_d$ matrix AA^*.

The generalized solution of the equation $Af = g$ for a compact linear operator can be written as

$$f^+ = \sum_{k=1}^{R} \frac{1}{\sigma_k} (g, v_k) u_k \tag{93}$$

For an infinite-rank CC operator, the generalized solution exists only for those data g such that

$$\sum_{k=1}^{\infty} \frac{1}{\sigma_k^2} |(g, v_k)|^2 < \infty \tag{94}$$

5.6 The Landweber Iterative Algorithm

Inverse problems are sometimes solved using iterative algorithms. We describe here the Landweber algorithm for the calculation of the generalized solution of an inverse problem $Af = g$ with a bounded linear operator A. Faster algorithms exist, but the Landweber algorithm is linear and numerically stable, and it gives a good illustration of properties common to other iterative methods.

The Landweber iteration produces successive estimates $f_0, f_1, \cdots, f_n, f_{n+1}, \cdots$ of the solution, defined by

$$f_0 = 0$$

Landweber iteration: $f_{n+1} = f_n + \tau A^*(g - Af_n)$ $n = 0, 1, 2, \cdots$ (95)

If the *relaxation parameter* τ satisfies $0 < \tau < 2/||A||^2$, the sequence of iterates converges to the generalized solution: $\lim_{n\to\infty} f_n = f^+$. If the initial estimate $f_0 \neq 0$, the iteration converges to the pseudo-solution that is closest (in the norm of H_o) to f_0.

Note from (95) that $f_n \in f_0 + \text{range}(A^*) = f_0 + \mathcal{N}(A)^{\perp}$. For a compact operator, each Landweber iterate can therefore be represented on the singular basis. Using the results in section 5.5, one checks that the coefficient of the n-th iterate on the singular vector u_k is (for $f_0 = 0$),

$$(f_n, u_k) = \frac{1 - (1 - \tau\sigma_k^2)^n}{\sigma_k} (g, v_k) \quad k = 1, \cdots, R \quad n = 1, 2, \cdots \quad (96)$$

When $0 < \tau < 2/\sigma_1^2 = 2/||A||^2$, this converges for each k to

$$\lim_{n \to \infty} (f_n, u_k) = \frac{1}{\sigma_k} (g, v_k) = (f^+, u_k) \quad (97)$$

As can be seen from (96), the convergence speed is governed by the quantity $|1 - \tau\sigma_k^2|$: convergence is fast when $|1 - \tau\sigma_k^2|$ is small, and slow if $|1 - \tau\sigma_k^2|$ approaches 1. If the rank R is finite, the following choice for the relaxation parameter,

$$\tau = \frac{2}{\sigma_1^2 + \sigma_R^2} \quad (98)$$

maximizes the speed of convergence of the slowest converging component. If the rank is infinite, $\sigma_k \to 0$ as $k \to \infty$ and $|1 - \tau\sigma_k^2|$ takes values arbitrarily close to 1. The singular components corresponding to small singular values are the slowest to converge. For most operators, these components correspond to rapidly oscillating singular vectors u_k, i.e. those components that convey information over the smallest details of the object.

6 Regularization of Linear Inverse Problems

The third criterion for well-posedness (section 5.2) is that the solution of the inverse problem should be stable in the presence of noise. Even after overcoming the problems related to the existence and uniqueness, stability remains a major issue because, for many infinite-dimensional problems, the generalized inverse operator A^+ is unbounded. A small error on the data g can then cause an arbitrarily large error on the generalized solution f^+. Notice that, for DD and CD problems, A^+ is always a bounded operator, with norm $||A^+|| = 1/\sigma_R$ where σ_R is the smallest singular value of A. However that bound can be very large, and a small error on the data can cause an unacceptably large error on the generalized solution. One speaks in this case of an *ill-conditioned* problem.

Regularization aims at restoring stability. This is done by replacing the generalized inverse A^+ by a modified operator $\mathcal{R}_\alpha : H_d \to H_o$, which depends on a *regularization parameter* $\alpha > 0$, and which satisfies the following conditions,

- The norm of $\mathcal{R}_\alpha$ is smaller than the norm of A^+, thereby improving stability.
- $\mathcal{R}_\alpha$ "resemble" A^+, in the sense that $\mathcal{R}_\alpha$ tends to A^+ when $\alpha \to 0$.

The regularized solution of the inverse problem is taken to be $f_\alpha = \mathcal{R}_\alpha g$. Consider some object f_{ex}, the corresponding ideal data $g_{ex} = A f_{ex}$, and measured data $g = g_{ex} + n_\varepsilon$ corrupted by measurement errors n_ε. Here $\varepsilon > 0$ represents the noise level, one may for instance assume that $||n_\varepsilon|| \leq \varepsilon$. Consider now the regularized solution $f_\alpha = \mathcal{R}_\alpha g$. The reconstruction error can be bounded using the triangular inequality as,

$$||f_\alpha - f_{ex}|| \leq ||A^+ A f_{ex} - f_{ex}|| + ||(\mathcal{R}_\alpha - A^+)g_{ex}|| + ||\mathcal{R}_\alpha n_\varepsilon|| \tag{99}$$

The three terms in the RHS can be interpreted as follows:

1. The first term is the norm of the component of f_{ex} that is lying within the null-space of A. In the absence of prior information, this transparent component is irremediably lost.
2. The second term is the *systematic error* caused by the replacement of the generalized inverse A^+ by the approximated, regularized, inverse $\mathcal{R}_\alpha$.
3. The third terms is the *statistical error* caused by the random measurement error n_ε.

As a rule, the systematic error decreases, and the statistical error increases, when the regularization parameter α decreases. Selecting the regularization parameter for a specific application then requires a compromise between these two terms. The selection of α is often done empirically, but more objective methods have also been proposed. One example is the method of Morozov, which is based on the observation ("discrepancy principle") that it makes little sense to fit data g with an accuracy better than ε if we know that these data are corrupted by measurement errors of magnitude $||g - g_{ex}|| = \varepsilon$. Therefore, given a family of regularized inverse operators $\mathcal{R}_\alpha$, Morozov's method selects the parameter α in such a way that

$$||g - A f_\alpha|| = ||g - A \mathcal{R}_\alpha g|| \simeq \varepsilon \tag{100}$$

This method provides only a reasonable order of magnitude for α. More sophisticated methods for selecting the regularization parameter have been proposed, see,[1] but in practice the value of α is often selected empirically.

6.1 Truncated and Filtered SVD Method

The generalized solution f^+ of an inverse problem with compact operator A can be written in terms of the singular vectors and singular values (93). When the operator A has an infinite rank (for a CC problem), or when the condition number σ_1/σ_R is large, the generalized solution does not depend on the data in a stable way, because the factor $1/\sigma_k$ in (93) strongly amplifies the measurement errors on the data components (g, v_k) corresponding to small singular values.

Stability can be restored by truncating the SVD series to the N_α first components. This defines the following regularized inverse:

$$\text{Truncated SVD:} \qquad f_\alpha = \mathcal{R}_\alpha g = \sum_{k=1}^{N_\alpha} \frac{1}{\sigma_k} (g, v_k) u_k \qquad (101)$$

with the condition that $\lim_{\alpha \to 0} N_\alpha = R$, which guarantees that $\lim_{\alpha \to 0} \mathcal{R}_\alpha = A^+$. The reconstruction error can then be bounded as (see (99)):

$$\|f_\alpha - f_{ex}\| \le \|A^+ A f_{ex} - f_{ex}\| + \left(\sum_{k=N_\alpha+1}^{R} |(f_{ex}, u_k)|^2 \right)^{1/2} + \frac{\varepsilon}{\sigma_{N_\alpha}} \qquad (102)$$

assuming a bound on the measurement error, $\|n_\varepsilon\| \le \varepsilon$. Mathematically, we say that the inverse problem has been regularized if N_α is chosen in function of the noise level ε in such a way that

$$\lim_{\varepsilon \to 0} \varepsilon / \sigma_{N_\alpha(\varepsilon)} = 0$$

$$\lim_{\varepsilon \to 0} N_\alpha(\varepsilon) = R$$

These two conditions guarantee that the statistical and systematic errors in (102) both tend to zero when the noise level ε tends to zero. In practice, one works with a fixed noise level and these asymptotic conditions have a limited relevance: more sophisticated – or then fully empirical – methods must be used to determine N_α. The above conditions nevertheless give insight into the expected dependence of N_α on ε.

The abrupt truncation of the higher SVD component in (101) often cause Gibbs-type artefacts near sharp transitions in the image, similar to those observed when applying an ideal low-pass filter. In practice, it is preferable to use a smoother transition, of the type

$$\text{Filtered SVD:} \qquad f_\alpha = \mathcal{R}_\alpha g = \sum_{k=1}^{R} \frac{F(\alpha, k)}{\sigma_k} (g, v_k) u_k \qquad (103)$$

where the digital filter $F(\alpha, k)$ is designed with a smooth transition between 1 for the first components (small k) and 0 for the largest components. A typical choice is $F(\alpha, k) = \sigma_k^2 / (\sigma_k^2 + \alpha)$.

Iterative algorithms can be regularized by stopping the iteration after a number of steps M_α that is chosen in function of the noise level. For a compact operator, the Landweber algorithm interrupted after M_α steps is equivalent (see (96)) to a filtered SVD with the digital filter

$$F(\alpha, k) = 1 - (1 - \tau \sigma_k^2)^{M_\alpha} \qquad (104)$$

6.2 Regularization with Variational Methods

The generalized solution f^+ of $Af = g$ is the element of minimum norm which best fits the data, in the sense that it minimizes the residual error $||Af - g||^2$. It makes little sense however, to fit noisy data with the highest possible accuracy. Many regularization methods are based on the replacement of the least-squares functional $||Af - g||^2$ by a penalized functional

$$\Phi_\alpha(f,g) = ||g - Af||^2 + \alpha U(f) \tag{105}$$

where the *penalty* $U : H_o \to \mathbb{R}$ is a functional defined on the object space H_o. The regularized solution is then defined as

$$f_\alpha = \arg \min_{f \in H_o} \Phi_\alpha(f,g) \tag{106}$$

assuming the minimum is unique. Uniqueness of the solution of the minimization problem (106) is guaranteed if U is a strictly convex functional, or if U is convex and $\mathcal{N}(A) = \{0\}$. Recall that a functional $U(f)$ is convex if

$$\forall f_1, f_2 \in H_o, \ \forall \lambda \in (0,1) \qquad U(\lambda f_1 + (1-\lambda) f_2) \le \lambda U(f_1) + (1-\lambda)U(f_2) \tag{107}$$

and strictly convex if the inequality is strict.

The penalty models prior knowledge on the solution: the functional is designed in such a way that $U(f)$ is large for those f that are deemed a priori unlikely. Standard examples include:

- The so-called *Tikhonov penalty* $U(f) = ||f||^2$ is proportional to the "energy" of the solution, and favors solutions which are not "too large". Because of the interaction with the properties of typical operators A, this penalty favors in fact smooth solutions.
- More general *quadratic penalties* of the type $U(f) = ||Bf||^2$, with $B : H_o \to H_o$ a linear operator with a bounded inverse B^{-1}, are used to enforce smooth solutions. The rationale is that ill-posedness is usually related to rapidly oscillating components, and penalizing these components stabilizes the solution. For a CC or a CD problem in $L^2(\mathbb{R}^d)$, a typical choice is a high-pass filter

$$(\mathcal{F}Bf)(\mathbf{v}) = (1 + |\mathbf{v}|^2)^{\beta/2}(\mathcal{F}f)(\mathbf{v}) \tag{108}$$

with some $\beta > 0$. For a DD problem, one typically penalizes large values of the differences between neighbouring voxels using for example the penalty

$$U(f) = ||Bf||^2 = \sum_{i=1}^{N_o} \sum_{j=1}^{N_o} c_{i,j}(f_i - f_j)^2 \tag{109}$$

with $c_{i,j} > 0$ if the voxels (or basis functions) i and j are in some sense neighbours, and $c_{i,j} = 0$ otherwise. In some cases, the neighbourhood of each voxel can be designed to reflect additional prior information on the solution, provided e.g. from another imaging modality.

- *Sparsity constraints.* Consider an orthonormal basis $\{\psi_\lambda, \ \lambda \in I\}$ of the object space, where I is a set of indices. The following penalty

$$U(f) = \sum_{\lambda \in I} |(f, \psi_\lambda)| \tag{110}$$

which represents the ℓ^1-norm of the sequence of the Fourier coefficients of f, favors solutions f having only few non-zero components on this basis. This approach can be used with a voxel basis to favor solutions that are zero except in a small number of voxels (e.g. for digital subtraction angiography). Wavelet basis have also been used for a ℓ^1 penalty, because some classes of "natural" images are known to have a sparse wavelet transform.

- Other non-quadratic constraints have also been used, which aim at enforcing solutions which are smooth except possibly at a limited number of boundaries between regions. Most priors are local in the sense that they penalize differences between neighbouring voxels, in the same way as in (109),

$$U(f) = \sum_{i=1}^{N_o} \sum_{j=1}^{N_o} c_{i,j} P(f_i - f_j) \tag{111}$$

but with a non-quadratic function $P(s)$ instead of $P(s) = s^2$.

6.3 Regularization with Variational Methods

For a quadratic penalty $U(f) = ||Bf||^2$ defined by a linear operator B with bounded inverse B^{-1}, the regularized solution can be obtained by solving the Euler-Lagrange equations for the minimization of $\Phi_\alpha(f,g) = ||g - Af||^2 + \alpha U(f)$:

$$f_\alpha = \mathcal{R}_\alpha g = (A^*A + \alpha B^*B)^{-1} A^* g \tag{112}$$

The inverse exists and is bounded even if the null-space of A is non-trivial, provided $\alpha > 0$ and B has a bounded inverse. In (112), note that the data g are first multiplied by the adjoint operator A^*. This automatically "kills" the inconsistent component of the data because $\mathcal{N}(A^*) = (\text{Range } A)^\perp$. Recall that in tomography, the adjoint of the Radon transform is the backprojection.

If A is a compact operator and B^*B commutes with A^*A, the regularized solution (112) is equivalent to a filtered SVD with the digital filter

$$F(\alpha,k) = \sigma_k^2 / (\sigma_k^2 + \alpha\beta_k^2) \tag{113}$$

where β_k^2 is the eigenvalue of B^*B corresponding to u_k, i.e. $B^*Bu_k = \beta_k^2 u_k$. In most applications however B^*B and A^*A do not commute because physically meaningful penalties are not easily expressed in terms of the singular basis of A. This limitation has motivated the use of alternative basis in object space, which only approximately diagonalize the operator A^*A, but are more appropriate to express regularizing constraints (see e.g.[11]).

If A is a convolution operator in $L^2(\mathbb{R}^d)$, with transfer function $\hat{h}(\mathbf{v})$ (see (39)) and if B is also a convolution operator with transfer function $\hat{b}(\mathbf{v})$, the regularized solution (112) is defined in terms of its Fourier transform as

$$(\mathcal{F}f_\alpha)(\mathbf{v}) = \frac{\hat{h}^*(\mathbf{v})}{|\hat{h}(\mathbf{v})|^2 + \alpha|\hat{b}(\mathbf{v})|^2} (\mathcal{F}g)(\mathbf{v}) \qquad (114)$$

If the inverse of $A^*A + \alpha B^*B$ cannot be calculated – or implemented – easily, the regularized solution (112) can be approached using a regularized Landweber iteration, with initial estimate $f_{\alpha,0} = 0$ and

$$f_{\alpha,n+1} = (I + \alpha B^*B)^{-1} (f_{\alpha,n} + \tau A^*(g - Af_{\alpha,n})) \qquad n = 0, 1, 2, \cdots \qquad (115)$$

and the condition $0 < \tau < 2/||A||^2$ for the relaxation parameter (section 5.6) is still sufficient to ensure convergence. Note the different roles of the two parameters: the relaxation parameter τ controls the convergence speed, whereas the regularization parameter α controls the stability for noise. More efficient iterative algorithms exist for minimizing a quadratic functional, such as the conjugate gradient algorithm.[26]

Let us remark that, as shown in,[8] another simple modification of the Landweber algorithm allows to cope with the case of ℓ^1-type penalty introduced in (110). The modification involves a thresholding (more precisely a soft-thresholding) of the coefficients (f_{n+1}, ψ_λ) at each iteration step.

6.4 Regularization with Variational Methods: Link with Stochastic Models

The functional $\Phi_\alpha(f, g)$, which defines the regularized solution in variational methods, can also be defined in a Bayesian framework, where both the object f and the data g are described as realizations of stochastic processes (for a continuous model) or of random vectors (for a discrete model).

We first describe the Bayesian approach for a DD problem. The data $g \in \mathbb{R}^{N_d}$ and the object $f \in \mathbb{R}^{N_0}$ are modelled as random vectors with specified probability distribution functions (PDF):

- $Prob(f)$ is the prior PDF (often simply called the *prior*) of the object, which models the prior knowledge about the properties of the solution.
- $Prob(g|f)$ is the conditional PDF of the data *given* a fixed object f. This PDF models both the system and the measurement noise, and it is also referred to as the *likelihood* function. The link between the likelihood and the operator A is defined by means of the average (expectation) value of the data, *given* f:

$$Af = E(g|f) = \int_{\mathbb{R}^{N_d}} g \, Prob(g|f) \, dg \qquad (116)$$

Using Bayes theorem, one can then calculate the *posterior probability distribution function*, also called the *posterior likelihood*:

$$L(f,g) = Prob(f|g) = \frac{Prob(g|f) \, Prob(f)}{Prob(g)} \qquad (117)$$

where

$$Prob(g) = \int_{\mathbb{R}^{N_o}} Prob(g|f) \, Prob(f) \, df \qquad (118)$$

The posterior likelihood is the probability of f *given* the data g, and it is then natural to define a bayesian (or "MAP") estimator of the solution as a maximizer of $L(f,g)$:[2,21]

$$f_{MAP} = \arg \min_{f \in \mathbb{R}^{N_o}} Prob(f|g) \qquad (119)$$

Maximizing $L(f,g)$ is equivalent to minimizing $-\log L(f,g)$. Up to factors which do not depend on f or can be absorbed in the regularization constant α, one easily checks that the variational method with the penalized least-square functional $\Phi_\alpha(f,g)$ in (105) is equivalent to a Bayesian approach with

- A prior PDF
$$Prob(f) = Z' \exp\{-U(f)\} \qquad (120)$$

- A Gaussian data model with covariance matrix proportional to the identity, and mean value $E(g|f) = Af$,

$$Prob(g|f) = Z \exp\{-||g - Af||^2/2\sigma^2\} \qquad (121)$$

where Z and Z' are two constants independent of f and g.

A second classical variational method derived from a stochastic description is the *Wiener filter* for convolution problems of the type

$$g(\mathbf{x}) = (Af)(\mathbf{x}) + n(\mathbf{x}) = \int_{\mathbb{R}^d} h(\mathbf{x} - \mathbf{x}_0) f(\mathbf{x}_0) \, d\mathbf{x}_0 + n(\mathbf{x}) \qquad (122)$$

where $f(\mathbf{x}) \in L^2(\mathbb{R}^d)$ and $g(\mathbf{x}) \in L^2(\mathbb{R}^d)$ are modeled as stochastic processes (the functional space equivalent of a random vectors) with known first and second moments. In the simplest case, the model specifies

- The expectation value of the object, $E\{f(\mathbf{x})\} = f_m(\mathbf{x})$.
- The autocorrelation function of the object, assumed stationary in the sense that its stochastic properties are invariant for translation,

$$E\{(f(\mathbf{x} + \mathbf{y}) - f_m(\mathbf{x} + \mathbf{y}))(f^*(\mathbf{x}) - f_m^*(\mathbf{x}))\} = \rho_{ff}(\mathbf{y}). \qquad (123)$$

The Fourier transform $(\mathcal{F}\rho_{ff})(\mathbf{v}) = \hat{\rho}_{ff}(\mathbf{v})$ of this autocorrelation function is the *object power spectrum*, and

$$E\{(\hat{f}(\mathbf{v}) - \hat{f}_m(\mathbf{v}))(\hat{f}^*(\mathbf{v}') - \hat{f}_m^*(\mathbf{v}'))\} = \hat{\rho}_{ff}(\mathbf{v})\,\delta(\mathbf{v} - \mathbf{v}'). \tag{124}$$

- The expectation value of the noise, usually assumed equal to zero, $E\{n(\mathbf{x})\} = 0$.
- The autocorrelation function of the noise, assumed stationary,

$$E\{n(\mathbf{x} + \mathbf{y})\,n^*(\mathbf{x})\} = \rho_{nn}(\mathbf{y}). \tag{125}$$

The Fourier transform $\hat{\rho}_{nn}(\mathbf{v})$ of this autocorrelation function is the *noise power spectrum*. White-noise is defined by a constant power spectrum, $\rho_{nn}(\mathbf{v}) = cst$.

- The cross-correlation between noise and object. These are usually assumed to be uncorrelated,

$$E\{(f(\mathbf{x}) - f_m(\mathbf{x}))n(\mathbf{y})\} = 0. \tag{126}$$

Consider restoration filters with impulse response $r(\mathbf{x})$ and transfer function $\hat{r}(\mathbf{v})$, which define approached solutions $f_r(\mathbf{x}) = f_m(\mathbf{x}) + (r*(g - f_m*h))(\mathbf{x})$ to the inverse problem $g = f * h$. The *Wiener filter* r_{opt} is the filter that minimizes the expectation value of the mean-square error:

$$r_{opt} = \arg\min_r E\left\{\int_{\mathbb{R}^d} |f_r(\mathbf{x}) - f(\mathbf{x})|^2\,d\mathbf{x}\right\} \tag{127}$$

Note that the expectation value involves averaging over all possible realizations of the object and over all possible realizations of the noise. The solution of this minimization problem yields the *Wiener filter*

$$\hat{r}_{opt}(\mathbf{v}) = \frac{\hat{h}^*(\mathbf{v})}{|\hat{h}(\mathbf{v})|^2 + \frac{\hat{\rho}_{nn}(\mathbf{v})}{\hat{\rho}_{ff}(\mathbf{v})}} \tag{128}$$

Note the similarity with the regularized deconvolution (114) derived in a non-stochastic framework. In particular Tikhonov regularization corresponds to a model in which both the noise and the object are "white" stochastic processes, with $\hat{\rho}_{nn}(\mathbf{v}) = \varepsilon^2$ and $\hat{\rho}_{ff}(\mathbf{v}) = E^2$. The Tikhonov regularization parameter is then given by the ratio $\alpha = \varepsilon^2/E^2$ between the expected energy of the noise and of the object.

In summary, the stochastic models provide a systematic way to define objective functions for variational regularization techniques. This is especially valuable when the prior knowledge on the solution of the inverse problem is sufficient to design a reliable prior probability distribution or object auto-correlation. Such information, however, is rarely available, and in most applications, one must satisfy oneself with a very empirical model.

References

1. Bertero M and Boccacci P 1998 Introduction to Inverse Problems in Imaging, Institute of Physics
2. Barrett H H and Myers K J 2004 Foundation of Image Science, Wiley, New York

3. Bertero M, De Mol C and Pike E R 1985 Linear inverse problems with discrete data. I: General formulation and singular system analysis *Inverse Probl.* **1** 301–330
4. Bertero M, De Mol C and Pike E R 1988 Linear inverse problems with discrete data. II: Stability and regularization *Inverse Probl.* **4** 573–594
5. Brankov J G, Yang Y and Wernick M N 2004 Tomographic image reconstruction based on a content-adaptive mesh model *IEEE Trans Med Imag* **23** 202–212
6. Champeney D C 1990 A Handbook of Fourier Theorems, Cambridge University Press, Cambridge
7. Daubechies I 1992 Ten Lectures on Wavelets, SIAM, Philadelphia, PA
8. Daubechies I, Defrise M and De Mol C 2004 An iterative thresholding algorithm for linear inverse problems with a sparsity constraint *Comm Pure Appl Math* **57** 1416–1457
9. de Boor C 1978 A Practical Guide to Splines, Springer, New York
10. Doyley M M, Meaney P M and Bamber J C 2000 Evaluation of an iterative reconstruction method for quantitative elastography *Phys Med Biol* **45** 1521–1540
11. Donoho D L 1995 Nonlinear solution of linear inverse problems by wavelet-vaguelette decomposition *Appl Comp Harmonic Anal* **2** 101–126
12. Hill D L G, Batchelor P G, Holden M and Hawkes D J 2001 Medical image registration *Phys Med Biol* **46** R1–R45
13. Khurd P and Gindi G 2005 Fast LROC analysis of Bayesian reconstructed emission tomographic images using model observers *Phys Med Biol* **50** 1519–1532
14. Lange K and Carson R 1984 EM reconstruction algorithms for emission and transmission tomography *J Comp Assist Tomo* **8** 306–316
15. Lazaro D, El Bitar Z, Breton V, Hill D and Buvat I 2005 Fully 3D Monte Carlo reconstruction in SPECT: a feasibility study *Phys Med Biol* **50** 3739–3754
16. Lee B I, Oh S H, Woo E Je, Lee S Y, Cho M H, Kwon O, Seo J K, Lee J-Y and Baek W S 2003 3D forward solver and its performance analysis for magnetic resonance electrical impedance tomography (MREIT) using recessed electrodes *Phys Med Biol* **48** 1971–1986
17. Lewitt R M 1992 Alternatives to voxels for image representation in iterative reconstruction algorithms *Phys Med Biol* **37** 705–716
18. Lewitt R M and Matej S 2003 Overview of methods for image reconstruction from projections in emission computed tomography *Proc IEEE* **91** 1588–1611
19. Mallat S 1999 A Wavelet Tour of Signal Processing, Academic, San Diego, CA
20. Metz C E and Goodenough D J 1972 Evaluation of receiver operating characteristic curves in terms of information theory *Phys Med Biol* **17** 872–873
21. Mumcuoglu E, Leahy R M and Cherry S R 1996 Bayesian reconstruction of PET images: methodology and performance analysis *Phys Med Biol* **41** 1777–1807
22. Natterer F 1986 The Mathematics of Computerized Tomography, Wiley, New York
23. Natterer F and Wubbeling F 2001 Mathematical Methods in Image Reconstruction, SIAM, Philadelphia, PA
24. Noo F, Clackdoyle R and Pack J D 2004 A two-step Hilbert transform method for 2D image reconstruction *Phys Med Biol* **49** 3903–3923
25. Patch S 2004 Thermoacoustic tomography: consistency conditions and the partial scan problem *Phys Med Biol* **49** 2305–2315
26. Press W H, Teukolsky S A, Vetterling W T and Flannery B P 1994 Numerical Recipes in C, Cambridge University Press, Cambridge
27. Strichartz R S 2003 A Guide to Distribution Theory and Fourier Transforms, World Scientific Publishing
28. Unser M 1999 Splines: a perfect fit for signal and image processing *IEEE Signal Proc Mag* **16** 22–38
29. Unser M 2000 Sampling - 50 years after Shannon Proc. IEEE **88** 569–587
30. Wang R K, Hebden J C and Tuchin V V 2004 Special issue on recent developments in biomedical optics *Phys Med Biol* **49** no 7 1085–1368
31. Webb S 1990 From the Watching of the Shadows, Adam Hilger, Bristol

32. Welch A, Campbell C, Clackdoyle R, Natterer F, Hudson M, Bromiley A, Mikecz P, Chillcot F, Dodd M, Hopwood P, Craib S, Gullberg G T and Sharp P 1998 Attenuation correction in PET using consistency information, Nuclear Science, IEEE Transactions on Volume 45, Issue 6, Part 2, Dec. 1998 Page(s):3134–3141

33. Wilson D W, Tsui B M W and Barrett H H 1994 Noise properties of the EM algorithm. II. Monte Carlo simulations *Phys Med Biol* **39** 847–871

34. Yendiki A and Fessler J A 2004 Comparison of rotation- and blob-based system models for 3D SPECT with depth-dependent detector response *Phys Med Biol* **49** 2157–2168

35. Zou Y and Pan X 2004 Exact image reconstruction on PI-lines from minimum data in helical cone-beam CT *Phys Med Biol* **49** 941–959

Exercises in PET Image Reconstruction

Oliver Nix

Abstract These exercises are complementary to the theoretical lectures about positron emission tomography (PET) image reconstruction. They aim at providing some hands on experience in PET image reconstruction and focus on demonstrating the different data preprocessing steps and reconstruction algorithms needed to obtain high quality PET images. Normalisation, geometric-, attenuation- and scatter correction are introduced. To explain the necessity of those some basics about PET scanner hardware, data acquisition and organisation are reviewed. During the course the students use a software application based on the STIR (software for tomographic image reconstruction) library [1,2] which allows them to dynamically select or deselect corrections and reconstruction methods as well as to modify their most important parameters. Following the guided tutorial, the students get an impression on the effect the individual data precorrections have on image quality and what happens if they are forgotten. Several data sets in sinogram format are provided, such as line source data, Jaszczak phantom data sets with high and low statistics and NEMA whole body phantom data. The two most frequently used reconstruction algorithms in PET image reconstruction, filtered back projection (FBP) and the iterative OSEM (ordered subset expectation maximation) approach are used to reconstruct images. The exercise should help the students gaining an understanding what the reasons for inferior image quality and artefacts are and how to improve quality by a clever choice of reconstruction parameters.

Keywords: PET · sinogram · corrections · tomographic reconstruction · FBP · OSEM

O. Nix
Deutsches Krebsforschungszentrum Heidelberg, Dep. for Medical Physics in Radiology, Im NeuenheimerFeld 28069120 Heidelberg, Germany
e-mail: O.Nix@dkfz-heidelberg.de

Y. Lemoigne, A. Caner (eds.) *Molecular Imaging: Computer Reconstruction and Practice*, 131
and Experiments,
© Springer Science+Business Media B.V., 2008.

1 Introduction

1.1 PET-Scanner Hardware

In PET (Positron Emission Tomography) two back to back emitted photons with an energy of 511 keV emerging from the annihilation of a positron originating from a β^+ decaying nucleus with an electron are detected. In most commercially available PET scanners they are detected using a scintillating material like BGO (bismuth germanate) or more recently GSO (gadolinium oxyorthosilicate) and LSO (lutetium oxyorthosilicate). Usually the block shaped scintillator material is segmented into a matrix of individual elements. The photon deposits energy, mainly by scattering, in the scintillator which is converted to scintillation light that is read out and converted to an electric signal using PMTs (photomultiplier tubes). More recently compact APDs (avalanche photo diode) are used, especially in dedicated PET scanner for small animal imaging. In a typical clinical PET scanner such as the Siemens/CTI EXACT HR+[3] the BGO scintillator material is segmented into arrays of 8×8 read out by four photomultiplier tubes. Due to light sharing between the individual photomultipliers the position of the impacting photon can be estimated using a center of gravity algorithm. The combination of scintillator and photomultiplier is called block detector. They are arranged as one or many rings around the subject. In case of the EXACT HR+ scanner four rings of block detectors exist, each equipped with 72 blocks. In total, this results in 576 channels for each of the 32 rings. In the early days of PET a lead shield, the septa, was positioned between detector rings to prevent the detection of coincident photons between different rings. With the septa inside the FoV (field of view) the data is reduced to a set of 2D planes. This constraint can be dropped for modern PET scanners due to increased computing power and storage capabilities together with the solution to the problem of integrating oblique slices in respect to the z-axis into the tomographic reconstruction process. Data acquisition in so called 3D mode became possible. The data used during the exercises were acquired in 3D mode which is today's standard.

1.2 Data Organisation in PET

The quantities measured in PET are photon energy, position and time. A line of response (LoR) is constructed if both photons have been detected in two different detectors within a certain time and energy window. The LoR can be considered a straight line between the two detectors located somewhere on the PET detector ring. The information content is that a PET event occurred somewhere along this line. Two approaches exist to store this data – listmode and matrix representation. In listmode format all detected events are written to an event list. In addition to the mentioned data additional information like gantry state or if available depth of interaction information can be stored. The listmode data format conserves all

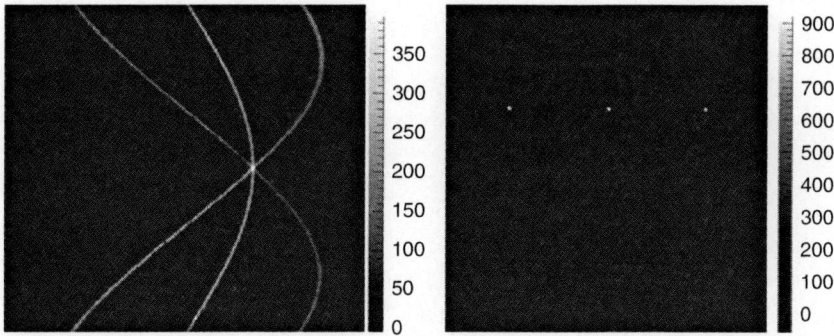

Fig. 1 Left hand side shows the sinogram resulting from three lines sources. Right hand side shows the image reconstructed from this sinogram

information available. It can be integrated in the reconstruction process to increase image quality. In matrix representation only a binary information about the event is kept, that is, that an event fulfilling the imposed criteria occurred in a given LoR. List mode format is mainly used in a research context and for experimental scanner systems while matrix representation is used in clinical routine. Any LoR within one plane can be defined by two coordinates – the tangent distance (x_r) from the center of a plane and an angle (ϕ). Matrix based image reconstruction nearly always starts from projections. A projection is defined by assembling all LoRs within one plane with a fixed angle ϕ. For a set of planes perpendicular to the z-axis the z-coordinate (z) is needed as additional coordinate to define the LoRs. This is the 2D PET case. In 3D PET planes with an arbitrary angle toward the z-axis are allowed. Therefore, a fourth coordinate is introduced to describe the LoRs. The obliqueness (δ) in respect to the z-axis. Planes with similar obliqueness are merged to reduce the data volume. This is effectively equivalent to an axial compression and the merged sinograms are sometimes called segments. Projections can be even if the center of the tomograph is located between two projection bins or odd otherwise. Frequently interleaving between two neighbouring projections is applied. That is merging one neighbouring odd and even projection thus halving the tangential sampling distance. Projections are assembled to sinograms by grouping the projections associated with one z-coordinate such that each row represents one projection. An example of the sinogram resulting from imaging a set of three line sources is shown in Fig. 1. The sinogram is the most widely used representation of PET data.

1.3 Data Acquired During a PET Scan

During a clinical PET scan a progression of emission and transmission scans is performed. During the emission scan the coincident photon events from the β^+ decaying radiotracer and the subsequent e^+e^- annihilation are recorded and stored in sinogram format. The activity distribution, that is the image, is reconstructed from this data. During the transmission scan a set of rod sources, usually ^{68}Ge, rotate around the subject. The transmission scan is similar to a CT scan except for using 511 keV photons and in comparison very low statistics. The transmission data is used to reconstruct an attenuation image or μ-map. No or only very little anatomical details can be seen on these images. Its sole purpose is to correct for photon attenuation inside the patient's body. Transmission data is again stored as sinograms. In combination with a blank scan which is a transmission scan acquired without any object inside the field of view, the attenuation image can be reconstructed. The blank scan is usually acquired over night as part of daily quality assurance procedures.

2 Corrections on PET Data

It is necessary to preprocess the data before running the actual reconstruction algorithm. A lack of these corrections results in deterioration of image quality and causes typical artefacts depending on the type of correction.

2.1 Normalisation

Normalisation is needed to correct for different efficiencies individual LoRs have. The total efficiency of a LoR is the product of detection efficiency, dead time efficiency and geometrical efficiency. In case of the EXACT HR+ scanner the efficiencies are represented as matrices of floating point numbers stored in a binary normalisation file. They are calculated from high statistic scans of a homogenous cylinder phantom or a line source during normalisation procedures performed approximately every two months. A copy of the normalisation data is added to the patient data record. The normalisation file provided for the exercises contain 32 * 576 scintillator efficiency factors and 63 * 288 geometrical efficiency factors. In addition crystal interference corrections need to be applied to take care for effects originating from the use of block detectors to construct the PET detector ring. A lack of normalisation can be detected easily by looking at the sinograms. In case of non normalised data the sinograms look coarse and channels with significant higher or lower efficiencies can be identified as lines with significantly different amplitudes or in graphical representation grey values. Normalised sinograms appear smoother. Figure 2 shows a non normalised sinogram acquired during a scan of the NEMA whole body phantom and the same data with normalisation applied. The image reconstructed from the non normalised sinogram shows strong ring like artefacts. These structures disappear if normalisation is applied.

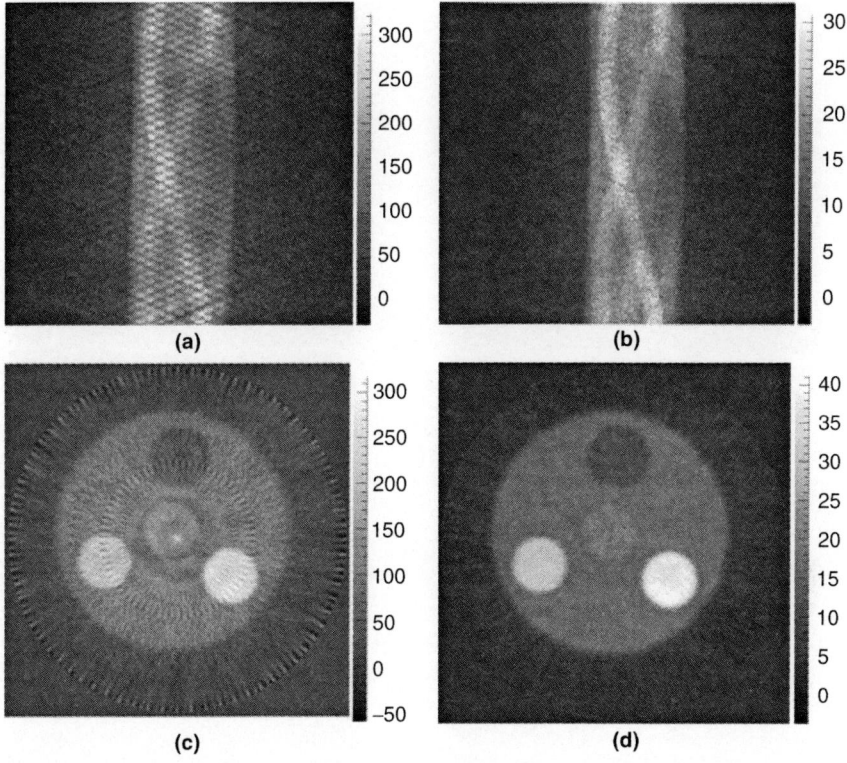

Fig. 2 Sinogram obtained from a NEMA whole body phantom scan. Upper left (**a**): no normalisation applied. Upper right (**b**): normalisation performed. Lower row: corresponding images reconstructed using a 3D filtered backprojection algorithm

2.2 Geometrical Corrections

Geometrical correction, also known as arc correction, is necessary to take into account the effectively different tangential bin sizes a given projection has. Different bin sizes result from the ring shaped arrangement of the detectors. The detectors at the edges of the projection appear smaller because their face side is inclined in respect to the projection axis. Variable tangential bin sizes are equivalent to different widths of the LoRs assembling the projection. This results in a non uniform sampling of the projection. The resolution at the edges of the field of view is deteriorated and artefacts occur. During arc correction the measured projections with non equal bin sizes are mapped onto projections with equal bin sizes by means of interpolation. Figure 3 shows the reconstructed image from a phantom consisting of 29 line sources homogenously distributed among the field of view. Close to the edges strong blurring and star like artefacts appear if no arc correction is applied.

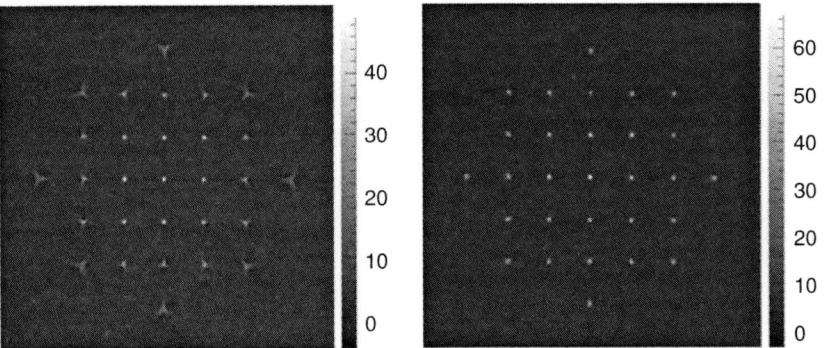

Fig. 3 Effects of arc correction. Left: no arc correction applied. Blurring and star like artefacts are clearly visible off z-axis. Right: same data and reconstruction with arc correction applied

2.3 Attenuation Correction

Photons travelling through matter can be absorbed. Absorption is a property of the media and the path length. In PET two photons are emitted back to back. Therefore, the total path length of the two photons is independent of the position on the LoR where the PET event occured. The probability that both photons reach the detectors is independent of their origin on the LoR. The attenuation correction factor for each projection bin or LoR can be estimated using transmission and blank scan. The correction factor for each detector pair is calculated by dividing blank through transmission scan. In practice the acquisition time of the transmission scan is in the order of a few minutes and therefore much shorter than the over night acquired blank scan. The different number of acquired events in each LoR needs to be considered. To avoid outlier the sinograms are smoothed, for example by median filtering. The EXACT HR+ scanner always acquires transmission and blank scan in 2D mode. For 3D data acquisition correction factors for oblique sinograms have to be derived from the 2D data sets. The used application does this by reconstructing attenuation images from the obtained attenuation correction sinograms using the 2D filtered backprojection algorithm. The resulting attenuation image is then forward projected into the oblique slices to get correction factors for the oblique sinograms. A calculated attenuation correction sinogram from a NEMA whole body phantom scan and the corresponding reconstructed attenuation image is shown in Fig. 4. In Fig. 5 images reconstructed from this data set with and without attenuation correction applied are displayed. It is clearly visible that the measured activity drops to around 50% of the expected activity towards the center of the phantom on the profiled line whereas a homogenous activity distribution is expected. The intensity drop in the reconstructed image disappears after applying attenuation correction to the emission sinograms.

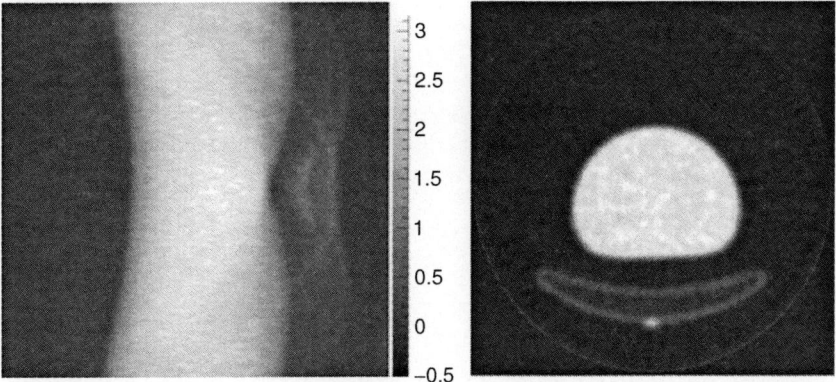

Fig. 4 Left: attenuation correction sinogram calculated from transmission and blank scan acquired during a scan of the NEMA whole body phantom. Right: reconstructed attenuation image

2.4 Scatter Correction

Photons travelling through tissue can not only be absorbed but also interact via scattering. A scattered photon will change its direction and thereby loose some of its energy depending on the scattering angle. Scattering will create false lines of response. A scattered photon can be discriminated because of its reduced energy. An energy window for photons is defined. Unfortunately, the energy resolution of the used scintillator materials is not good enough to suppress scatter completely. Scattered events cause background in the reconstructed image reducing the image contrast. Especially in quantitative PET scatter is a problem because the calculated absolute activities or standard uptake values (SUV) are falsified. In 3D PET up to 30% of the measured activity originates from scatter if no correction is applied[3]. A correction for scatter can be applied during the exercises by performing an object based scatter estimation using the SSS (single scatter simulation)-algorithm first introduced by Watson et al. in 1996.[4] The basic idea behind the algorithm is to get an estimate of the volume where scatter mainly occurs by reconstructing an attenuation image first. A certain number of elementary scatter volumes is distributed randomly inside the scatter volume.

The probability that a photon is scattered inside this volume, redirected and detected in another detector resulting in a false LOR is calculated using the Klein-Nishina formula. The amount of scatter in each LoR is estimated by summing up the contributions to each LoR from all elementary scatter volumes.[5] Figure 6 shows the image reconstructed from a Jaszczak phantom scan. No scatter correction is applied in the left image. An increasing background towards the center of the phantom is visible. After correcting the emission sinogram for scatter, the background is clearly reduced. The object or image based scatter correction can not eliminate scatter contributions originating from outside the FoV.

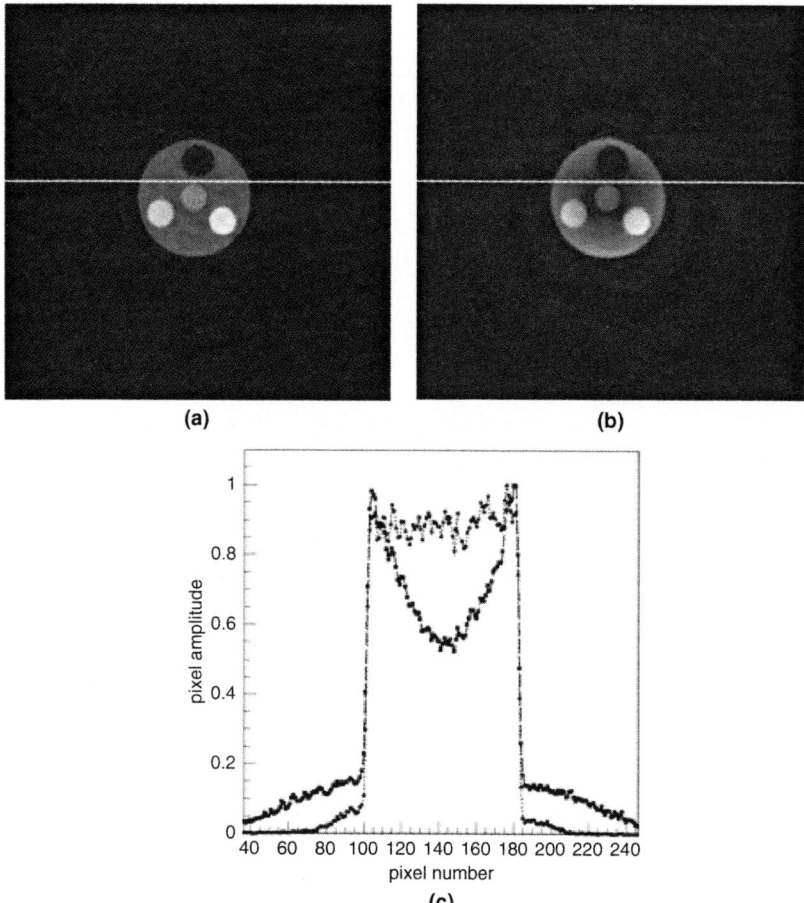

Fig. 5 Effect of attenuation correction. Along the profiled line the expected activity is constant. In case attenuation correction is performed the observed activity is nearly constant (**a**). If no attenuation correction is applied to the data the intensity drops visibly towards the center of the phantom (**b**). On the profiled line the activity drop is close to 50% (**c**)

3 Image Reconstruction

Two classes of reconstruction algorithms exist, analytical and iterative ones. The most popular analytical reconstruction method is the filtered backprojection (FBP) algorithm. The OSEM (ordered subset expectation maximisation) algorithm is the most popular iterative method used in clinical routine. Reconstruction algorithms were discussed in more detail in the lecture on PET image reconstruction preceding the exercises. All images shown in this chapter were acquired with a deluxe Jaszczak phantom with hot spot insert that was scanned with a Siemens/CTI EXACT HR+ scanner. The rods of the hot spot insert have diameters of 4.8, 6.4, 7.9, 9.5, 11.1,

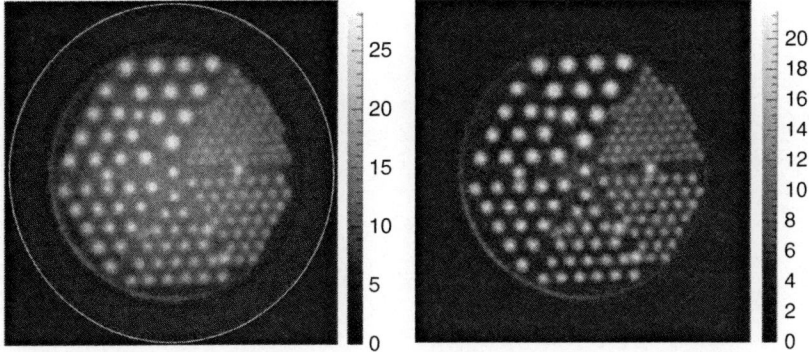

Fig. 6 Model based scatter correction using the SSS (single scatter simulation) algorithm. Left: no scatter correction applied. Right: scatter correction applied

12.7 mm. The phantom was filled with an aqueous solution of ^{18}F-FDG with an initial activity of 2.05 mCi. The data were acquired in standard 3D mode until 2.0 $* 10^7$ (low statistic acquisition) and $60 * 10^7$ (high statistic acquisition) counts were recorded.

3.1 Filtered Back Projection

Filtered back projection (FBP) is, as its name suggests, a combination of a filtering step performed on the projection data and a backprojection step. For 2D data all projections belonging to one z position are fourier transformed in respect to x_r using a FFT (fast fourier transform) algorithm. This data is then multiplied with a ramp filter and inverse fourier transformed back into projection space. The filtered projection is then backprojected onto the image. For more details see Defrise[6] and references therein. For 3D data sets the situation is more difficult due to the limited axial extend of the scanner. This causes incomplete data sets because of missing projections. This problem is overcome by the 3D reprojection algorithm (3DRP).[7] Missing projections are estimated from the activity distribution obtained by reconstructing the direct non oblique sinograms (a subset of the 3D data set). This estimation is then forward projected into those projections which do not exist in the real scanner. Therefore, this algorithm is also referred to as PROMIS (PROject MISsing data). The importance of the filtering step is demonstrated in Fig. 7.

Prerequisite for good image quality is a sufficient number of recorded PET events. Although this is a matter of course it is the first thing to look at if the image looks noisy and quality is in general bad with non of the artefacts visible described in the previous chapter. See Fig. 8.

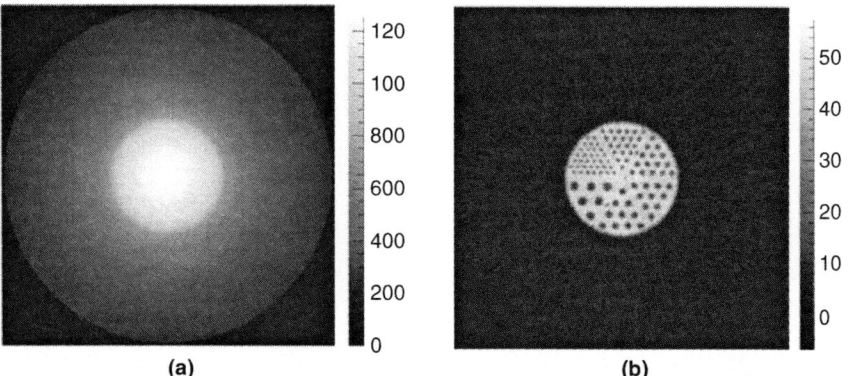

Fig. 7 Effects of the filtering step in 2D filtered back projection. (**a**) no ramp filter is applied to the projection data, simple backprojection. (**b**) ramp filtering of the projections before backprojection

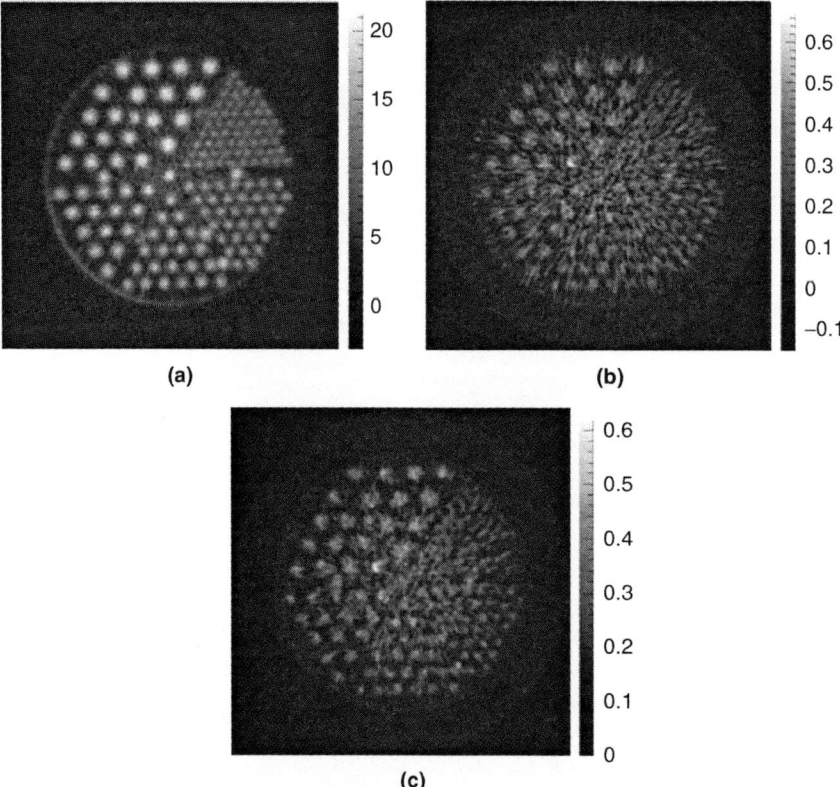

Fig. 8 (**a**) Image reconstructed from the high statistic scan using the 3DRP algorithm. Even the rods with the smallest diameter of 4.8 mm can be resolved. (**b**) the same for the scan with only a small number of events acquired using the FBP3D algorithm. Image (**c**) is reconstructed from the low statistic data set using the OSEM algorithm (12 subsets, 5 full iterations). Image quality is clearly dominated by scan statistics. In case of low statistics the iterative OSEM approach leads to better results due to the consideration of the statistical nature of the measured data in the reconstruction model compared to the analytical FBP3D algorithm

3.2 Iterative Approaches

In iterative image reconstruction the true activity distribution $f(x, y, z)$ is found by iteratively approximating the measured data with data derived from an estimated image. The initial estimation is usually a flat activity distribution normalised to one. The initial image is forward projected into the projections to get an estimation of what would have been measured by the scanner if this image had been the real one. Forward projection is the inverse operation of backprojection. It is done by summing up all amplitudes along a path forming a LoR. The LoRs are assembled to a sinogram which can be directly compared to the measured sinogram. Based on this comparison the estimated image is updated. Many approaches exist on how the comparison of the estimation and the measured data can be done and how to update the image estimate. The maximum likelyhood expectation maximization (MLEM) algorithm first proposed by Shepp and Vardi[8] is one of the most popular ones. It considers the statistical nature of PET data acquisition by means of Poisson statistics and by assigning different weights depending on the number of counts in a LoR. In MLEM each iteration requires forward projection and backprojection of the complete data set.

This is computationally expensive and convergence is slow. Ten to 100 iterations are required. Hudson and Larkin[9] developed an accelerated version of the EM algorithm based on the concept of ordered subsets (OSEM). The full data set containing all projections is split up in a certain number of subsets. These subsets are reconstructed successively and the image update step is done after each processed subset. The speed of convergence is greatly increased while the image quality remains similar to a MLEM reconstruction. The convergence behaviour as a function of the number of full iterations can be seen in Fig. 9. A typical choice for the number of subsets is between 2 and 20. Figure 10 shows the convergence of the OSEM algorithm with a fixed number of iterations as a function of the number of subsets chosen.

4 Conclusions

Several considerations are required to obtain high quality PET images. The PET scanner itself has to be set up properly and the individual detector responses have to be determined and quantified correctly. Before applying any reconstruction algorithm, the acquired data has to be normalised and arc corrected. If transmission scans were performed attenuation correction factors can be calculated. Especially in quantitative PET a correction for scatter is important. As far as possible this is done during the data acquisition itself by defining an energy window on the detected photons. In addition object based scatter corrections can be calculated offline. The image quality strongly depends on the number of acquired PET events. Low statistics always results in bad image quality. Statistics is the prevailing factor to image quality in clinical PET due to limited acquisition times and activities.

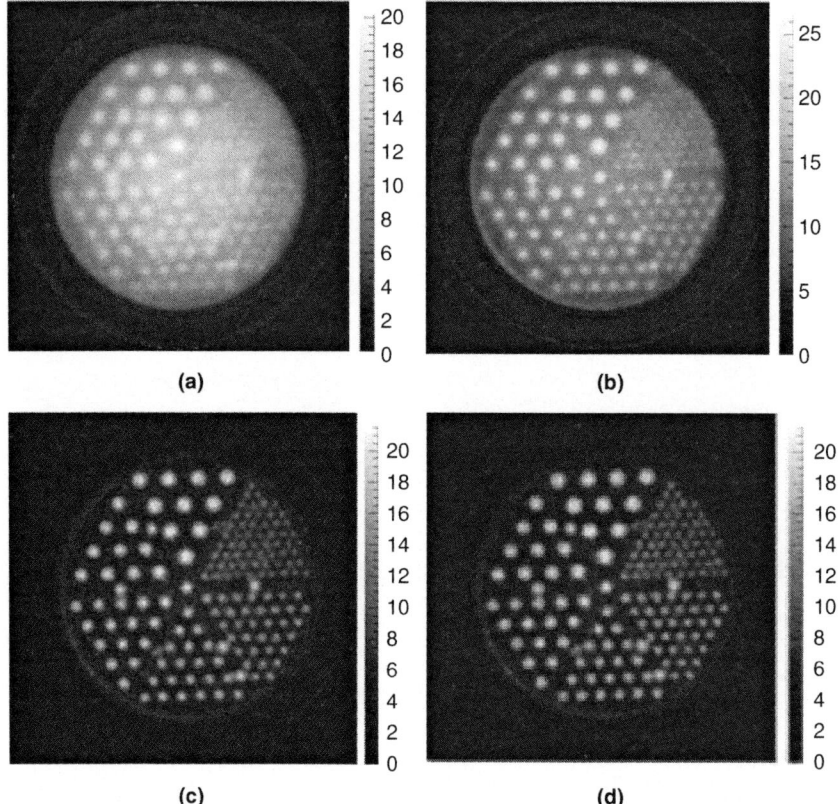

Fig. 9 Images reconstructed using the OSEM algorithm as a function of the number of full iterations performed. Number of subsets is 12. (**a**) 1 iteration, (**b**) 3 iterations, (**c**) 6 iterations and (**d**) 10 iterations

A clever choice of the reconstruction algorithm and its parameters is therefore essential. Iterative algorithms, like OSEM, are the preferred choice in case of limited statistics. The most important parameters in iterative reconstruction are the number of iterations and the choice of the number of subsets. If the number of iterations performed is too small, the image will look blurred and constrast is low. If too many iterations are performed, noise will be amplified. The optimum number of iterations in OSEM depends on the number of subsets chosen. The smaller the number of subsets, the more iterations are needed. The exercises were intended to mediate a first impression on all of the mentioned aspects. Without considering them properly, the resulting PET images will be suboptimal.

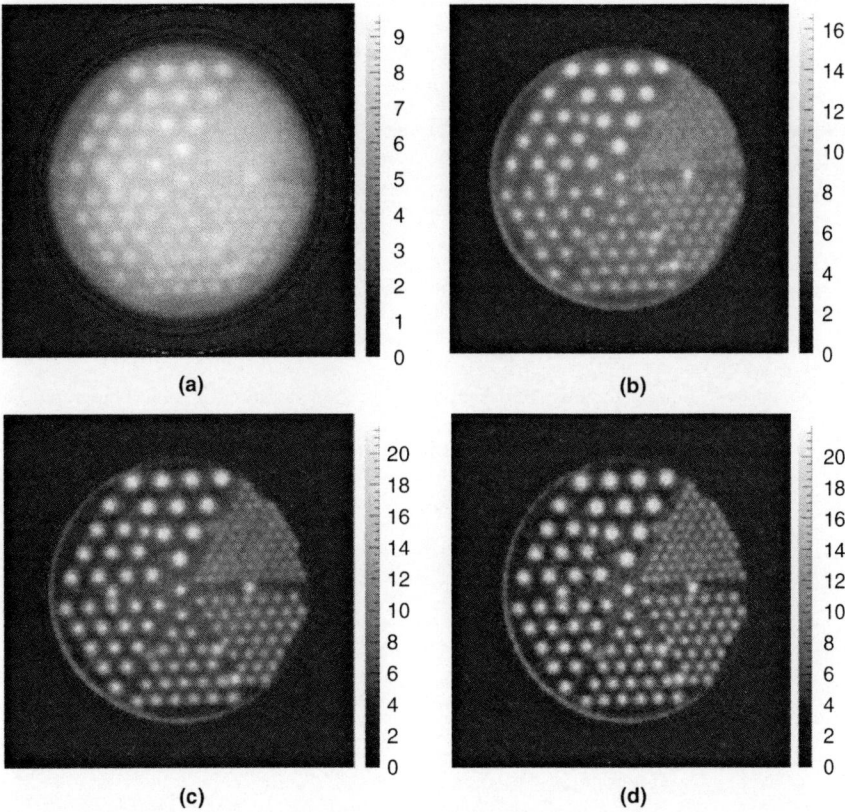

Fig. 10 Convergence of the OSEM algorithm as a function of the number of subsets used. The number of performed iterations is 5. (**a**) MLEM (1 subset), (**b**) 4 subsets, (**c**) 9 subsets and (**d**) 18 subsets. The larger the number of subsets the faster the convergence. The number subsets chosen and the number of full iterations performed must be balanced. If the number of subsets and iterations is too small the algorithm will not have converged resulting in a low contrast blurred image. See (a), (b). If the number of subsets is too large in respect to the number of full iterations (or vice versa) noise will be enhanced and the image will be increasingly grainy. See (d)

Acknowledgements The author likes to thank Dr. J. Doll for his help in acquiring the phantom data used during the exercises.

References

1. Labbé, C., Thielemans, K., Belluzzo, D. et al.: An Object-Oriented Library for 3D PET Reconstruction Using Parallel Computing. In: Evers, H., Glombitza, G., Lehmann, T., Meinzer, H.-P. (Eds.):*Informatik aktuell*, Springer (1999) pp. 268–272
2. Hammersmith Imanet, "STIR Software for Tomographic Image Reconstruction", Last update, 01/26/06. Available from: http://stir.sourceforge.net/

3. Adam, L.E., Zaers, J., Ostertag, H., et al.: Performance Evaluation of the whole-body pet scanner ECAT EXACT HR+ following the IEC standard. *IEEE Trans Nucl Sci* **44** (1997) 1172–1179

4. Watson, C.C., Newport, D., Casey, M.E., deKemp, R.A., Beanlands, R.S., Schmand, M.: Evaluation of simulation-based scatter correction for 3-D PET cardiac imaging.*IEEE Trans Nucl Sci* **44** (1997) 90–97

5. Werling, A., Bublitz, O., Doll, J., Adam, L.-E., Brix, G.: Fast implementation of the single scatter simulation algorithm and its use in iterative image reconstruction of PET data. *Phys Med Biol* **47** (2002) 2947–2960

6. Defrise, M., Kinahan, P.E.: Data Acquisition and Image Reconstruction for 3D PET. In: Bendriem, B., Townsend, D.W. (Eds.): *The Theory and Practice of 3D PET*. Kluwer, Dordrecht, The Netherlands/Boston, MA/London (1998)

7. Kinahan, P.E., Rogers, J.G.: Analytic 3D image reconstruction using all detected events. *IEEE Trans Nucl Sci* **36** (1989) 964–968

8. Shepp, L.A., Vardi, Y.: Maximum likelihood reconstruction for emission tomography.*IEEE Trans Med Imag* **MI-1** (1982) 113–122

9. Hudson, H.M., Larkin, R.S.: Accelerated image reconstruction using ordered subsets of projection data.*IEEE Trans Med Imag* **13** (1994) 601–609

Simulation of Detectors for Biomedical Imaging

Paolo Russo

Abstract The goal of this lecture is to provide basic information on the role of computer simulations in the investigation of the properties of semiconductor radiation detectors, when evaluating the predicted response of biomedical imaging systems under development. It will be presented some examples on the simulation of the response of room-temperature radiation detectors (Si, GaAs, CdTe) for medical imaging, and their comparison with experimental data. Simulations with the EGS4 Monte Carlo code will include pencil X-ray beam irradiation of thick semiconductors substrates in the 10–100 keV diagnostic energy range, in order to extract information on the intrinsic spectroscopic and spatial resolution performance of detectors for digital radiography. The problem of the simulation of the electric field internal to biased compound semiconductor detectors will be addressed. Examples of Monte Carlo simulations for detector response characterization in nuclear medicine (PET, scintigraphy) will also be shown.

1 Introduction

In this lecture will be shown some elementary examples of computer simulations performed in order to predict the response of semiconductor radiation detectors for biomedical imaging applications. These application fields include digital radiography, gamma-ray scintigraphy, positron emission tomography. In all cases, the EGS4 Monte Carlo code system [1] was used for the simulations. Rather than showing *how* these simulations have been done, the approach here is to show *why* they have been carried out and the final information that they can provide to the researcher.

P. Russo
Università di Napoli Federico II and INFN Napoli, Italy
e-mail: Paolo.Russo@na.infn.it

Y. Lemoigne, A. Caner (eds.) *Molecular Imaging: Computer Reconstruction and Practice*,   145
and Experiments,

2 Room Temperature Semiconductor Detectors

Radiation detectors based on semiconductor crystal substrates operating at room temperature have long been investigated for use in medical imaging. In digital radiography, they were studied as an alternative detector with respect to conventional film-screen combinations.

In nuclear medicine, they have been studied as alternative to scintillator based devices for the realization of gamma cameras with high energy resolution and high spatial resolution; they were initially investigated also for positron emission tomography (PET). The structure of a semiconductor detector is illustrated in Fig. 1 in its various geometries, from a single pixel to double-sided microstrip to pixel detectors.

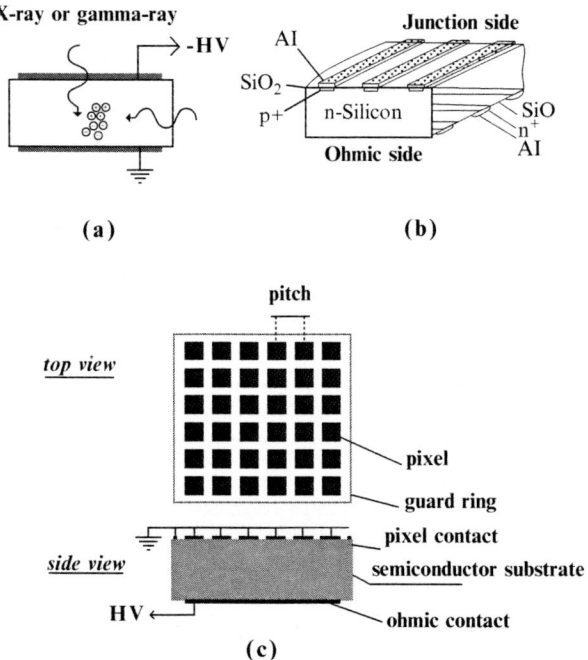

Fig. 1 (**a**) A single pixel (pad) detector made by a semiconductor substrate equipped with electrical contacts connected to a voltage bias supply. The incident (from the top or from the side) radiation produces electron-hole pairs in the detector, which drift by means of the applied electric field to the corresponding collecting electrodes, where they are amplified. (**b**) Scheme of a double-sided microstrip detector based on a silicon crystal, where strip junctions are realized on both sides in orthogonally to each other, so as to provide the X-Y coordinates of interacting photons by time coincidence of signals from top and bottoms collecting strips. (**c**) Scheme of a pixel detector, in which the collecting electrode on one side of the semiconductor substrate is segmented in a matrix array of square pixels: identification of the signal of the collection of the radioinduced charge on one of the pixels gives directly the coordinates of the interaction of the X-ray or gamma ray in the substrate

In its basic setup (Fig. 1a), the semiconductor substrate is used as a photoconductor slab equipped with electrical contacts on both sides: when the incident photon (X-ray or gamma-ray) radiation of energy E interacts in the substrate (either front-on or by the side), it generates a number of electron-hole (e-h) pairs $N = E/w$, where w is a (almost energy-independent) quantity specific of each semiconductor (for example, $w = 3.6\,eV/e$-h pairs for silicon at room temperature). These charges are separated by the applied electric field obtained by an external high voltage power supply, and they drift toward corresponding collection electrodes. During this process, a charge is induced in the external circuit connected to the electrodes: for example, a negative charge due to drifting electrons is induced at the anode electrode and this charge can be amplified and converted to a proportional voltage level. The dark (i.e., with no incident radiation) current of the detector can be reduced if a surface barrier junction can be realized on the doped semiconductor substrate. For example, on an n-type silicon substrate, a p-type junction can be realized; with gallium arsenide (GaAs) substrates, either semiconductor or semi-insulating, Schottky barrier junctions can be realized on one side and ohmic (non-inverting) contacts on the other side, and metal-semiconductor junctions can be realized as well on cadmium telluride (CdTe) substrates. In this case, the detector can be reverse biased so that a very low leakage current results, thus reducing the detector noise.

In order to realize a position sensitive detector useful for imaging, i.e. one able to provide X-Y coordinates of the 2D position of the photon interaction in the detector, two geometries can be setup. In Fig. 1b is shown a double-sided microstrip silicon detector, where junctions are realized on both detector substrate sides in the form of thin (e.g., 12–25 µm) parallel strips with a fine pitch (e.g., 50–100 µm).

With a suitable high voltage reverse bias, the diode detector can be completely depleted so that the detector active length is the full substrate thickness Δx (e.g., 300 µm). Alternatively, one can segment the contact on one side of the semiconductor detector substrate in the form of pixels and make a uniform contact on the opposite side (Fig. 1c): this can be done either in the photoconduction mode or by realizing junctions under each pixel electrode.

These detectors are normally operated in the photon counting mode, i.e. a photon is detected and counted if it deposits in the detector a charge greater than a given threshold. Alternatively, a photoconductor device can be operated in energy-integrating mode, i.e. it generates a photocurrent whose intensity is proportional to the total amount of energy deposited in the detector element during the X-ray exposure. The quantum efficiency $q(E)$ of such a semiconductor detector at photon energy E is given by $q(E) = [1 - \exp(-\mu(E)\Delta x)]$, where $\mu(E)$ is the linear attenuation coefficient of the substrate material at energy E. In order to estimate the intrinsic detection efficiency $\eta(E)$ – the fraction of incident photons detected – for such a detector, for any given detection threshold, one can perform a simulation with, e.g., a Monte Carlo computer code. This is a very basic task that allows to predict fundamental characteristics as detection efficiency in the energy region of interest and to compare it with measured values, in order to assess the detector performance. In Fig. 2 is shown the result of a simple Monte Carlo simulation carried out with the EGS4 code system[1] for the calculation of detector intrinsic efficiency. In this

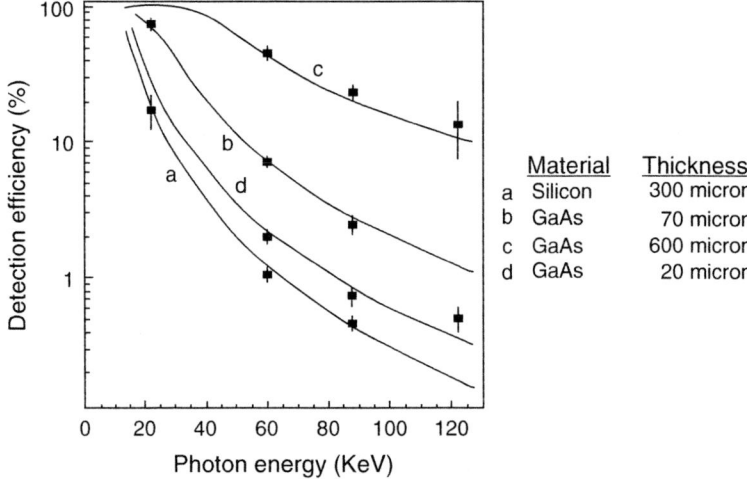

Fig. 2 Detection efficiency in the diagnostic energy range of Si and GaAs detectors of various thicknesses, obtained after EGS4 Monte Carlo simulation. In the simulation, for being detected, a photon has to release at least 10 keV (detection threshold) in the substrate, equivalent to about 2,800 electrons in Si and 2,400 electrons in GaAs. Simulations (continuous curves) are compared with experimental measurements (filled squares) (Data of University of Pisa Medical Physics group)

simulation, a large number of photons is directed normally to the surface of a detector made with a Si or GaAs semiconductor substrates of various thicknesses, and their energy release in the detector is scored; a detection threshold of 10 keV equivalent energy was set, i.e. a photon is detected if it deposits in the substrate an energy greater than this value. This calculated data are then compared with measurements taken in the laboratory with gamma emitting reference sources and prototype single-pixel (pad) detectors. This elementary simulation was the first task in a research project aiming at realizing semiconductor detectors for digital radiography, i.e. with photons in the so-called diagnostic energy range (10–100 keV). Fig. 2 clearly shows that apart from energies useful for X-ray mammography (around 20 keV mean energy), a 300-μm thick Si detector or a 20- μm thick GaAs detector do not provide significant absorption of X-ray photons, but a few hundred μm thick GaAs detector has enough efficiency at mammographic energies. In a parallel simulation aiming at characterizing thick GaAs detectors for nuclear medicine gamma-ray imaging at 140 keV (Tc-99m radiotracer main emission), an EGS4 simulation provided an estimate of 13% intrinsic detection efficiency for a 1-mm thick detector (Fig. 3). Such detectors were actually realized on a semi-insulating GaAs substrate equipped with Schottly contact(s) on one side and a ohmic contact on the other side, in the form of single pad detectors up to 1-mm thickness and in the form of pixel detectors up to 0.6 mm thick (see below).

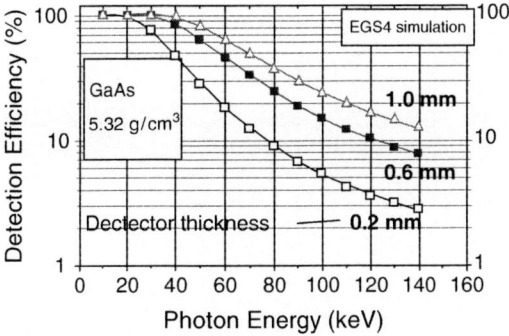

Fig. 3 Detection efficiency of GaAs detectors of various thicknesses, obtained after EGS4 Monte Carlo simulation. A detection threshold of 10 keV was assumed. At 140 keV (gamma-ray emission of Tc-99m radiotracer used in nuclear medicine) a detector as thick as 1 mm has 13% detection efficiency

3 Digital Radiography[2]

Digital radiography and digital mammography are research areas where semiconductor X-ray detectors have been extensively investigated. The main limitation of these detectors is related to the difficulty in realizing large sensitive areas (e.g., $18 \times 24\,cm^2$ for mammography) with minimum or no dead space between single detector units. The present status of commercial availability favors the so-called Digital Radiography (DR) imaging devices, that make use of flat panel detectors based on a matrix array of thin film transistors and a matrix array of a:Si:H photodiodes coupled to a scintillator layer (indirect detection devices) or a a:Se photoconductor pixel detector (direct detection devices). Both are energy-integrating devices. In mammography, however, at least one commercial system exists based on silicon microstrip detectors irradiated sideways and working in single photon counting.

In order to show some basic simulations carried out in this field with Si detectors, let us consider the problems of the limiting spatial resolution attainable and of their low intrinsic detection efficiency. Figure 4 shows the results of an EGS4 simulation aimed at estimating the lateral spread of the charge deposit in 300 μm thick silicon detectors. Once a photoelectron or a Compton electron has been generated in the Si substrate by the interaction of an X-ray photon, it loses kinetic energy in the medium along a path where it is multiply scattered: a detailed simulation of the particle track allows to determine the average lateral extent of the energy deposit in the detector, giving an estimate of the limiting spatial resolution attainable as a function of the photon energy. Figure 4 shows the result of a detailed EGS4 Monte Carlo simulation of monoenergetic X-ray photons (20, 60, 100 keV) incident normally to the 300-μm thick Si detector slab. At these low energies, electrons are generated almost orthogonally to the photon direction, and the lateral extent of their tracks is increasing at increasing photon energy (due to a larger electron range at higher energies), being a few μm at 20 keV and in the order of ∼30 μm at 60 keV (Fig. 4).

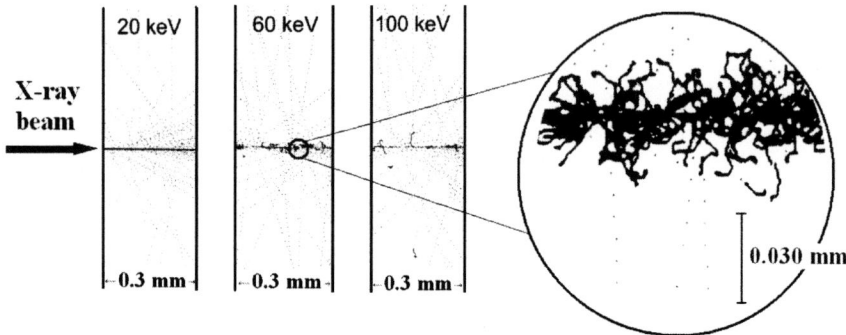

Fig. 4 In an EGS4 simulation, a monoenergetic pencil beam of 100,000 photons of 20, 60 or 100 keV energy was directed normally onto the surface of a Si slab 300 μm thick. The plots show the side views of the trajectories of only 5,000 tracks produced by secondary radiation (photons, dotted lines; electrons, continuous lines) generated in the detector. Most of the energy is deposited along the incident direction, but the lateral spread of the charge deposited increases with increasing photon energy. The rightmost plot is the zoomed view of the photoelectron tracks, Compton electron tracks and Compton scattered photon trajectories at 60 keV, occurring mostly at 90 deg with respect to the incident direction: in this zoom, the vertical bar corresponds to 0.030 mm

Hence, for a Si detector, a pitch in the order of 50 μm will be large enough to avoid firing more than one pixel or microstrip due to charge spread, up to about 60 keV.

In a second simulation concerning detection efficiency, we investigated via EGS4 Monte Carlo simulation a detector geometry made by a stack of several (microstrip) Si detectors, each 0.3 mm thick. While at 30 keV photon energy (e.g., mammography energy range) a stack of five detector slabs (1.5 mm total detector thickness) resulted enough to produce a total detection efficiency greater than 50% (Fig. 5c), another way was explored via computer simulation to increase efficiency at higher photon energies, by interposing a metal (molybdenum) thin slab in between two Si detector slabs (Fig. 5a).

The Mo layer has high attenuation for photons and performs as a converter layer producing secondary electrons which are efficiently detected by the Si detector both behind (i.e., upstream) and in front of the metal layer, but it should be thin enough to allow these secondary electrons to exit the Mo slab with sufficient kinetic energy. In this setup, the choice of the Mo layer thickness would be critical, and a simulation helps the researcher in predicting the response of detector geometries he would even never test experimentally, if not necessary. In Fig. 5b is shown the detection efficiency of the front and back Si detectors, as a function of the Mo slab thickness, at 60 keV primary photon energy. It is seen that while without the Mo layer the 0.3 mm thick Si detectors have roughly 1.5% detection efficiency, their efficiency increases up to about 2.5–3% with increasing Mo thickness up to about 50 μm. Then, with further increase of the Mo thickness, while the detection efficiency of the front detector does not increase the back detector sees less electrons and its efficiency declines (Fig. 5b). As a result of this simulation, a Mo converter layer of 50 μm thickness was found adequate to almost double the detection efficiency of the stack of 0.3 mm

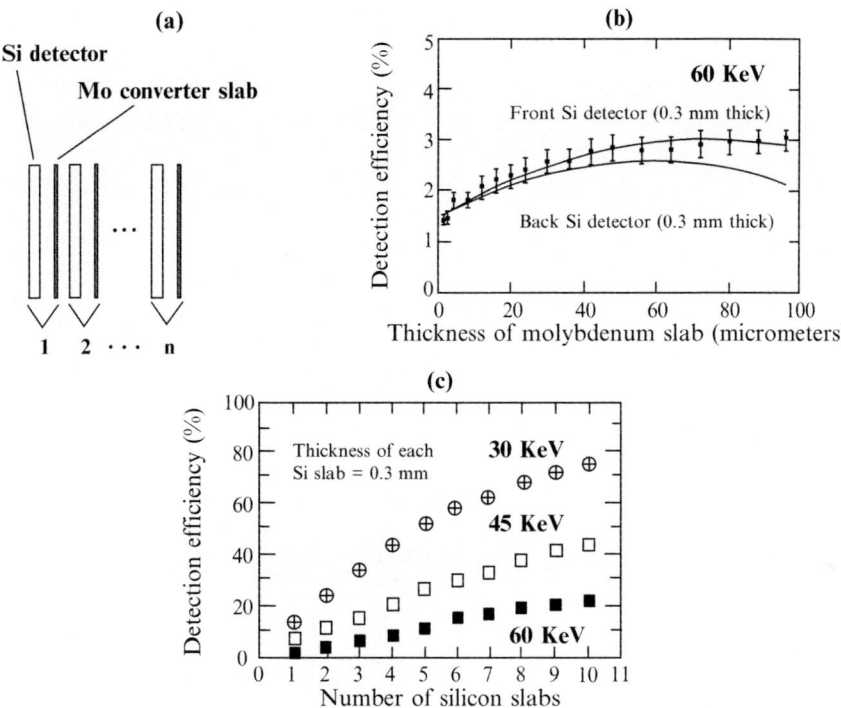

Fig. 5 (**a**) Scheme of a detector sandwich of n units made by silicon microstrip detector and a molybdenum thin slab used a converter for increasing the photon attenuation. (**b**) Simulation (made with EGS4 Monte Carlo code) of the detection efficiency of the sandwich detector (1 Si detector 300 μm thick, 1 Mo slab, 1 Si detector) as a function of the thickness of the Mo slab. Curves are parabolic best fit to the simulation data. (**c**) Simulated detection efficiency (without Mo converter plates) of a stack of 300 μm thick Si detectors, at three photon energies

thick Si detectors at 60 keV, and an arrangement of, e.g., 10 detectors in cascade with interposed Mo converter is able to determine an overall detection efficiency of about 40% at 60 keV (Fig. 5b, c).

In these simulations, a threshold of 2,000 e-h pairs (i.e., 7.2 keV) was set for determination of detection efficiency.

A more complex imaging task was simulated with an EGS4 Monte Carlo code for the double-sided microstrip detector for photon counting digital mammography: a parallel beam of 50,000 photons (whose energy was sampled from a Mo-target X-ray tube peaked at 24 keV) was directed onto a breast phantom consisting of a 60-mm thick water block (maximum diameter of compressed breast in the mammographic clinical procedure) containing at its center a microcalcification (a cube of 0.5 mm side). A stack of two double-sided Si microstrip detectors (each 0.5-mm thick, with 200 μm pitch microstrips) was used as X-ray detector behind the phantom (Fig. 6a). In the detectors, energy deposit was scored in equivalent pixels of 200 × 200 μm area. A line profile of the pixel counts (above a 7.2 keV detection

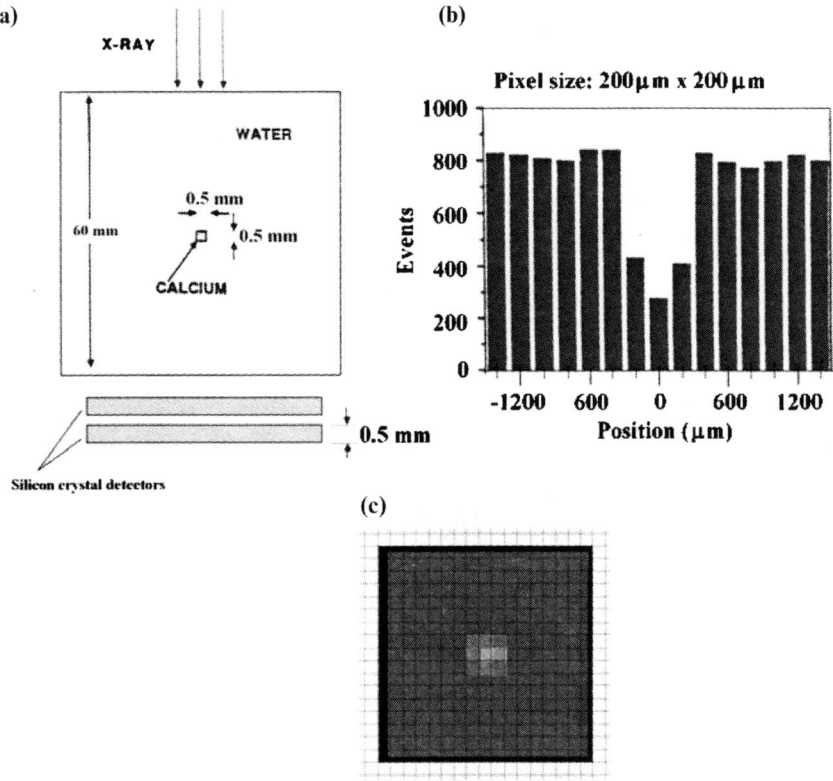

Fig. 6 (**a**) For simulating with an EGS4 Monte Carlo code a digital mammography system based on two stacked double-sided Si microstrip detectors 0.5 mm thick, an X-ray beam extracted from the emission spectrum of a Mo anode X-ray tube was directed on a 60-mm thick water slab containing a 0.5 mm microcalcification. (**b**) The line profile of the detector counts clearly reveals the presence of the detail with high contrast. (**c**) simulated image of the microcalcification detail as seen by the photon counting 2D Si microstrip detector

threshold) indicates that the microcalcification is clearly visible, extending over 3 pixels, with a very high image contrast as determined by the average counts inside the detail and in the background (Figs. 6b, c).

4 Positron Emission Tomography[3]

Semiconductor detectors have been proposed also for detection of high energy gamma rays in Positron Emission Tomography (PET), where photons with up to 511 keV energy must be detected: this task is normally performed by thick scintillator detectors (e.g., BGO, density = 7.13g/cm^3, effective atomic number $Z_{\text{eff}} = 74$) readout by photomultiplier tubes. Semiconductor detectors operating

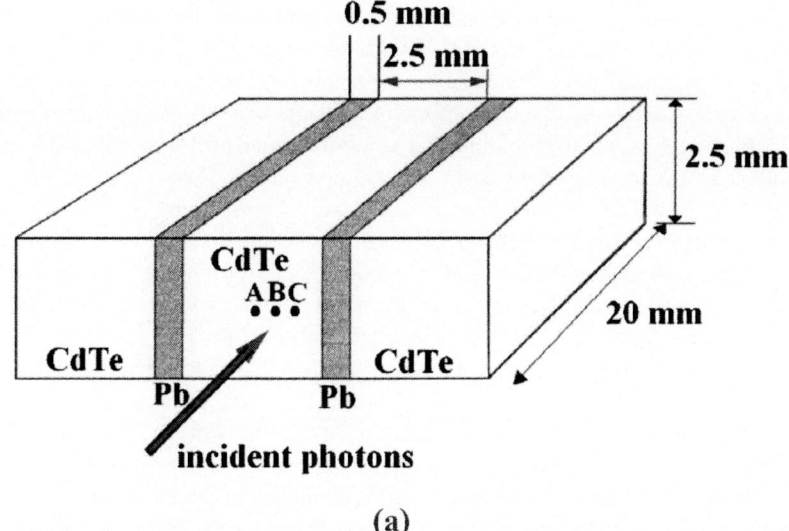

(a)

	Position	Efficiency (%)	Lateral contamination (%)	Cross talk probability (%)
Noise threshold (9.4 keV)	A	63.2 ± 1.1	0.02 ± 0.02	1.2 ± 0.1
	B	62.9 ± 1.1	0.07 ± 0.03	1.5 ± 0.2
	C	62.9 ± 1.1	0.09 ± 0.04	1.5 ± 0.2
Low threshold (200 keV)	A	38.9 ± 0.8	0.45 ± 0.09	0.18 ± 0.06
	B	38.4 ± 0.8	0.48 ± 0.09	0.24 ± 0.07
	C	37.6 ± 0.8	0.53 ± 0.10	0.32 ± 0.08
High threshold (350 keV)	A	18.1 ± 0.6	0.26 ± 0.07	0.0
	B	17.8 ± 0.6	0.22 ± 0.07	0.0
	C	17.3 ± 0.6	0.26 ± 0.07	0.0

(b)

Fig. 7 (**a**) Geometry for the simulation of the response of CdTe crystal detectors ($2.5 \times 2.5 \times 20\,mm^3$) for 511 keV photon imaging in PET. The single crystals are separated by a 0.5 mm thick lead septum. Position of incidence of the beam (10,000 photons) are A (at 1.25 mm from the septum), B (at 0.75 mm from the septum), C (at 0.25 mm from the septum). (**b**) Results of the Monte Carlo simulation in terms of detection efficiency, lateral contamination and cross talk probability, for three detection thresholds at each incidence position

at room temperature have some potential interest in PET detector research; cadmium telluride (CdTe) based detectors could be a choice, due to its high density ($6.2\,g/cm^3$) and high effective atomic number $Z_{eff} = 50$. In Fig. 7a is shown a possible geometry for this detector application, in which finger-like CdTe crystals of

$2.5 \times 2.5 \times 20 \, \text{mm}^3$ are oriented with their long axis in the direction of the incident photons. The $2.5 \times 2.5 \, \text{mm}^2$ sensitive area determines the pixel size and pixels are separated from each other by Pb septa of 0.5 mm thickness, to prevent cross talk from adjacent detectors disposed in a ring arrangement. A Monte Carlo simulation with EGS4 was performed in order to predict the detector response in various geometry and readout configurations, to ten thousand 511-keV photons impinging along the crystal axis onto three entrance positions (A, B, C, in Fig. 7) at 1.25, 0.75 and 0.25 mm from the Pb septum, respectively. The parameters investigated were the detection efficiency (fraction of incident photons producing a number of e-h pairs above threshold in the central pixel), lateral contamination (fraction of incident photons producing a number of e-h pairs above threshold in one of the two lateral crystals) and cross talk probability (fraction of incident photons producing a number of e-h pairs above threshold in both the central pixel and one of the two lateral crystals). The simulation was repeated for three different detection thresholds (just above noise at 9.4, at 200 and at 350 keV). High energy thresholds in the 200-350 keV range are used in PET systems in order to reject most of the photons deriving from Compton interactions in the biological target. The result of this simulation is shown in tabular form in Fig. 7b. Increasing the detection threshold from 200 to 350 keV reduces the detection efficiency from about 38% to about 18%, but the cross talk probability then drops from 0.1–0.3% to zero. Correspondingly, the probability of lateral contamination almost halves, and its value of about 0.3% is small enough to allow for a very limited spatial resolution degradation, due to the spread of the charge in adjacent pixels.

5 Scintigraphy[4]

While – as seen above – the detection efficiency of semiconductor detectors for 511 keV photons is still low (e.g., <40%) and not comparable with that attainable with thick scintillator crystals used in PET systems, semiconductor detectors can approach the efficiency of scintillator based systems for low energy gamma-ray imaging, and active research is being conducted for the realization of imagers based on semiconductor detectors that could offer improved performance in specific imaging tasks in nuclear medicine where compactness, high energy resolution and high spatial resolution are of fundamental importance.

The basic detector used in planar gamma-ray imaging in nuclear medicine is the gamma camera, a detector based on a large scintillator crystal (e.g., NaI:Tl) readout by an 2D array of photomultiplier tubes and coupled to a lead collimator that selects the direction of the incident photons (Fig. 8).

The spatial resolution R of such a detector depends on the scintillator crystal intrinsic resolution R_{scint} and on the collimator resolution R_{coll}, according to $R = \sqrt{[(R_{\text{scint}})^2 + (R_{\text{coll}})^2]}$. In turn, the collimator resolution depends on the source-collimator distance z and on collimator hole diameter d and length L, according to $R_{\text{coll}} = d(z+L)/L$. The spatial resolution of clinical gamma cameras is in the order

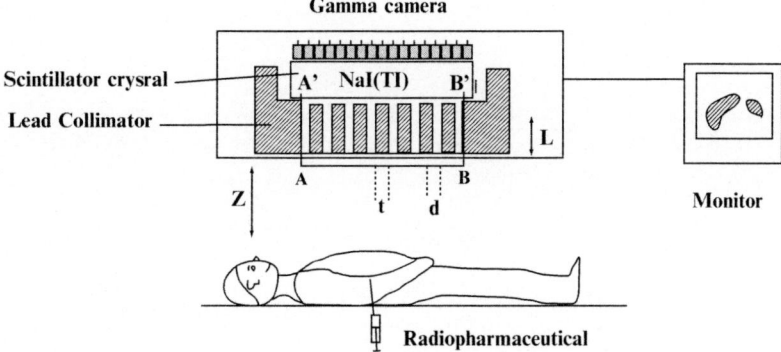

Fig. 8 Basic setup of a planar (scintigraphic) gamma-ray imaging system in nuclear medicine, employing a scintillator (e.g., NaI:Tl) based gamma camera equipped with a lead collimator in order to image (with spatial resolution in the order of several mm) the radioactive gamma-ray emission from injected radiotracers (e.g., Tc-99m, 140 keV photon energy). A parallel hole collimator is shown, having holes of size d, septa of thickness t and hole length L

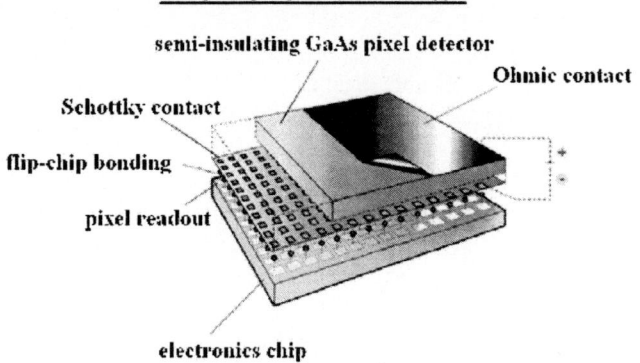

Fig. 9 Scheme of a photon counting hybrid pixel detector (here based on a semi-insulating GaAs substrate) consisting of a pixel detector bump-bonded pixel-by-pixel to the microelectronic circuit containing pixel amplification, threshold discrimination and counting[4]

of several mm, and it is ultimately limited by the intrinsic resolution of the scintillator crystal and its readout system. Analogously, the energy resolution (used for rejecting Compton scattered photons in the tissue) is not better than 10% at 140 keV photon energy. A direct-detection semiconductor pixel detector could improve the intrinsic spatial resolution by offering sub-millimeter pixel sizes and the energy resolution as well, which in the case of CdTe and CdZnTe semiconductor detectors a few mm thick can reach $\sim$5%. A semiconductor pixel detector proposed for Tc-99m photon counting gamma-ray imaging in compact gamma cameras is based on the so-called hybrid technology, in which the pixel detector is readout pixel-by-pixel by a separated electronic integrated circuit by means of micro-bondings (Fig. 9).

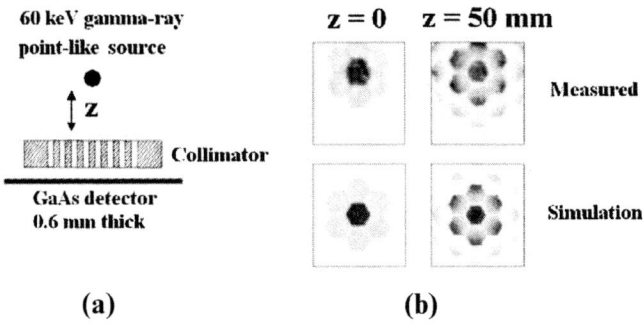

Fig. 10 (a) Setup for the measurement and the simulation of the imaging performance (at 60 keV photon energy) of a photon counting GaAs hybrid pixel detector, 0.6 mm thick, having 64 × 64 square pixels of 170 μm pitch. The Low Energy High Resolution parallel hole lead collimator of a clinical gamma camera was used: it is 40 mm thick and has hexagonal holes of 1.8 mm with septa of 0.2 mm. (b) Data of University of Pisa Medical Physics group

Substrates used for this hybrid technology include Si, semi-insulating GaAs and CdTe, with thicknesses up to 1–1.5 mm. An EGS4 Monte Carlo simulation was setup in order to predict the spatial resolution performance of such a detector, based on a 0.6 mm thick GaAs substrate, having 170 × 170 μm pixels. A source emitting 60 keV gamma-rays was used, either in contact or at 50 mm distance from the parallel hole collimator (Fig. 10a).

This last was a Low Energy High Resolution collimator of a clinical gamma camera, whose resolution is matched to that typical of NaI(Tl) based detectors. The photon counting GaAs pixel detector has an intrinsic resolution of 0.34 mm (twice the pixel pitch), so that the collimator resolution (1.8 mm at $z = 0$ and 4 mm at $z = 50$ mm source-collimator distance) should dominate the overall spatial resolution. This is confirmed by both simulation and experimental measurements (Fig. 10b), thus proving that semiconductor hybrid pixel detectors with sub-millimeter pitch have potential to produce ultra high spatial resolution for gamma-ray imaging.

6 Charge Transport in Semiconductor Radiation Detectors[5,6]

Detector material, substrate thickness, dark current and pixel pitch are not the only parameters that determine the final spectroscopic and imaging performance of a semiconductor detector. As a matter of fact, the specific, detail mechanisms of electron and hole transport in the semiconductor bulk are the limiting factors that allow to obtain an X-ray and gamma-ray detector with high detection efficiency, high energy and spatial resolution, and low noise.

A key role in the design and detector performance evaluation is then played by computer simulations that take into account not only the physics of photon interaction, charge slowing-down and energy deposit in the detector bulk, but also the drift of the radiation-generated electrons and holes under the internal electric field determined by the application of an external voltage bias. In this drift process toward the collecting electrodes, material impurities in the semiconductor substrate and local reduction of the electric field may represent traps for electrons and holes and low-drift-velocity regions, respectively, so that short electron and hole lifetimes and inefficiencies in the collection of charge at the electrodes take place. A reduced collected charge with respect to the photo-generated one determines a degradation of the energy information, a lower energy resolution and an increase of the noise level. Moreover, applying a constant voltage across the opposite sides of a semiconductor pixel detector determines a complex structure of the electric field lines in the substrate – possibly complicated by the presence of sub-surface junctions – so that radiation-induced charge can diffuse and/or be transported under adjacent pixels, thus lowering the spatial resolution. Computer programs can help analyze these complicated phenomena, by adding electrostatics and electrical charge transport simulation tools to the Monte Carlo code for photon interaction description. A number of simulation packages are available commercially for description of charge transport in biased semiconductor substrates.

In order to show an example illustrating this use of computer simulations, Fig. 11 shows the case of a $1 - mm^2$ semi-insulating GaAs detector equipped with a Schottky contact on one side and an ohmic contact on the opposite side. The detector substrate is 0.6 mm thick and the single pad has a diameter that was set in the range $70 - 520 \mu m$. The following parameters were set for the charge transport simulation: electron lifetime 3 ns, hole lifetime 20 ns, electric-field dependent electron and hole mobilities, no detrapping and no charge diffusion mechanisms included. EGS4 was used for the Monte Carlo generation of photon interactions in the substrate, with a simulated electronic noise of 1,000 electrons (4.2 keV) added to the statistical charge generation spread. The results of the simulation show the large variation in electric field intensity (Fig. 12a) and potential (Fig. 12b) inside the detector, both laterally (i.e., in a direction parallel to the detector surface) and in the detector depth below the front contact. Low field regions affect negatively

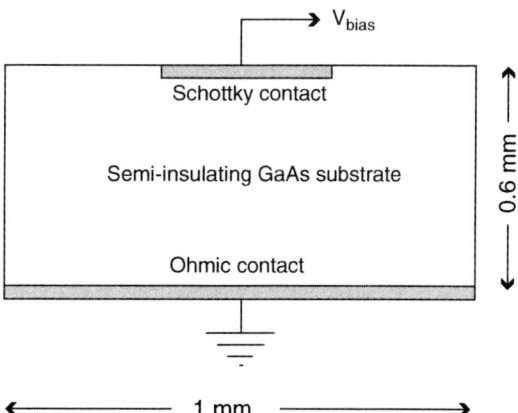

Fig. 11 GaAs pad detector used in the simulation. The front circular contact has a diameter varied between 70 and 520 μm. The reverse bias voltage was set at −600 V

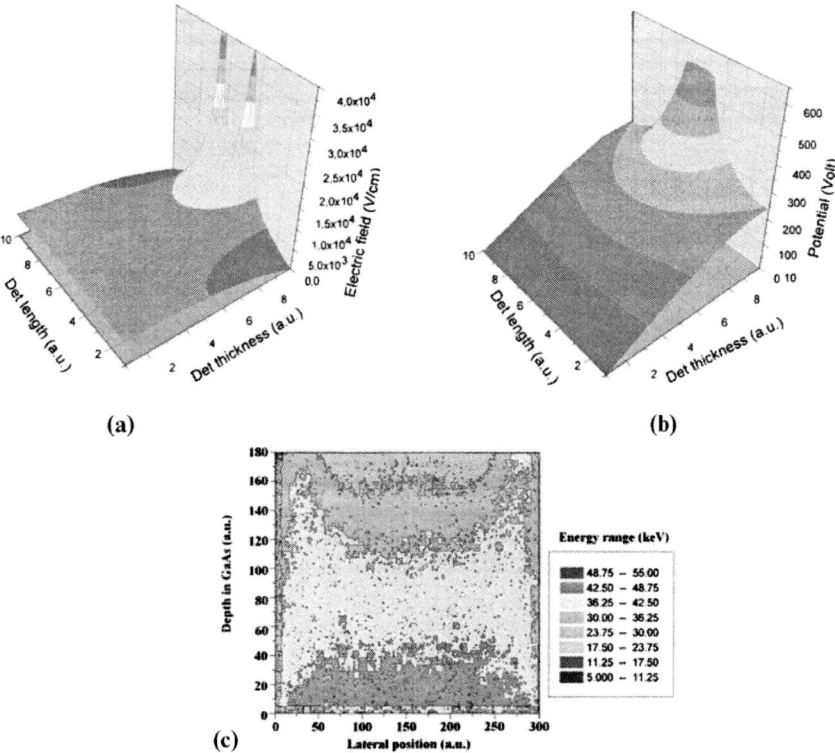

(a)   (b)

(c)

Fig. 12 Plot of the electric field (**a**) and electrical potential (**b**) distribution inside the biased GaAs detector, having a contact pad of 220 μm diameter, under 600 V reverse bias. Please note the large variation in electric field intensity from below the front pad to the ohmic side of the detector. The field also reduces rapidly laterally with respect to the pixel, so that a high field region is present only below the front contact. Detector thickness is 0.6 mm and detector length is 1 mm. (**c**) Scatter plot of the charge collection efficiency (expressed as fraction of 60 keV primary photon energy locally deposited in the substrate volume) under -600 V reverse bias applied to the pixel contact, for a GaAs detector, having a contact pad of 520 μm (By courtesy of Dr. M. G. Bisogni, University of Pisa Medical Physics Group)

the charge transport and collection properties, and it is seen that in regions far away from the front contact a significant reduction in electric field occurs, and that the potential drops rapidly at the sides of the front (pixel) contact, reducing the lateral charge collection. As regards the efficiency ε of charge collection at the electrodes, one can express this quantity by the energy deposited locally in the detector substrate volume after uniform irradiation of the detector with photons of fixed energy E (i.e., by representing the quantity $\varepsilon \cdot E$), generated by Monte Carlo simulation of the interaction in the GaAs substrate.

This plot is presented in Fig. 12c for 60 keV photons and shows (i) an increased charge collection efficiency sideways to the pixel contact (due to a higher field), (ii) a low collection efficiency below the pixel contact due to high charge trapping in a high electric field, and (iii) a high collection efficiency region close to the ohmic contact, where electron collection is favored.

As a final example of the application of a mixed Monte Carlo-electrical simulation, an investigation is presented where a semi-insulating GaAs pixel detector having square pixels of 40μm side and 10μm interpixel gap (i.e., 50μm pitch), 200μm thick, is studied at 40keV photon energy: this total energy deposit would correspond to about 10,000 e-h pairs generated in the detector active area.

A reverse voltage bias of 200 V is applied to the pixellated side with respect to the ohmic contact on the back, and the photon interaction is supposed to occur just in the middle of the interpixel zone, 25μm below the detector surface. Hole end electron lifetimes were set at 5 or and 3 ns, respectively, and no charge carrier trapping or detrapping was simulated. The electron and hole mobilities were field-independently and fixed to 1,000 and 350cm^2/Vs, respectively. The 3D semiconductor simulation tool used here is a commercial code (Davinci) that simultaneously solve Poisson's equation for the electrostatic potential and current continuity equations for electrons and holes, with suitable stability conditions. Lateral charge diffusion was also taken into account. The simulation can show the time-dependent evolution of the charge drift toward respective electrodes from the initial time $t = t_0$ of e-h charge pairs creation. Figure 13 shows (superimposed and slightly offset laterally for clarity) the drift and diffusion process of the charge clouds of electrons and holes, at various time delays.

The analysis of this plots suggests that some sharing of the charge (holes) between the two neighboring pixels occurs at collection times greater than 0.7 ns: this charge sharing effect reduces the total signal to pixels so that full energy deposition under a pixel area is not guaranteed, and may produce a lowering of the spatial resolution. In photon counting systems, where the pixel is considered hit if the charge deposit is above a given charge threshold, this charge sharing process for photons interacting in the interpixel area may determine a loss of the photon count or a double count in two adjacent pixels, a phenomenon which has to be seriously taken into account in designing and realizing semiconductor pixel detectors for X-ray and gamma-ray medical imaging applications.

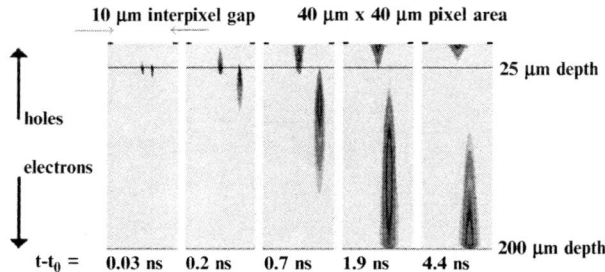

Fig. 13 Temporal evolution at different times of the ten thousand electrons (downgoing) and holes (upgoing) charge clouds created in a 200 m thick GaAs pixel detector at t = t$_0$ in the middle of the region between two adjacent pixels, 25 m below the detector surface. For clarity, in this plot the electron cloud was slightly shifted laterally, thus avoiding overlap with the hole cloud. The simulation allows to show the lateral diffusion of the charge, the different mobilities of the charge carriers which induce differences in the charge collection times at the electrodes (Data by courtesy of Dr. M. Chmeissani, IFAE, Barcelona, Spain)

References

1. W. R. Nelson, H. Hirayama, D. W. O. Rogers. The EGS4 code system. SLAC report 265 (1985).
2. Bencivelli W. et al. Use of EGS4 for the evaluation of the performance of a silicon detector for X-ray digital radiography. *Nucl. Instrum. Methods* A **305**, 574–580 (1991).
3. M. Conti et al. Use of the EGS4 Monte Carlo code to evaluate the response of HgI2 and CdTe detectors for photons in the diagnostic energy range. *Nucl. Instrum. Methods* A **322**, 591–595 (1992).
4. Medipix Collaboration website: www.cern.ch/medipix
5. M. G. Bisogni, A. Cola, M. E. Fantacci. Simulated and experimental spectroscopic performance of GaAs X-ray pixel detectors. *Nucl. Instrum. Methods* A **466**, 188–193 (2001).
6. M. Chmeissani, B. Mikulec. Performance limits of a single photon counting pixel system. *Nucl. Instrum. Methods* A **460** 81–90 (2001).

Computer Aided Detection (CAD) Systems for Mammography and the Use of GRID in Medicine

Cad for Mammography

Adele Lauria

Abstract It is well known that the most effective way to defeat breast cancer is early detection, as surgery and medical therapies are more efficient when the disease is diagnosed at an early stage. The principal diagnostic technique for breast cancer detection is X-ray mammography. Screening programs have been introduced in many European countries to invite women to have periodic radiological breast examinations. In such screenings, radiologists are often required to examine large numbers of mammograms with a double reading, that is, two radiologists examine the images independently and then compare their results. In this way an increment in sensitivity (the rate of correctly identified images with a lesion) of up to 15% is obtained. [1,2] In most radiological centres, it is a rarity to find two radiologists to examine each report. In recent years different Computer Aided Detection (CAD) systems have been developed as a support to radiologists working in mammography: one may hope that the "second opinion" provided by CAD might represent a lower cost alternative to improve the diagnosis. At present, four CAD systems have obtained the FDA approval in the USA.[†] Studies[3,4] show an increment in sensitivity when CAD systems are used. Freer and Ulissey in 2001 5 demonstrated that the use of a commercial CAD system (ImageChecker M1000, R2 Technology) increases the number of cancers detected up to 19.5% with little increment in recall rate. Ciatto et al.,[5] in a study simulating a double reading with a commercial CAD system (SecondLook[‡]), showed a moderate increment in sensitivity while reducing specificity (the rate of correctly identified images without a lesion). Notwithstanding

A. Lauria
Università di Napoli Federico II and INFN Napoli, Italy
e-mail: alauria@na.infn.it.

[†] URL: http://www.icadmed.com/. Accessed March 2006.
URL: http://www.r2tech.com/main/home/index.php. Accessed March 2006.
URL: http://www.kodak.com/global/en/health/productsByType/medFilmSys/eqp/system/mamCad .jhtml. Accessed march 2006.
URL: http://imaginis.com/breasthealth/news/news6.03.03.asp. Accessed March 2006.

[‡] CADx Systems Inc., now owned by iCAD, Nashua, NH, USA.

Y. Lemoigne, A. Caner (eds.) *Molecular Imaging: Computer Reconstruction and Practice,*   161
and Experiments,
© Springer Science+Business Media B.V., 2008.

these optimistic results, there is an ongoing debate to define the advantages of the use of CAD as second reader: the main limits underlined, e.g., by Nishikawa[6] are that retrospective studies are considered much too optimistic and that clinical studies must be performed to demonstrate a statistically significant benefit from the use of CAD.

The implementation of a screening program generates huge amount of data (images and programs) that are distributed over a large area in different hospitals and mammographic centres. At present, it is not possible to share data and images among different centres. The GRID technology allows distributed database data to be shared as easily as they are on a local database.

1 Breast Cancer: Occurrence and Radiological Diagnosis

Every year in Europe about 75,000 women die for breast cancer.[7,8] Early detection is the best way to fight breast cancer because therapies are much more effective if the cancer is at an early stage.[1,2] Mammography is the most efficient diagnostic technique in detecting the signs of breast cancer. Mammography systems generate a gray-level analogical or digital image, the intensity levels of the gray levels are proportional to the X-ray transmission through tissue. The intrinsic limit of the mammography technique is due to the fact that the absorption coefficient of the tumorous lesions of 20 KeV X-rays is not very different from surrounding tissue, so it is not always possible to distinguish the cancers signs. Moreover, some studies show that diagnostic errors are due not only to the imaging modality, but can be attributed to radiologists' differing visual perceptions, or even to the variation of reading conditions. Many studies have been performed in last decades to reduce these kinds of misdiagnoses.

2 Screening Mammography and Double Reading

The screening is the periodical recall (every year or two) of mammographic checks of women from 49 to 69 years old. A recent study[9] shows the efficacy of the screening in terms of survival. The screening offers the possibility to immediately operate on asymptomatic women in which the disease is not yet manifest. That is surely advantageous but it still remains the problem to prevent the reading errors from radiologists. To this aim it was very advantageous the double reading by two radiologist independent each other. In the European countries where both these modalities (screening and double reading) have been adopted a decrease in the mortality rate was obtained.

3 Mammographic CAD Systems

The first study in which the advantages resulting from the use of a computer in mammography was published in 1967.[10] It showed that the digitalization and the computerized analysis of the radiographic image reveal mammographic anomalies. Since then, the development of the CAD system has been parallel with the development of technology, increasing computing capacities and improving mammographic images. In the area of digital radiology, the study of CAD systems has yielded significant results. The interest in CAD systems continues to grow.

A CAD system, using trained algorithms to recognise suspicious pathologies in a digitalised mammogram, provides a way to capture the experience of a specialist and to code it in a computerized way. *Computer Aided Detection* is a support to diagnosis by a radiologist as an aid to locate lesions. In 1995, for the first time ever, a CAD system was introduced – in the clinical practice in the University of Chicago, where screening mammograms were digitised and analysed to look for massive and microcalcification lesions.[11,12] The studies of the last years show that the computerised research of the lesion in mammography significantly increases the diagnostic capability of the radiologist.[3,4,13–17] Other studies on the search of massive lesions showed that a computerized system is able to identify suspicious lesions that were missed by radiologists. Several different approaches have appeared in the literature, but it is possible to identify two major research lines: one for the microcalcification lesions and the other for the massive ones. These lesions have different morphologies. Microcalcifications are very small (≤ 0.5 mm diameter) accumulation of calcium, deposited in breast tissue. They are usually of varying shapes and frequently occur in tight groups (clusters). Sometimes these calcifications, seen on a mammogram, occur with an associated mass. When their form is not regular they could be malignant. The analysis of the microcalcification has two main problems: the localization and the classification. The former is, principally, due to the size of the calcifications (from 100 micron to few millimetres). The morphology and the way they are distributed inside the tissue are two essential parameters in the recognition and in the evaluation of the microcalcifications. In Fig. 1 a scheme of different types of microcalcifications is shown. Some parameters useful to determine if the microcalcification clusters are malignant or not, are:

1. The shape of the cluster (rounding and regular shape is associated with to benign lesions)
2. Different density and irregular margins of microcalcifications
3. Total numbers of microcalcifications (superior to 15–20 is indicative of malignant lesions)
4. Density of the microcalcifications (superior to 10 in an area of $2 \, cm^2$ is indicative of malignant lesions)
5. Presence of multiple clusters of microcalcifications

What a radiologist does, is to associate all of the morphological characteristics of the microcalcifications (or of a cluster of them) to a grade of malignity, even if this

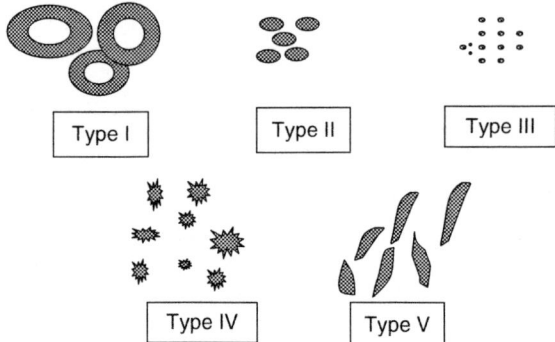

Fig. 1 Different kind of microcalcifications: (I) "teacup" calcifications, always benign; (II) eggshell calcifications, benign in 80% of the cases; (III) round calcifications, malignant in 40% of the cases; (IV) irregular calcifications, malignant in 70% of the cases; (V) linear calcifications, always malignant[18]

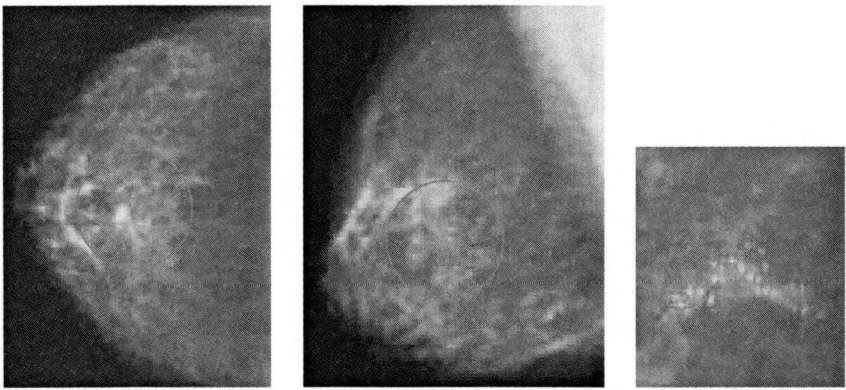

Fig. 2 Example of malignant microcalcifications

association is depends to the surrounding tissue. In Fig. 2 examples of malignant lesions are shown.

Massive lesions are groups of cells that are more dense than the surrounding tissue; their shape, margin and density characterize them are used by radiologists for their classification.

In order to classify a massive lesion it is necessary to consider its shape, margins, presence of halos, density and size. In Fig. 3 a scheme of different shapes is shown.

Even for this kind of lesion the irregularity of the shape reflects the grade of malignity of the lesion. Star-shaped lesions (Fig. 4) are generally malignant. Although it is important to emphasize that the human process of reading, analysing and evaluating the radiological image can not be replaced by a computer, the benefits of the elaboration and analysis process with computerized support cannot be overlooked. To this aim, in the last years different algorithms were developed to optimize the informative content of the image.[19–21]

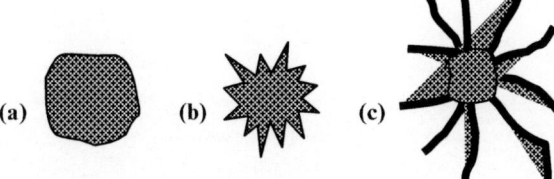

Fig. 3 Massive lesion with regular (**a**), star-shaped (**b**) and star shaped with spike lines radiating in all directions from a central mass (**c**)

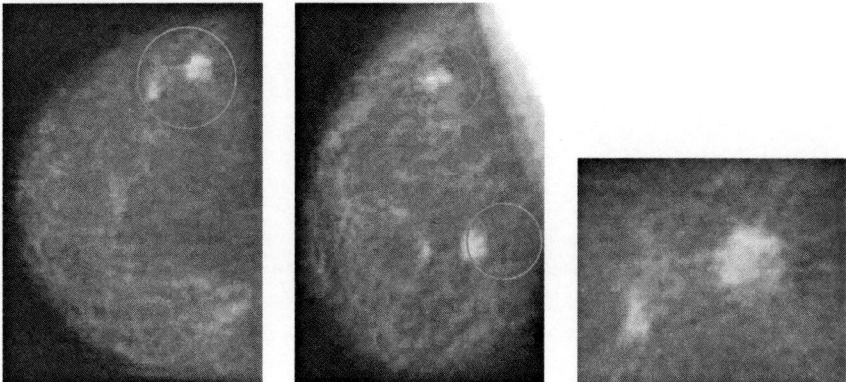

Fig. 4 Example of star lesions

4 Architecture of CAD Systems

The purpose of a CAD system is to extrapolate and to analyse the characteristics of malignant and benign lesions in a mammogram. This assists radiologists with their diagnostic accuracy, reducing, in this way, the total number of false positive cases, and the number of unnecessary biopsies.[22] These systems process images, previously digitised, searching them for signs of breast neoplasies. Once the mammogram is in a digital form, the recognisability of the lesions from an expert system happen in three steps:

1. Preprocessing of the image through digital technique to eliminate the parts of the imagecontaining meaningless information, e.g. background and noise (*image preprocessing*).
2. Preprocessing of the image through digital technique to eliminate the parts of the imagecontaining meaningless information, e.g. background and noise (*image preprocessing*).
3. Analysis of the image and extraction of features to classify the images (*feature extraction*). In this step, using algorithms to elaborate the information contained in the pixels of the image through specific algorithms such as contrast enhancement, low-pass or high pass algorithms, Region Of Interests (ROI) containing

Fig. 5 Scheme of a computerized automatic system for
the detection of lesions5

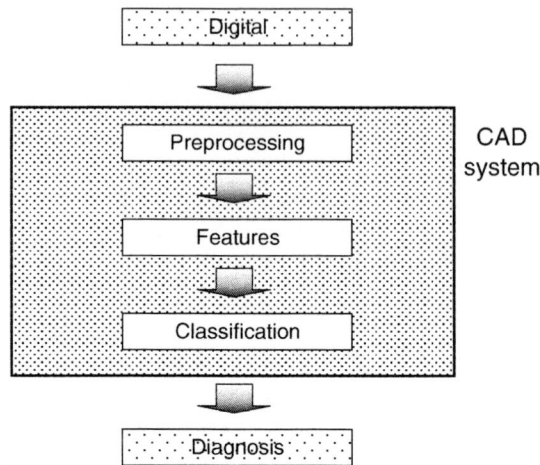

suspicious lesions were found. From these ROIs characteristic parameters are extracted that will be evaluated in the following step for the classification.

4. Evaluation and classification, performed by a Neural Network, of the features extracted from the ROIs in the previous step (*data classification*).

In Fig. 5 is schematized the workflow of an automatic system.

4.1 Neural Network System

The choice of an appropriate and well tested Neural Network (NN) plays a fundamental role in determining the effectiveness of a CAD system. An artificial NN is made of interconnected elements called neurons; their distinctive feature is their ability to learn and to classify from given patterns of data, in a manner that mimics in some elementary way the mechanisms of *natural* networks of brain neurons. [23]

Neurons are computational elements acting as integrators of their weighted inputs. In an artificial NN there can be n input (each of them can be weighted by a parameter w_i) that will be summed and then passed through an activation function. The output y of the NN depends on the value of the activation function. This simple structure becomes highly complex when many neurons are organised in a parallel way. The choice of the inputs, weights, activation function, output threshold and their architecture, determine the ability of the NN to learn and to classify.

A typical topology adopted for static conditions is the so-called *feedforward* scheme. In such a scheme the NN is composed of i input nodes, each one connected to the h nodes of a second layer, the *hidden layer*. These are connected to the nodes of a third output layer, which generates the output. The number of neurons in the hidden layers can be determined in empirically, by a trial and error process; the number of hidden neurons has an upper limit of $i - 1$.

The learning process takes place through the training phase, in which the weights of the interconnections between nodes are progressively updated toward the values required to obtain a desired output. In supervised learning approach (as opposed to unsupervised learning) the training process is carried out using a, in which a certain input vector will be mapped to a desired output vector, followed by a proper computation of the weights according to a learning algorithm. Once the NN has completed the training phase, it will be able to recognise input patterns with high confidence; the more input-output pairs the NN will be trained on, the higher the confidence will be. To test its ability, the trained NN undergoes a testing phase, in which unknown patterns are given as input and the NN ability in classifying them is evaluated. Several learning schemes exist for the training of the network. An often adopted scheme is the *backpropagation* algorithm, where the error is backpropagated layer by layer and results are used to update the weights of the network. NNs have been applied to medical applications extensively over the last few years, and in particular in mammography.[24]

5 Evaluation Tests for a CAD System

The output given by a CAD system, generally, is not a dicotomic value (positive/negative), but expresses a probability that a disease could exist or not.The parameters used to evaluate of the diagnostic effectiveness of a system (a human or a machine) are:

1. True Positives (TP): number of images of ill patients correctly recognised
2. False Positives (FP): number of images of ill patients uncorrectly recognised as healthy
3. True Negatives (TN): number of images of healthy patients correctly recognised
4. False Negatives (FN): number of images of healthy patients uncorrectly recognised as ill

Starting from these parameters the following factors of merit were defined:

1. Sensitivity = the rate of correctly identified images with a lesion, that is: TP/(TP + FN)
2. Specificity = the rate of correctly identified images without a lesion, that is: TN/(TN + FP)
3. The area under the ROC curve. The ROC (*Receiver Operating Characteristics*) curve is obtained reporting sensitivity vs. (1 - specificity):

$$(1 - \text{specificity}) = FP/(TN + FP)$$

Each one of these parameters has a specific meaning. The sensitivity measures the capability of a diagnostic system to recognise images with a lesion. Specificity measures the capacity to correctly recognise images of healthy people. These two parameters (that are defined as rates, ranging from 0 to 1) are completely independent from each other, so they do not point out the overall capacity of a diagnostic system to correctly recognise both healthy or not healthy cases. Paradoxicallly a diagnostic system could have a very high capacity to recognise images with pathologies (so, it will have high sensitivity) while at the same time, it may not able to recognise healthy cases, pointing out them as ill (in this case, it will be low specificity), It could also have the opposite attributes (low sensitivity and high specificity). From an economic and social point of view, these parameters are both important because a highly sensitive system will not lose ill cases, that probably will be recognised after other medical examinations, in a more advanced state of the disease, with a minor possibility of success of the therapies. At the same time a highly specific system will avoid useless and expensive (economically and temporally) medical examinations due to false alarms.

To evaluate the system in the further depth, another parameter was introduced, that is the area under the ROC curve (A_z), that represents the overall diagnostic capability of a system. To build the ROC curve, a rating method is generally adopted[11]: it is used to classify each image in one of five (or more) classes: (1) surely negative, (2) probably negative, (3) perhaps positive, (4) probably positive and (4) surely positive. The A_z value ranges from 0 to 1 (ideal case). The more the value is near to 1, the better the diagnostic capability of a system will be. In mammography A_z ranges from 0.80 to 0.90.[25]

6 The CALMA Experiment

A prototype of CAD has been developed in Italy in the research frame of the project CALMA (Computer Aided Library for Mammography).[26] It is an example of a CAD system in the field of mammography. It was installed and tested in different Italian hospitals.[27]

To evaluate the diagnostic capability of a radiologist with and without the support of a CAD system the following study was performed. Three radiologists with different experience in reading mammograms were asked to read 190 mammographic images in a conventional way (the images are on films and they must be viewed with a diaphanoscope). In a following step, the same radiologists read the same images with the CAD support. At the end, the obtained diagnoses were compared. To enlarge the study and to compare a CAD developed in a research frame to a CAD that was already available in the market, a different CAD system was considered: *SecondLook* (CADx, Medical System). The results obtained in this study are in line with the results reported in literature. They can be summarized as follows. The increase in sensitivity is counterbalanced by a decrease in specificity, as expected. However the decrease is smaller in absolute value than the increase and the net result is an improvement of diagnostic accuracy. Specifically, the values A_z of the areas under the ROC curves increased by 0.03 on average when radiologists were supported by CAD ($P < 0.05$). This value provides a quantitative assessment of the general increment in diagnostic accuracy observed when radiologists are supported by CAD as a second reader. It was not possible to establish a strong dependence on the skill of the readers, but when sensitivity is concerned, we observed that the youngest one is the most conditioned by CAD. Moreover it was not possible to compare the two CAD systems in terms of sensitivity and specificity values, because it would have been necessary to collect a significantly larger number of images to obtain a statistically significant difference, as reported by Hoffmeister.[28]

7 The Use of GRID in a Screening Environment

The amount of data generated by a screening program is so large that it can't be managed by a single computing centre. In addition, data are generated according to an intrinsically distributed pattern: any hospital participating to the program will collect a fraction of the total dataset. Still, that amount will increase linearly with time and, if fully transferred over the network to diagnostic centres, would be large enough to saturate the available connections. However, the availability of the whole database to a radiologist, regardless of the data distribution, would provide several advantages:

- The CAD algorithms could be trained on a much larger data sample, with an improvement of their performance, in terms of both sensitivity and specificity.
- The CAD algorithms could be used as real time selectors of images with high breast cancer probability: radiologists would be able to prioritize their work, with a remarkable reduction of the delay between the data acquisition and the human diagnosis.
- Data associated to the images and stored on the distributed system would be available to select the proper input for epidemiology studies or for the training of young radiologists.

A grid approach to a screening environment consists of a distributed computing system and the associated data management: the key concept is the Virtual Organisation (VO), a group of distributed users with a common goal and the will to share their resources. The development of a Grid-enabled version of CALMA (MAGIC-5 project) is in progress: the collaboration can be seen as a Virtual Organisation.[29,30] In each hospital digital images and data describing the mammograms, known as *metadata*, are stored in the local database and registered to a common server (*Data Catalogue*). The algorithms for the image analysis will be sent to the remote sites where images are stored, rather than moving them to the radiologist's sites. A preliminary selection of cancer candidates will be quickly performed and only mammograms with cancer probabilities higher than a selected threshold are transferred to the diagnostic sites and interactively analysed by radiologists. The workflow is schematically shown in Fig. 6.

At the present, a designated server http://gpcalma.to.infn.it based on the *AliEn* technology http://alien.cern.ch, has been configured by the MAGIC-5 collaboration, with a central Server running common services. The server connects configured

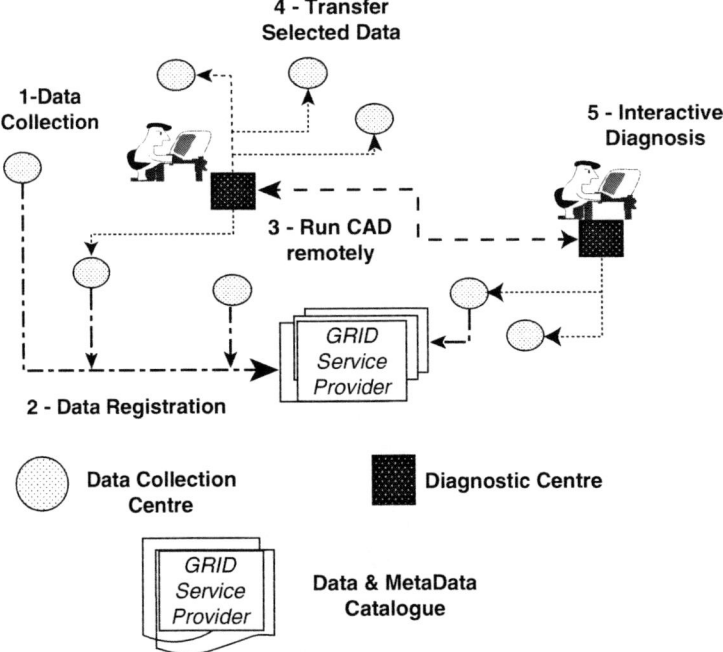

Fig. 6 Schematic representation of the workflow of a tele-diagnosis for breast cancer screening: (1) Data is collected from the Diagnostic Centres to the Data Collection Centres. (2) Data is registered in the Data Catalogue. (3) CAD runs remotely. (4) Selected data are transferred back to Data Collection Centres. (5) Interactive diagnosis between the Diagnostic Centres is completed

clients in several sites of the MAGIC-5 project (Lecce, Napoli, Palermo, Sassari, Torino). The server and the clients make up a prototype in which different utilities are available. It is possible to register new patients from remote sites, attributing them a Logical Name (that can be different from the name on local hard drive where it was stored), that will be associated to their physical location. It is then possible to run CAD algorithms, from remote sites, by selecting images from a list of Logical Names. For this purpose the PROOF (Parallel ROOt Facility, http://root/cern.ch) system was installed on designated personal computers providing the functionality required to run interactive parallel processes on a distributed cluster of computers. The output of the CAD is a list of coordinates and probabilities related to the ROIs identified by the algorithms.

A Graphic User Interface (Fig. 7) drives the execution of three basic functionalities related to the *Data Catalogue*:

- Registration of a new patient that is based on the generation of a unique identifier, which could be easily replaced by the identification code
- Registration of a new exam, associated to an existing patient
- Query of the *Data Catalogue* to retrieve all the physical file names of exams associated to a given patient and eventually analyse them

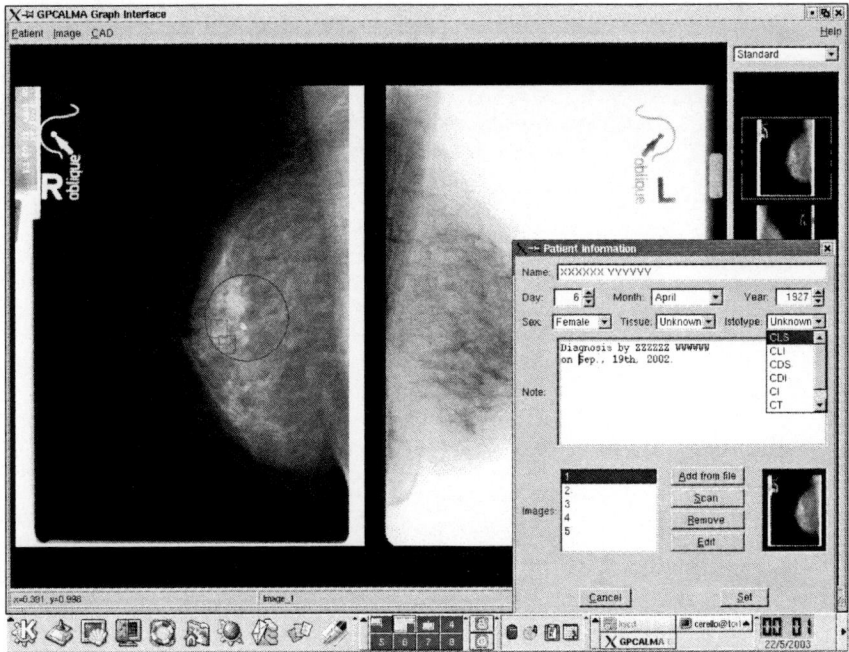

Fig. 7 The graphical interface of MAGIC-5. It is possible to introduce new patients (data and images), modify data, to run the CAD algorithms, and to visualize the radiologist's diagnosis. In addition, it is possible to elaborate the image to optimize contrast and luminosity, or to zoom in on it

The GRID approach to mammographic screening is very promising: it would allow the availability of a distributed database of mammograms from any node (i.e., hospital) in a *Virtual Organisation*. It would also allow for the division of the data collection and data analysis functions: data could be collected without the presence of radiologists, who would take advantage of the CAD results in order to select the sub-sample of mammograms with highest cancer probability in order to prioritize their work. In principle, only mammograms considered positive by the CAD algorithm would have to be moved over the network to a different site, where a radiologist could analyze them, almost in real time.

8 Future of CAD Systems

Many authors maintain that the actual benefit due to the adoption of the CAD systems in mammography remains to be fully assessed.[6,31] Future studies must be done in order to obtain an improvement of the algorithms for the search of microcalcification clusters and massive lesions, to reach results in terms of sensitivity and specificity near to 100%. Presently biomedical applications are under development in the CAD field, such as lung CAD and Alzheimer's disease diagnosis 30.

References

1. E. Thurfjell, K. Anders Lernevall and A. Taube, Benefit of independent double reading in a population-based mammography screening program, Radiology **191** pp. 241–244 (1994).
2. I. Anttinen, M. Pamilo, M. Soiva and M. Roiha, Double reading of mammography screening films-One radiologist or two?, Clin. Radiol. **48** pp. 414–421 (1993).
3. V.T. Freer and M.J. Ulissey, Screening mammograpy with computer-aided detection: prospective study of 12,860 patients in a community breast center, Radiology **220** p. 781 (2001).
4. G.M. Brake, N. Karssemeijer and J. Hendriks, Automated detection of breast carcinomas not detected in a screening program, Radiology **207** p. 465 (1998).
5. S. Ciatto, M. Rosselli Del Turco, G. Risso, S. Catarzi, R. Bonari, V. Viterbo, P. Gnutti, B. Guglielmoni, L. Pinelli, A. Pandiscia, F. Navarra, A. Lauria, R. Palmiero and P.L. Indovina, Comparison of standard reading and computer aided diagnosis (CAD) on a national proficiency test of screening mammography, Eur. J. Rad. **45** pp. 135–138 (2003).
6. R.M. Nishikawa and M. Kallergy, Computer-Aided detection, in its present form, is not an effective aid for screening mammography, Med. Phys. **33**(4) pp. 811–814 (2006).
7. F. Bray, P. McCarron and D. Maxwell Parkin, The changing global patterns of female breast cancer incidence and mortality, Breast Cancer Res. **6** pp. 229–239 (2004).
8. F. Levi, F. Lucchini, E. Negri and C. La Vecchia, Trends in mortality from major cancers in the European Union, including acceding countries, in 2004, Cancer **101** pp. 2843–2850 (2004).
9. N. Bjurstam, L. Bjorneld, et al., The Gothemburg breast screening trial. First results on mortality, incidence and mode of detection for women ages 39–49 years at randomization, Cancer **80** p. 2091 (1997).
10. F. Winsberg, M. Elkin, J. Macy, et al., Detection of radiographic abnormalities in mammograms by means of optical scanning and computer analysis, Radiology **89** p. 211 (1967).
11. M.L. Giger, K. Doi, H. MacMahon, et al., An "intelligent" workstation for computer-aided detection, RadioGraphics **13** p. 647 (1993).
12. R.M. Nishikawa, M.L. Giger, D.E. Wolverton, et al., Prospective testing of a clinical mammography workstation for CAD: analysis of the first 10,000 cases. In: N. Karssemeijer et al. (Eds). Digital Mammography, pp. 401–406, Kluwer, Nijmegen (1998).
13. W.P. Kegelmeyer, J.M. Pruneda, P.D. Bourland, A. Hillis, M.W. Riggs and M.L. Nipper, Computer-aided mammographic screening for spiculated lesions, Radiology **191** p. 331 (1994).
14. Z. Huo, M.L. Giger, C.J. Vyborny, U. Bick, P. Lu, D.E. Wolverton and R.A. Schmidt, Analysis of spiculation in the computerized classification of mammographic masses, Med. Phys. **22** p. 1569 (1995).
15. R.L. Birdwell, D.M. Ikeda, K. O'Shaughnessy and E. Sickles, Mammographic characteristics of 115 missed cancers later detected with screening mammography and the potential utility of computer-aided detection, Radiology **219** p. 192 (2001).
16. L.J. Burhenne, S.A. Wood, C.D'Orsi, S. Feig, D.B. Kopans, K. O'Shaughnessy, E.A. Sickles, L. Tabar, C.J. Vyborny and R.A. Castellino, Potential contribution of computer-aided detection to the sensitivity of screening mammography, Radiology **215** p. 554 (2000).
17. R. Taft and A. Taylor, Potential improvement in breast cancer detection with a novel computer-aided detection system, Appl. Radiol. **12** p. 25 (2001).
18. R. Passariello, RADIOLOGIA elementi di tecnologia, IDELSON-GNOCCHI, Roma, terza edizione (1999).
19. N. Petrick, H. Chan, B. Sahineret, et al., Combined adaptive enhancement and region-growing segmentation of breast masses on digitized mammograms, Med. Phys. **26** p. 1642 (1999).
20. S.P. Keshavmurthy, M.M. Goodsitt, H.P. Chan, et al., Design and evaluation of an external filter technique for exposure equalization in mammography, Med. Phys. **26** p. 1655 (1999).
21. N. Petrick, H. Chan, D. Wei, et al., Automated detection of breast masses on mammograms using adaptive contrast enhancement and texture classification, Med. Phys. **23** p. 1685 (1996).

22. M.L. Giger, Computer-Aided Diagnosis RSNA Categorical Course in Breast Imaging, pp. 249–272 (1999).
23. The Biomedical Engineering Handbook, editor in chief Joseph D. Bronzino, CRC Press/IEEE Press, BocaRaton, FL (1995).
24. J.M. Boone, G.W. Gross and V. Greco-Hunt Neural Networks in radiologic diagnosis I: introduction and illustration, Invest. Radiol. **25** pp. 1012–1016 (1990).
25. J.A. Swets, Measuring the accuracy of diagnostic systems, Science **249** pp. 1285–1293 (1988).
26. S. Amendolia, M.G. Bisogni, U. Bottigli, A. Ceccopieri, P. Delogu, A. Marchi, V.M. Marzulli, R. Palmiero and S. Stumbo, The CALMA project: a CAD tool in breast radiography, Nucl. Instr. Meth. A **460** p. 107 (2001).
27. M. Bazzocchi, I. Facecchia, C. Zuiani, V. Londero, S. Smania, U. Bottigli and P. Delogu, Applicazioni di un sistema di Computer Aided Detection (CAD) a mammografie digitalizzate nell'individuazione di microcalcificazioni, Radiol. Med. **101** p. 334 (2001).
28. J.W. Hoffmeister, S.K. Rogers, M.P. DeSimio and R.F. Brem, Determining efficacy of mammographic CAD systems, J. Digit. Imaging **15** (suppl 1) pp. 198–200 (2002).
29. S. Bagnasco, U. Bottigli, P. Cerello, S. Cheran, P. Delogu, M.E. Fantacci, F. Fauci, G. Forni, A. Lauria, E. Lopez Torres, R. Magro, G.L. Masala, P. Oliva, R. Palmiero, L. Ramello, G. Raso, A. Retico, M. Sitta, S. Stumbo, S. Tangaro and E. Zanon, GPCALMA: a GRID based tool for mammographic screening, Methods of Inform. Med. **2** pp. 244–248 (2005).
30. R. Bellotti, P. Cerello, V. Bevilacqua, M. Castellano, G. Mastronrdi, S. Tangaro. F. De Carlo, S. Bagnasco, U. Bottigli, R. Cataldo, E. Catanzariti, S.C. Cheran, P. Delogu, I. De Mitri, G. De Nunzio, M.E. Fantacci, F. Fauci, G. Gargano, B. Golosio, P.L. Indovina, A. Lauria, E. Lopez Torres, R. Magro, G.L. Masala, R. Massafra, P. Oliva, A. Preite Martinez, M. Quarta, G. Raso, A. Retico, M. Sitta, S. Stumbo and A. Tata, Distributed medical image analysis on a GRID infrastructure, Future Gener. Comp. Sy., **23**(3) pp. 475–484 (2006).
31. M. Bazzocchi, F. Mazzarella, C. Del Frate, F. Girometti and C. Zuiani, CAD systems for mammography: a real opportunity? A review of the literature, La Radiol Med. **112** pp. 329–353 (2007).

A Short Introduction to Computer Networks
Introduction to Computer Networks

Ulrich Fuchs

Abstract Computer Networks have become an essential tool in many aspects: human communication, gathering, exchange and sharing of information, distributed work environments, access to remote resources (data and computing power) and many more. Starting from an historical overview, this paper will give an introduction to the underlying ideas and technologies. The second half will concentrate on the most commonly used network technology today (Ethernet and TCP/IP) and give an introduction to the communication mechanisms used.

Keywords: Computer · network · Ethernet · TCP · IP · history

1 Definition – A Try

The definition of the term "computer network" is widely discussed, people try to pin down various numbers and types of equipment that is needed. Also the point of view changes: telecommunication carriers have a different view of networks than hardware or software producers have.

Personally I prefer to stay with the following definition, which will guide us through this presentation:

A network is a system of *data sources* and *data receivers* that are connected by certain *media*, *transmission* and *switching* equipment or other networks.

This definition is recursive and at a first glance doesn't seem conclusive but we will see that it accurately describes our day-by-day experience of networked computers.

U. Fuchs
CERN, PH Dept., CH-1211 Geneva 23, Switzerland
e-mail: Ulrich.Fuchs@cern.ch

Y. Lemoigne, A. Caner (eds.) *Molecular Imaging: Computer Reconstruction and Practice*, 175
and Experiments,

The idea behind networks, however, is quite clear: they allow people to share information and resources. And this very need led to their invention, already quite some time ago.

2 An Eye on History

2.1 The Early Times

It is popular misconception that networks came up with the early internet or at least long after the first computers were invented. We have proof that the first networks existed already 1,000 years BC.

2.1.1 Taking the First Step Towards the Information Highway

In order to transmit information it was written down and slaves were sent running between senders and destinations. The "Shrine of the book" museum in Jerusalem displays a piece of papyrus that is clearly a message from a general to its officers in the field. Today, we consider these early messengers a carrier network, just like computer networks today, we only replaced running slaves by copper cables and we are using electric signals instead of papyrus.

To increase the speed of communication, the running slaves ($\sim$5 km/h) were soon replaced by riders (a horse reaches $\sim$20 km/h) and carrier pigeons ($\sim$60 km/h).

2.1.2 The Idea with the Fire

Though the speed from slaves to birds increased 12 fold it was still not sufficient for the receiver to be able to react in case there was imminent danger. To warn people from attacking enemy ships, a chain of fire towers was built along coast lines (e.g. in Israel and Italy). A fire lit on top of the tower could easily be seen from the next tower (5–20 km away) whose guards, in a turn, also lit a fire. One could also change the color of the fire by adding different chemical substances, giving different meanings to the different colors.

Speaking today's terms this was the first wireless network with wavelength-multiplexing. The throughput can be estimated to be about 0.02 bytes/sec. This is not much but a lot faster than a running slave or riders.

Another example are smoke signals used by the native north American Indian.

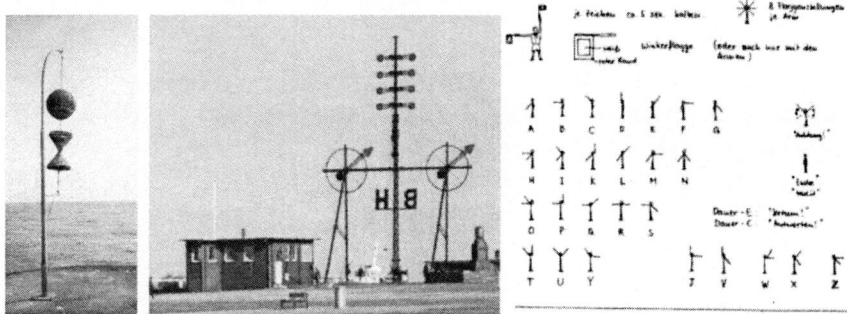

Fig. 1 Examples for Semaphores

2.1.3 Hold it Right There

Invented in France under King Louis XIV the first *Semaphore* networks were installed around the year 1783. A Semaphore is a visual system for transmitting information (Fig. 1). Combined with the same idea already used by the fire towers, observe the others and repeat, information (including the complete alphabet) could be transmitted at speeds of about one character per second (1 byte/sec).

Semaphores were used in nautics until the 1950s to indicate static messages like wind or water levels. The sign language covering the complete alphabet is still in use.

2.1.4 The First Electric Network

In 1845 the painter Samuel Morse invented the first electric network pulling wires between Washington and Baltimore, all necessary equipment and the 'code' to translate characters into electrical signals, the famous Morse Alphabet. Though limited in its abilities (38 characters, maximum speed about 30 characters per second) his invention paved the road for electric networks and signal transmission.

Officially abandoned in 2001, Morse signals are still in use today.

2.1.5 The First Speech Transmission

In 1876 Graham Alexander Bell invented an apparatus for transmitting speech over a wire pair, this was the birth of the telephone (Fig. 2).

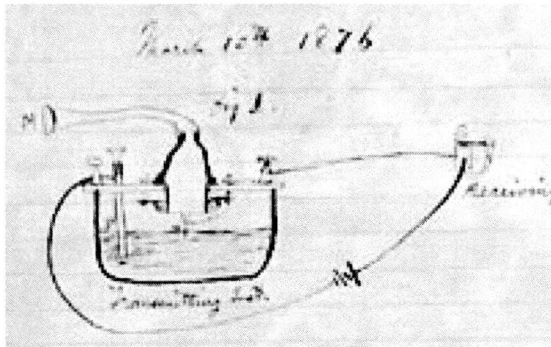

Fig. 2 Graham Alexander Bell and the original drawing of the first telephone

2.1.6 RS232 and ASCII

The quality of telephones and connections improved with time but it was not until the Second World War that there was a need for fast transmission of complex (e.g. encrypted) information. A typewriter (Flexowriter) modified with a relay bank and a serial connection was used, giving birth to two important standards:

- To connect two Flexowriters, the serial connection had to be standardized, in terms of plugs, voltages and pin layouts. The standard developed is called *RS232* and is still in use today for all "serial" equipment.
- To make sure information typed on one side is correctly transmitted to the other needed the definition of a *protocol*, i.e. a set of rules, how to convert letters into electric signals. The translation of the character set worked out was called *ASCII*, the American Standard Code for Information Interchange.

The idea of the Flexowriter (~10 characters per sec) developed and resulted in a commercial product: the Teletype (early version of Telex) in the early 1960s (~30 characters/sec).

2.1.7 Connecting the First Computers

In the early 1970s computer scientists in Stanford built a special device to translate characters into sounds (modulator) and back from sounds to characters (demodulator), the combined device was called *MoDem*. Connected to a computer these modems could transmit data over normal telephone lines literally around the globe. Soon special telephones with built-in modes were available, boosting the transmission speeds to 120 bytes/sec (i.e. 120 characters/sec).

Based on pure modem point-to-point connections, a network of computers was built allowing its users to exchange email and documents. This network called BITNET was widely used by scientists and officially switched off in the end-1970s.

2.2 Giving Birth to the Internet

2.2.1 ARPAnet

In 1969 the American Department of Defense launched a networking project with the goal to assure communication between two machines, even if part of the infrastructure is inoperative. There should be no point to point connections or hierarchy but rather a decentralized structure allowing multiple paths between senders and receivers over geographically different communication paths or other meshed networks. The network that originally connected six universities was called ARPAnet (Advanced Research Projects Agency Network) (Fig. 3).

Started in 1969 the network counted 13 connected machines in 1971, 60 machines in 1977 and 10,000 machines in 1980. The initial speed was 2.4 kbit/sec, later the speed was increased to 50 kbit/sec.

Part of project was to develop a new communication protocol that was needed for addressing networks consisting of other networks, "inter-net"s. This communication protocol was a suite of two independent parts, the *Interface Protocol* (IP) and the *Network Control Protocol* (NCP). Later, after many enhancements, NCP was renamed to *Transmission Control Protocol* (TCP). Today, the protocol stack "TCP/IP" is used by most computers around the world.

The ideas, structure and protocols developed for the ARPAnet made it ideal for further expansion and are considered the early Internet as we know it today.

2.2.2 The Internet in the USA (1980–1989)

The Internet grew remarkably within the next years, many universities connected their campus network and added their connections to other universities. To recruit students and offer information also industrial servers joined. To allow their users to work from home universities and companies offered dial-in lines to connect to their networks.

The usage of the new infrastructure was as roughly: 60% remote access to company/university computers (telnet), 30% file transfer (ftp), 5% discussion forums and information exchange (news), 2–4% email and 1% other.

2.3 The Internet in Europe (1972–1989)

The development of the Internet in Europe had a different origin and therefore took a completely different direction. It started in particle physics at CERN.

In the 1960s there were basically only point to point connection between mainframe computers and terminals. 1971 the computer scientist Ben Segal at CERN connected the PDP11 in the library to the main CERN computer, the CDC6600 in the Computing Centre via a home-made cable. This was the first network Europe. The data ports he used on the two computers were not made for fast data

INTERNET MAP

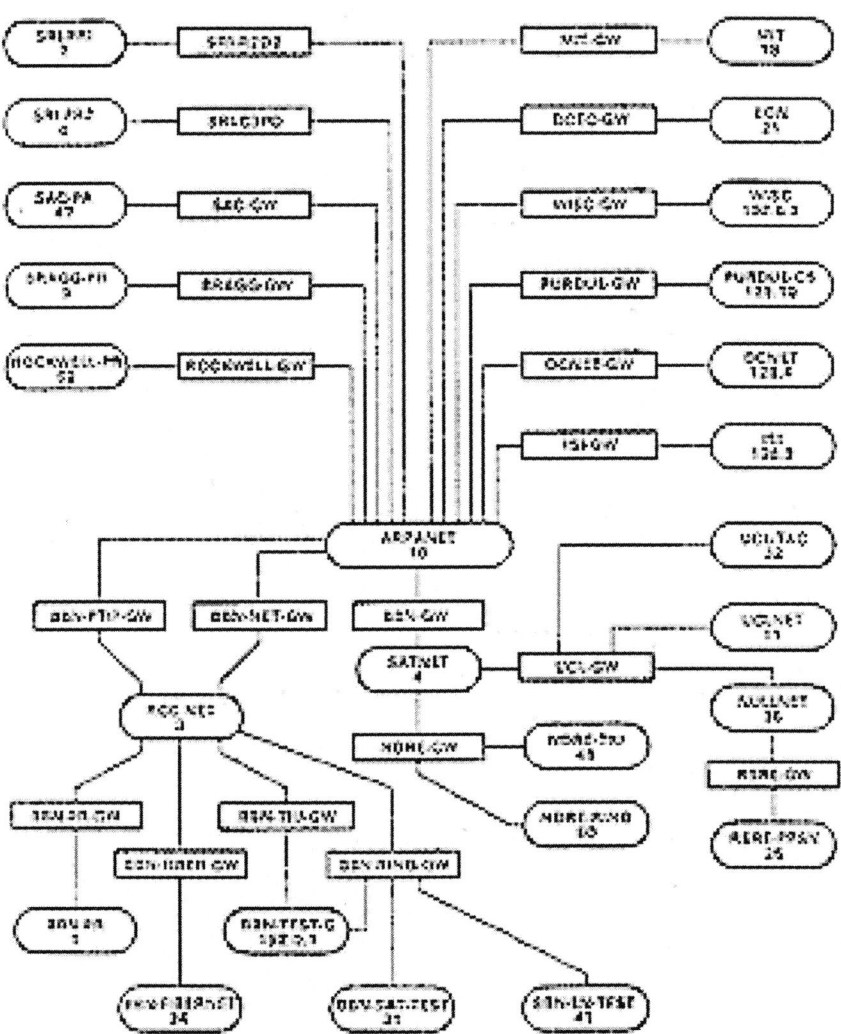

Fig. 3 A complete map of the ARPAnet 1978

communication, so the connection was slow, but it worked and proved a principle: to transfer data from one machine to another it was no longer necessary to copy the files to tape, carry them over to the other machine and read the files back. One could simply copy them between the two machines.

1978 the companies DEC, INTEL and XEROX developed a networking standard called "ETHERNET". It allowed up to 254 machines to be connected together and communicate at high speed (2 Mbit/sec).

A year later in 1979 Ben Segal wanted to install an Ethernet at CERN and failed at first because network interface cards were very expensive ($\sim$$1 Mio). But he convinced the concerned companies that this test would be to the benefit for all and got them for free. In the beginning this network was only used within the computing center but later all over CERN and allowed physicists to access the central machines from terminals in their buildings, instead of having to drive to the computing center. This is considered to be the first Ethernet in Europe.

In 1981 Ben Segal heard about TCP/IP and immediately deployed it at CERN, as it allowed connecting almost any equipment from any vendor. It turned out to be a big success. Physicists could have terminals in their offices, access their data remotely and run analyses and write programs without standing in line at the Computer operator's desk with a stack of punch-cards.

Because of these advantages many universities connected their campus network to CERN, forming a network that was called HEPnet (High Energy Physics network). Today we consider this the birth of the Internet in Europe.

2.3.1 The World Wide Web

In 1989 the young computer scientists Tim Bernes Lee and Robert Callieau at CERN had an idea for a new network service: instead of central machines offering all the data, many computers could 'offer' something and hypertext system providing links from one information source to others could guide the user. They wrote two programs: one running on computers all the time offering data to the network, a client to retrieve and interpret the data and they developed a simple transport protocol and description language for the communication between server, client and the user.

This work was the birth of the World Wide Web: the http protocol, web servers (CERNserv, later APACHE) and web clients (CERNcli, then MOSAIK, now NETSCAPE).

The initial idea was to install one server at CERN with one or two pages being the CERN telephone book. The success was overwhelming, within weeks universities around the world started creating web pages and publishing files. In 1990 CERN was the biggest web site with several thousand pages, articles and diagrams. Slowly the web became an ingredient of every scientists work and in 1993 the CERN management agreed to donate the protocols and program code to public domain. This was the take off for the World Wide Web. Consisting of just several thousand machines in 1993, the growth rate today is more than four machines per second.

Today the usage of the internet changed significantly: 1% Remote access, 8% Email, 1% other activities and 90% Web. That's why today the two terms "internet" and "World Wide Web" are used synonymously though basically they have nothing to do with each other.

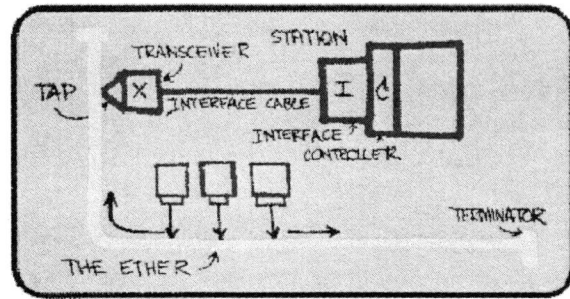

Fig. 4 The first drawing of an ETHERNET network

3 A Run through Technology

3.1 Ethernet

On May 22nd 1973 a member of XEROX research staff called Robert Metcalfe wrote a memo to his bosses on the enormous potential of a new network technology he called ETHERNET (Fig. 4), suggesting that a group of computers could share a common communication path.

He successfully convinced Digital Equipment, Intel, and Xerox Corporations to work together to promote Ethernet as a standard. Today an international computer industry standard, Ethernet is the most widely installed LAN protocol.

The main difference to existing networks up to that time was the common communication path for all machines (the ether), which in reality was a coax cable maximal 300 m long. The speed was defined to be 3 Mbit/sec. Because most of the cables were about 1 cm in diameter and yellow, this network technology got the nick name "yellow sausage".

Later the speed was increased to 10 Mbit/sec and the old yellow cable was replaced by cheaper, normal 5 mm coax cable, which made installations easier and cheaper, thus the nick name "cheapernet". The technical name of this network standard, however, was *10Base2*.

With the introduction of Hubs and Switches the cable specification changed again and became a telephone-like cable with RJ45 plugs (*10baseT*).

By improving the cable quality and timing in controller cards, the speed is still rising today:

- 1995: Better cables allow a higher throughput of 100 Mbit/sec: *100baseT*
- 1996: Duplex mode (send and receive simultaneously): *100BaseTx*
- 1998: Gigabit Ethernet at 1 Gbit/sec: *1000BaseTx*
- 2001: Ten Gigabit Ethernet, 10 Gbit/sec, on optical fibers: *10000BaseF*
- 2004: Ten Gigabit Ethernet, 10 Gbit/sec, again on copper: *10000BaseTwX*
- 2008?: Thirty Gigabit Ethernet, 30 GBit/sec
- 2009? Rumor: 100 Gbit Ethernet, 100 Gbit/sec

3.2 Layered Protocol

To send a present by parcel post requires some thinking on wrapping, correct addressing and choosing a carrier. Sending Data over a network requires exactly the same steps.

As an example, let's assume we want to transmit a line of text (think: send a book by postal mail). We have go through the following four steps, one by one:

	Name	Network	Parcel post
1	Data service	Encode data to machine readable format (Bytes)	How to wrap and pack the present?
2	Transport service	Transport data packets between different software applications	Who is the receiver and how do we make sure he gets it the parcel?
3	Network service	Transfer of data between different machines	By Train? Airplane? Truck? Bike? Ship?
4	Link service	Exchange data between different hardware components	How do we lift/handle the packet?

Let's go over each layer and discuss some of their technical details.

3.2.1 Data Service

Today we are mostly using three different "conversion systems" to convert human-readable text into computer-readable numbers:

- ASCII
  Still around from the Flexwriter in World War 2, this 95 character set got upgraded to 256 characters, comprising some national special characters, picture elements and control codes.
- ISO98859
  Consisting of 11 national character sets (English, French, German, Greek, Russian ...) and their special characters. Updated to the standard ISO98859-15 it now contains the Euro symbol also.
- UNICODE
  Mostly used by IBM machines this encoding consists of 256 pages with 256 characters each (in total 65,536).

Now having converted a line of text into a string of numbers, they can be passed to the next layer.

3.2.2 Transport Service

The transport service is responsible for breaking the long string of numbers it gets from the Data Service up into smaller groups, so-called *packets*, that can be handled by the next layer. It adds sequence numbers and error correction mechanisms to each packet. Every packet is then passed to the next layer and from now on treated independently from all other packets.

3.2.3 Network Service

The network service takes the data packets from the transport service and packs them again in a data structure that, apart from more error correction and integrity checks, for the first time also contains a destination address. This data structure is called a *frame*.

The network service is responsible for packing and addressing the data packets in frames and to assure that they arrive on the receiving machine.

3.2.4 The Link Service

The link service is responsible for actually getting the network frame to the other machine. It converts frames into a sequence of electric signals which can be sent over a wire or in light pulses for fiber communication or.

It therefore defines cables, plugs, electric signals and more. Only on this level we distinguish different kinds of connections, like Ethernet, telephone modem, DSL or wireless networks.

3.2.5 Summary

By going top-down through the four layers, the source computer converts a text into numbers, breaks them up in packets, packs the packets in frames and sends them (Fig. 5).

The receiving destination computer goes through the same steps but bottom-up: receive some signals, form frames, unpack the frame to get the packet, unpack the packet to get numbers, convert numbers back to text. It should be noted that this is a simplified view of all the layers and their protocols and not at all complete.

3.3 The Trick with TCP/IP Inter-Nets

3.3.1 The IP Addressing Scheme

Just like telephone numbers, every computer is assigned a world-wide unique number, the so-called *IP-Address* (e.g. "192.168.233.167"). This number contains all information needed for a frame to find its way from any source to the destination, it contains the exact continent, country and network owner (company) and depending on the setup even building, floor and finally the machine itself.

Every section of the IP-Address is a number ranging from 1 to 254 which gives a total of about four billion different addresses. In 1980 this was believed to be enough to connect all computers in the world but today we suffer from an address shortage. This problem will be solved by the new IPv6 standard which will provide millions times more addresses but is still in development.

In addition to the IP address of the destination the frame also contains the IP address of the source for confirmations and replies but also a very interesting field, the so-called time-to-live counter TTL. This counter defines how many times a network frame is allowed to be passed-on from one system to another before it expires. The main reason for the existence of this counter was to avoid network flooding due to loops in the infrastructure but it can also be an interesting tool for measuring network connections between two end points, namely answering the question: how many times did the frame get passed on? To send a frame around the world in 1995 it was necessary to allow a TTL from 128 to 250. In 2005 the number reduced to 16 to 32 (the default). This means that we have more and more direct connections to all parts of the world.

3.3.2 What About TCP?

Because one computer can run multiple applications simultaneously that are all sending and receiving data from the network, it is necessary to have unique identifiers for all of them. To talk to the network layers they are using defined internal connections, so-called *ports*. The combination of source port, source address, destination port, destination address is called a *socket* and unique throughout the whole internet. By connecting to a specific socket, an application on one computer knows

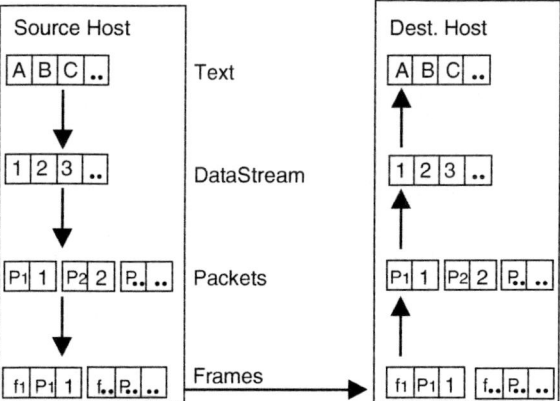

Fig. 5 Simple network communication

exactly whom it is talking to at the other end, even if there are multiple applications running on the destination machine.

The TCP suite is responsible for establishing connections between sockets and handling them. It breaks data up into packets, re-assembles data from packets and does error correction and integrity checks on it. Another important task in this level is flow control, i.e. suppress duplicate transmissions and re-transmit or request the re-transmission of lost or damaged packets. Clearly this connection handling adds some overhead to the network traffic, in the case of TCP it's about 20%.

3.3.3 Network Structure

Because networks span geographic distances, the Ethernet addressing scheme is no longer appropriate. In addition the increased network traffic makes it necessary to break a network up into "connected islands" with local traffic confined to one island. Therefore a new network component is needed: a *Router*. This device acts as relay station between several networks and re-directs the traffic between them according to what they believe to be the "best way". However, it is the administrator who defines the meaning of "best" in this context. Having to choose between several connections that all lead from A to B, the "best way" could be fastest way, but it could also be the cheapest or most reliable connection.

Because now the immediate 'destination' of a frame is no longer the target machine but the next router on the path, the frame might change on its way (Fig. 6). The contained packet, however, stays untouched.

Most routers on the internet talk to each other and exchange their knowledge on possible connections and what is "the best" at the moment. This information is therefore time dependant and can change rapidly.

See Fig. 7 for an example of time dependency of the "best way".

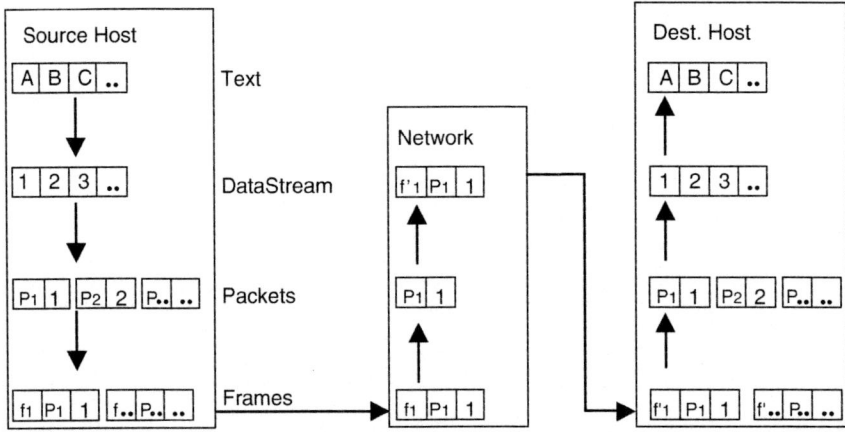

Fig. 6 Network transmission with change of frame

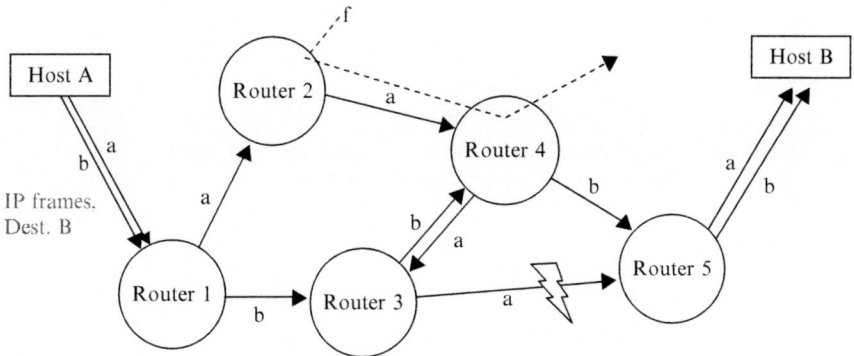

Fig. 7 Time dependant routing of frames

Host computer A sends the first frame "a" which travels the path Router 1, 2, 4, 3, 5 to host computer B.

Due to an incident the connection between Router 3 and 5 becomes unavailable. Both Routers, number 3 and 5, re-calculate the best route between them and exchange this information with Router 4. In addition a high data flow "f" starts between Router 2 and Router 4 due to other activity on the internet. Router 2 notifies Router 1 that the best way to reach Router 4 is no longer via itself due to high load. Router 1 therefore concludes that the best way is now to go to Router 3.

Now host computer A sends the second frame "b" and it travels what is the best way at this moment, namely along the path Router 1, 3, 4, 5 to host computer B.

3.3.4 DNS – An Important IP Service

As mentioned before, every computer in a TCP/IP network has a unique IP address used for communication. For humans, however, it is quite inconvenient to deal with numbers all the time. We are used to address things by name. Right from the start all IP implementations have foresaw a kind of "number lookup service". First implemented as a file on each machine this concept soon developed into a global service: the *Domain Name Service* (DNS). This service acts as directory service to translate names into IP addresses and back. The syntax for names is

- For machines: *host.domain*
- For users: *user@host.domain*

The name space is hierarchical from so-called *top-level* domains for countries (fr,ch,de,es,it,..com,org,...) down to companies, departments and machines. Names are to be read right-to left, e.g.:

- *steve@pcabco04.cern.ch*
  In Switzerland, at CERN, there is a machine called pcabco04 on which there is a user called 'steve'.

- *kelly@mail8.mail.cern.ch*
  In Switzerland, at CERN, in the mail service domain on the machine mail8 there
  is a user called 'kelly'.
- *connect pc18.epp.physics.uni-munich.de*
  We want to connect to a machine: in Germany, at the University of Munich in the
  Physics department in the 'epp' group to a machine called pc18.

3.3.5 A Word on Organization

How is the internet organized? Many people contribute to standards and enhancement of the internet community. This is done by the principle "do it if it doesn't
disturb". New ideas are published as *Request for Comments* (RFC). If the author
doesn't receive comments on his plans within a certain period of time, he can go
ahead.

This scheme of "freedom of mind" already led to remarkable developments and
is a constant source of inspiration for new programs and services.

4 Conclusions

Networks already exist for a very long time and have always been far from trivial.

Acknowledgements I would like to thank ESI and CERN for giving me the opportunity to participate in their collaborative projects and especially all the ESI staff for their support and good
working spirit.

PET Pharmacokinetic Modelling

Wolfgang Müller-Schauenburg* and Matthias Reimold

Abstract Positron Emission Tomography is a well-established technique that allows imaging and quantification of tissue properties in-vivo. The goal of pharmacokinetic modelling is to estimate physiological parameters, e.g. perfusion or receptor density from the measured time course of a radiotracer. After a brief overview of clinical application of PET, we summarize the fundamentals of modelling: distribution volume, Fick's principle of local balancing, extraction and perfusion, and how to calculate equilibrium data from measurements after bolus injection. Three fundamental models are considered: (i) the 1-tissue compartment model, e.g. for regional cerebral blood flow (rCBF) with the short-lived tracer [^{15}O]water, (ii) the 2-tissue compartment model accounting for trapping (one exponential + constant), e.g. for glucose metabolism with [^{18}F]FDG, (iii) the reversible 2-tissue compartment model (two exponentials), e.g. for receptor binding. Arterial blood sampling is required for classical PET modelling, but can often be avoided by comparing regions with specific binding with so called reference regions with negligible specific uptake, e.g. in receptor imaging. To estimate the model parameters, non-linear least square fits are the standard. Various linearizations have been proposed for rapid parameter estimation, e.g. on a pixel-by-pixel basis, for the prize of a bias. Such linear approaches exist for all three models; e.g. the PATLAK-plot for trapping substances like FDG, and the LOGAN-plot to obtain distribution volumes for reversibly binding tracers. The description of receptor modelling is dedicated to the approaches of the subsequent lecture (chapter) of Millet, who works in the tradition of Delforge with multiple-injection investigations.

Keywords: PET · positron emission tomography · pharmacokinetics · receptors · modelling

W. Müller-Schauenburg* and M. Reimold
Dept. of Nuclear Medicine, Radiological Clinic, University Clinic of Tuebingen, Hausserstr. 142, D-72076 Tübingen, Germany
e-mail: wolfgang.mueller-schauenburg@uni-tuebingen.de

Y. Lemoigne, A. Caner (eds.) *Molecular Imaging: Computer Reconstruction and Practice, and Experiments,*
© Springer Science+Business Media B.V., 2008.

1 Introduction

Tracer kinetic modelling addresses two different topics: Classical kinetic modelling uses blood data to describe the time courses of whole body tracer (or drug, resp.) distribution or whole organ balances in order to obtain parameters like whole organ blood flow or kidney clearance. In contrast, PET primarily measures local 3D distribution of tracers in time and their modelling yields local parameter distributions either averaged over a Region-of-Interest or on a pixel-by-pixel basis. The corresponding balances are local ones.

The current paper focuses on PET modelling, using the classical whole body or whole organ kinetic modelling as a reference for understanding the principles and approaches.

This article on PET kinetic modelling has developed as introduction to various abbreviated forms of the PET pharmacokinetic course which was turned to an international course by Nico Leenders and his co-workers.[1] Exercises and dialogue have ever been an important part of this course. They could not be transferred to this chapter.

1.1 Target Parameters of Kinetic Modelling

Usually, the aim of kinetic modelling is to obtain one of the following parameters:

1. Tracer delivery (clearance from blood), usually denoted K_1. For freely diffusible tracers, for example, $[^{15}O]H_2O$, this measure reflects tissue perfusion in $ml_{BLOOD}/(min \times ml_{TISSUE})$.
2. Trapping rate, such as $rCMR_{GLU}$ (regional cerebral metabolic rate of glucose) or $[^{18}F]F$-DOPA influx.
3. Binding potential, a measure proportional to the density of binding sites such as neuroreceptors. To measure receptor density (usually denoted B_{max}) independent from the affinity with which the tracer binds to the receptor, more complicated experimental settings (e.g. multi-injection approach) are required.

1.2 Kinetic Modelling Versus Simplified Clinical Applications

Patients get intra-venous bolus injections of positron emitting radioactive tracers. One frequently used tracer is $[^{18}F]FDG$, which reflects sugar metabolism for getting functional data from the brain or from the heart. The currently most important clinical application of $[^{18}F]FDG$ is to detect and characterize malignant tumors or evaluate their response to treatment since many tumors exhibit increased glucose metabolism.

In contrast to glucose, the tracer FDG is trapped in most organs (i.e. *not* metabolized to CO_2) and is subject to renal excretion. Trapping simplifies kinetic analysis and target to background ratio improves over time, so that a single late static image contains all clinically relevant information. This is an essential prerequisite for clinical application in oncology, for which *whole body* imaging is widely available since the mid 1990.

A PET scanner acquires data from a field-of-view of about 15 m in the longitudinal direction of the patient. For whole body imaging, e.g. with FDG, multiple fields-of-view are combined. For kinetic analysis, ideally, a full time course of radioactivity concentration ("time-activity-curve") is available, which means that imaging is restricted to one 15 m field-of-view (Fig. 1).

Classical kinetic modelling also requires time-activity curves from arterial plasma. These time activity curves may vary between different vessels and organs (delay and dispersion), however, measured arterial plasma concentration (samples taken e.g. from the forearm) more or less reflects a common input to all local tissue areas.

Quantification of PET data requires some sort of reference. For whole body imaging with FDG, tracer uptake is usually expressed as "SUV" (standardized uptake

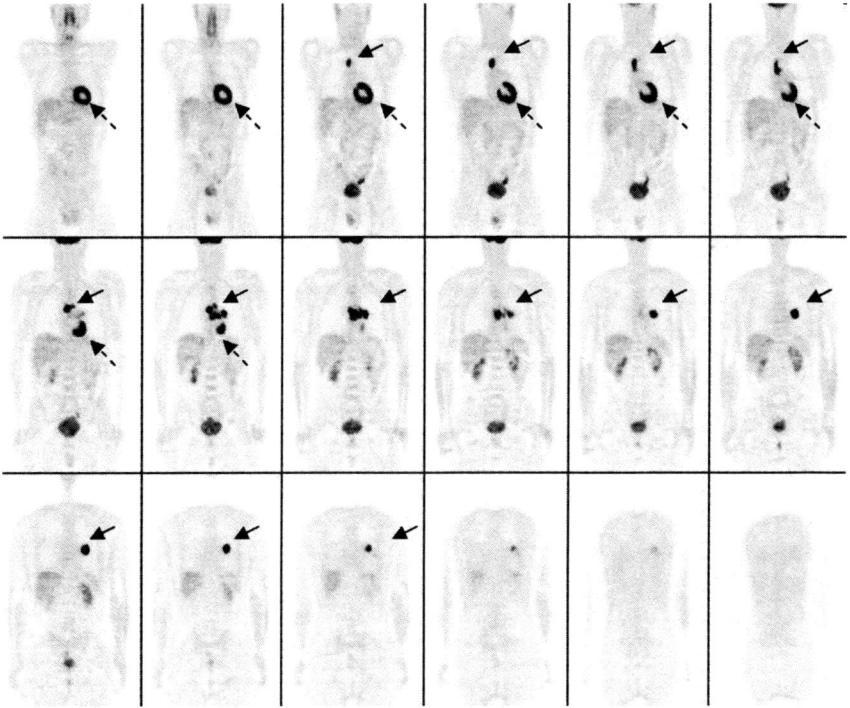

Fig. 1 [^{18}F]FDG-PET: sugar metabolism, "whole body" tomograms identifies tumor (full line arrow), brain (most is cut off), heart (dashed arrow). PET receptor images are shown in the PET methodology chapter of Morel and Millet

value), that is, measured radioactivity concentration in $[\mathrm{kBq/cm^3}]$ (corrected for decay) normalized to injected radioactivity [kBq/g patient weight]. For a hypothetical homogenous distribution, the SUV is 1 g/cm³. Even for full kinetic modeling (i.e. when arterial data are available), a reference tissue may be required as described below (section 6.2). An important side-effect of such a reference region is that arterial sampling may be avoided, if both the target and the reference region are modelled simultaneously and if certain assumptions are made.

2 Concepts

2.1 Volume of Distribution

The idea of a *volume of distribution* will be important, for example, when we look for the density of specific binding sites, that is, neuroreceptors in the brain.

The definition of a *volume of distribution* addresses an equilibrium state. In systems with tracer excretion (e.g. urinary), such equilibrium is experimentally obtained by constant infusion over a certain period of time (this period may be shortened by an optional initial bolus). As described below (section 2.4) a constant infusion may be simulated by mathematical integration over time of bolus injection data.

The concept underlying the term *distribution volume* is a dilution process in a homogeneous medium:

$$\text{Concentration} = \frac{\text{dose}}{\text{volume}} \rightarrow \text{volume} = \frac{\text{dose}}{\text{concentration}}$$

This term may be generalized to a *virtual* distribution volume that links a measured dose (in tissue) with a measured concentration (in the blood). It can be understood as a hypothetical tissue volume that would be needed to obtain the observed concentration if the tracer was homogeneously distributed (same concentration in blood *and* tissue at equilibrium). It should be noted that the typical reference is plasma rather than whole blood concentration.

$$\text{Distribution volume in tissue} = \frac{\text{dose in tissue}}{\text{blood concentration}}$$

This concept is also used to describe a whole body distribution:

$$\text{Whole body distribution volume} = \frac{\text{whole body dose}}{\text{blood concentration}}$$

Since the average amount of tracer per milliliter tissue may be much higher than the corresponding blood concentration, e.g. for substances incorporated in bone, the whole body volume of distribution may be larger than the physical body volume.

In contrast to the classical distribution volume described above, the distribution volume in PET modelling is normalized to (cm^3 tissue) or to (cm^3 scanner volume):

$$\text{PET distribution volume} = \text{distribution volume}/cm^3\text{tissue} = \frac{\text{dose}/cm^3 \text{ tissue}}{\text{blood concentration}}$$

Therefore the PET distribution volume corresponds to what is classically denoted *partition coefficient* (here: tissue-to-plasma). It is unitless and reflects the ratio of concentrations at equilibrium.

Often, scanner volume and tissue volume are used as synonymes, however, they have to be distinguished e.g. when correcting for a fractional intravascular volume that is subtracted from scanner volume to obtain the (remaining) tissue volume.

One should keep in mind that concentrations measured in PET do not address real biochemical concentrations. It is rather like measuring by PET the sugar concentration in a kitchen (pixel) instead of the true chemical concentration in the cup of coffee on the kitchen table (homogenous subspace).

2.2 $K_1 = \{Perfusion\} \times \{Extraction\ Fraction\}$

The formula in the subtitle connects the following three variables

- Perfusion = flow per tissue volume (ml blood/min/cm^3 tissue)
- K_1 "blood clearance to tissue" (ml blood cleared/[min $\times cm^3$ tissue])
- Extraction fraction (often "extraction") (unitless), related to *permeability* and *contact time* in the capillary. *Cave: extraction depends on perfusion (contact time ~1/perfusion)*

Perfusion reflects the function of blood to nourish tissue and to clear e.g. CO_2 and metabolites. This aspect explains its units, which may be not straightforward, especially for a physicist. In physics flow is primarily a volume per time and per area, e.g. describing blood flow in a single vessel. The reference "per cm^3 tissue" in the denominator of perfusion may be understood from the nourishing function of blood and the related branching of vessels. The total flow (volume/min) per total organ volume subdivides e.g. into two halves of that flow supplying the two halves of the organ, therefore the ratio in each half remains the same, as long as the tissue is nourished homogeneously. The process of subdividing organ volumes ends at the capillary level.

K_1 is the input constant into tissue. Traditionally it is written as capital letter in contrast to the other transport constants k_2, k_3, k_4. The units of the latter are [1/min], while K_1 has an additional ratio (ml blood/cm^3 tissue) because K_1 couples to the tracer concentration in *blood* (and not in *tissue*).

As a consequence,

$$\text{Perfusion} \times \{\text{arterial concentration}\}$$

describes the amount of tracer that's offered to tissue per time, while

$$K_1 \quad \times \{\text{arterial concentration}\}$$

refers to the part of it that actually reaches the tissue. The factor that links K_1 and perfusion, the unitless "extraction fraction", depends on flow. This may be understood from looking at the contact time: at a low flow, the contact time is long and all tracer may leave the capillary vessel during capillary transit. Extraction approaches 1. In contrast to that, at very high flows, only a small fraction of tracer leaves the vessel during one capillary transit. Extraction decreases with increasing flow, approaching a state where the increase in flow is counterbalanced by a decrease in extraction, resulting in a nearly flow-independent K_1. The most common description of this flow dependence of the extraction is the Renkin-Crone-formula. [2,3]

2.3 Fick's Equation

Fick's equation reflects the conservation of "mass" or tracer (corrected for decay). It describes the transport as a balance of amounts, not as equilibration of concentrations. In classical kinetic modelling this refers to whole organs or large subspaces like the central compartment, the blood. In PET it is a local balance.

A short form Fick's equation reads as

$$\text{change} = \text{in} - \text{out}$$

We mention here one classical application of Fick's equation to a whole organ balance, the *Kety-Schmidt* approach to measure whole brain blood flow, since this has been adapted to PET blood flow measurement. The balance is, in short, the integral of the equation above:

$$\text{accumulated} = \int \text{in} - \int \text{out}$$
$$V \times \quad C_{\text{TIS}}(\text{at late times}) = \text{flow} \times \quad \int (C_{\text{ART}} - C_{\text{VEN}})$$

From this we get flow/V, if we assume at late times (at the end of the measurement, which is the upper limit of the integrals)

$$C_{\text{TIS}} \sim C_{\text{VEN}}$$

The relation between C_{TIS} and C_{VEN} is used in PET the other way round than described above: In PET we measure the tissue concentration C_{TIS}, but we do not know the venous concentration C_{VEN} at the different pixels, while Kety and Schmidt measured the whole brain venous concentration, but they did not have any direct information on the tissue concentration in pre-PET time.

There is another point worth mentioning: The proportionality of venous and tissue concentration was used by Kety and Schmidt *only at late times* (nearly in equilibration). In PET the proportionality is assumed *for all times*. This is an example for an obviously wrong assumption being used successfully, because an averaging evaluation gives sufficient stable results. The assumption of proportionality between venous and tissue concentration is obviously wrong at early times, when the tracer has just entered tissue at the arterial end of the capillary passage (assuming a high clearance rate from blood to tissue for a good blood flow tracer), while the venous blood does *not* reflect at all average tissue concentration at early times.

The assumed proportionality of venous and tissue concentration is not described as identity because of differences in solubility yields stable ratios which do not approach equal concentrations.

2.4 Getting Infusion Data from Bolus Injection Data

We now describe a "trick" or tool to obtain equilibrium data, like e.g. the *volume of distribution,* from data measured after bolus injection. This way, we will understand why formulas for the *volume of distribution* often contain integrals.

The only prerequisite is a linear relation between the input to the system (i.e. tracer in blood plasma) and the system output (i.e. tracer in tissue), which is given when the amount of administered (unlabelled) tracer is low enough to assume that the system is not altered by the tracer ("tracer principle"). Linearity also means that different portions of tracer arriving locally over time do not interact.

Under this assumption, a constant input into a system, an infusion (Fig. 2, 2nd column, bottom), may be subdivided in time into many identical discrete bolus injections. Since all these bolus injections result in identical time courses of concentration in the system, only shifted in time, the sum of all these contributions to

Getting infusion data from injection data		
	Input "tracer application"	Output "measurement in tissue"
Injection: (easy to perform)		
↓ *integration* ↑*derivation*	↓ *integration*↓	↓ *integration*↓
Infusion: (equilibrium data for distribution volume)		

Fig. 2 Schematic, table-like deduction of the "trick" to convert measurements after single bolus injections into data related to infusion, using system linearity and integrating "input" and "output"

the system output corresponds to the integral over time from a single discrete bolus. We thus simulate infusion by integrating the data obtained after bolus injection over time.

3 Three Basic Models

Figure 3 draws the structure of three main examples and applications of PET kinetic modelling: blood flow, FDG and receptors.

There is one closed rectangle for each tissue compartment, corresponding to a local balance and a related differential equation describing the change in the compartment as the sum of all input to that compartment minus the sum of all output from it.

The PET measurement cannot distinguish between the signals from the different compartments and those from the part of the blood that is included in the PET pixel (fractional blood volume). Therefore the time courses of all tissue compartments and that of the fractional blood volume are added up to one single PET measurement.

The main reason to draw an open rectangle for the arterial input is that there is *no local balance* for it. The change of arterial concentration is *not* given by the local efflux to tissue and by the local backflow from tissue. It results from *all* venous backflows and the efflux to *all* tissue parts of the whole body. A minor additional point of view for the open rectangle is the above mentioned aspect that only a part of the arterial blood space, a fraction of it, is included as "fractional blood volume" in the local PET pixel reflected by the model drawings in Fig. 3.

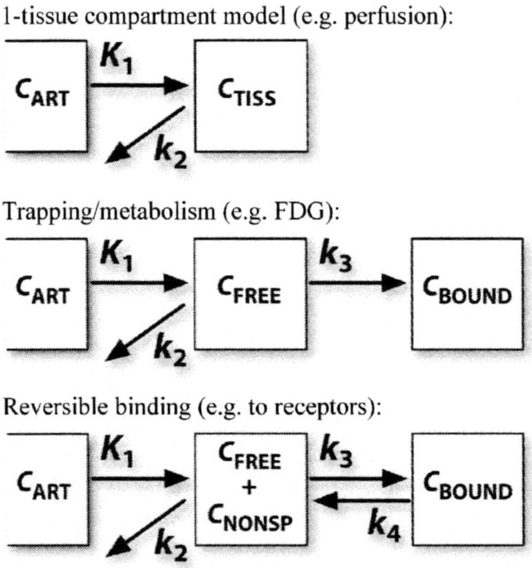

1-tissue compartment model (e.g. perfusion):

Trapping/metabolism (e.g. FDG):

Reversible binding (e.g. to receptors):

Fig. 3 Three basic PET kinetic models C_{ART} = concentration of nonmetabolized free tracer in arterial plasma

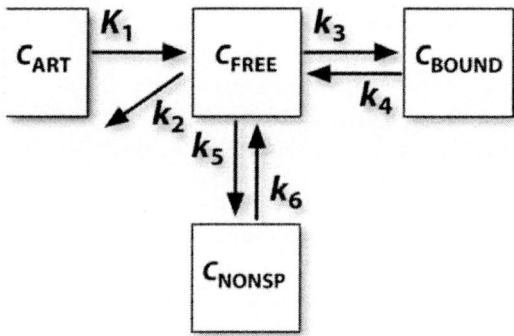

Fig. 4 Receptor model including nonspecific binding

The 2-tissue-compartment system is based upon an underlying 3-tissue-compartment system where non-specific binding has its own compartment (Fig. 4). Within the 2-tissue-compartment system it is assumed that C_{FREE} is a constant fraction f_{ND} of $C_{FREE+NS}$.

3.1 Basics on the Time Courses in the different Compartments

PET measurements, which add signals from all different compartments and from the fractional blood volume, can be simulated after *calculating* the different time courses *separately*.

As described above, there is no local balance for the arterial concentration. The arterial concentration is assumed to be given and has to be measured. Later we will consider the special approach to replace the measurement of arterial concentration by including measurements from a so-called *reference tissue*.

Since there is one balance per compartment, the number of the differential equations corresponds to the number of compartments. Solving the differential equations leads us to exponentials and to the simple rule that the number of exponentials in the basic solution corresponds to the number of equations, i.e. to the number of (tissue) compartments. (It should be noted that this does not imply that exponentials and compartments can be attributed pairwise).

The structure behind that rule is the following: if we insert an exponential function as solution, the differentials become multiplications with that exponent. We thus get a set of ordinary linear equations, from which we get one equation for the exponent as condition of solubility, the degree being the number of compartments or equations.

There is another way to consider the behavior of the compartment systems of Fig. 3: the mathematical approach to ordinary differential equation is to look first for a special solution with a δ-input and then to express the general solution as convolution of the arterial input function with the special solution found before. To look for this special solution, we suppose an amount of tracer given by K_1 being initially present in the first tissue compartment, and set the influx from plasma to zero for the rest of the time course. This solution represents the "answer" of the system to a δ-input, and is called *impulse response function (IRF)*.

For the flow model this *impulse response function* is simply

$$IRF = K_1 e^{-k_2 t}$$

and the full solution (omitting the fractional blood volume) reads as

$$C_{TIS} = C_{ART} \otimes K_1 e^{-k_2 t}$$

The FDG model is somehow special within the scheme on the number of exponentials: It has only one "real exponential", but in accordance with the rule there is an additional constant, which is the special case of an exponential with a time constant zero. Therefore the rule is still working.

The one "real exponential" of the FDG-model is easily seen from Fig. 3: The starting amount given by K_1 will clear from the first tissue compartment by the time constant $k_2 + k_3$:

$$IRF_{FREE} = K_1 e^{-(k_2+k_3)t}$$

The second tissue compartment C_{BOUND} integrates C_{FREE}. This introduces a constant but no new exponential. The complete IRF can be seen directly by subdividing the tracer that enters the first compartment into a reversible part $\sim k_2$ that flows back to the venous blood and a part that will be trapped $\sim k_3$. Since the reversible part can only be cleared with the exponential described above, we get

$$IRF = K_1 \frac{k_2}{k_2 + k_3} e^{-(k_2+k_3)t} + K_1 \frac{k_3}{k_2 + k_3} \tag{1}$$

This equation may also be obtained from $IRF = IRF_{FREE} + IRF_{BOUND}$ with

$$IRF_{BOUND} = \int k_3 \times IRF_{FREE}$$

by reordering the terms.

3.2 Distribution Volumes Related to the Three Basic Models

The equilibrium conditions necessary to calculate the distribution volumes can be directly seen from Fig. 3: According to section 2.1 we express the distribution volume of each tissue compartment as the ratio of the respective concentration over (arterial) blood (plasma) concentration in equilibrium. The total distribution volume is obtained as the sum of the separate volumes.

In equilibrium we have constant concentrations and all fluxes into and from each compartment compensate. From these conditions of compensation we get the necessary ratios, which need to be combined to get for each compartment the ratio to the blood concentration.

3.2.1 1-Tissue Compartment Model

$$k_2 \times C_{\text{TIS}} = K_1 \times C_{\text{ART}}$$
$$\frac{C_{\text{TIS}}}{C_{\text{ART}}} = \frac{K_1}{k_2}$$
$$V = \frac{K_1}{k_2}$$

3.2.2 2-Tissue Compartment Model

$$k_2 \times C_{\text{F+NS}} = K_1 \times C_{\text{ART}} \quad k_2 \times C_{\text{BOUND}} = k_3 \times C_{\text{FREE}}$$
$$\frac{C_{\text{F+NS}}}{C_{\text{ART}}} = \frac{K_1}{k_2} \quad \frac{C_{\text{BOUND}}}{C_{\text{F+NS}}} = \frac{k_3}{k_4}$$

$$V_{\text{F+NS}} = \frac{C_{\text{F+NS}}}{C_{\text{ART}}} = \frac{K_1}{k_2} \quad V_{\text{T}} = \frac{C_{\text{BOUND}}}{C_{\text{F+NS}}} = \frac{k_3}{k_4}$$
$$V = V_{\text{F+NS}} + V_{\text{T}} = \frac{K_1}{k_2}\left(1 + \frac{k_3}{k_4}\right)$$

3.2.3 3-Tissue Compartment Model

The calculation of the volume of distribution extends further from the 2-tissue compartment model to the three compartments of Fig. 4.

$$V = \frac{K_1}{k_2}\left(1 + \frac{k_3}{k_4} \frac{k_5}{k_6}\right)$$

4 Simulations and Evaluations

When we considered above the time courses in the different compartments for given model parameters and given arterial input, we did a simulation. Evaluation of patient data works the other way round: From measured time courses of the arterial input and of the total tissue PET signal (on pixel by pixel basis or on region-of-interest basis), we want to obtain either all parameters of the model or at least important and stable combinations like the metabolic rate constant for glucose or distribution volumes of receptors.

The *metabolic rate of glucose* is e.g. obtained from FDG measurements in a two step procedure: First, the *rate constant* is measured or estimated for FDG, corresponding to the expression described above

$$\frac{K_1 k_3}{k_2 + k_3} \tag{2}$$

In a second normalisation step this constant, related to FDG, is converted to the corresponding constant related to glucose, by dividing it by an organ dependent so-called *lumped constant*, and by multiplication with the *blood concentration of (non-radioactive) glucose* to get the *metabolic rate of glucose* from the *glucose rate constant*.

The process of parameter estimation or parameter measurement reverses the convolution procedure described above. Primarily this is the field of non-linear fitting e.g. according to the simplex algorithm[4,5] or the Marquardt-Levenberg algorithm.[6,7]

A point that becomes less important with increasing computer power, is the computational speed. Especially if parameters are calculated on a pixel-by-pixel basis this has been the field of linear approaches. It is a common property of linear approaches that the variables used are not statistically independent over time like a concentration and its integral, which leads to a bias for the estimated parameters.

5 Linearizations

5.1 Flow

Integrating the differential equation from the flow model in Fig. 3

$$\frac{dC_{TIS}}{dt} = K_1 \times C_{ART} - k_2 \times C_{TIS}$$

yields

$$C_{TIS} = K_1 \times \int C_{ART} - k_2 \times \int C_{TIS} \tag{3}$$

or, after dividing by the integral of the arterial concentration

$$\frac{C_{TIS}}{\int C_{ART}} = K_1 - k_2 \times \frac{\int C_{TIS}}{\int C_{ART}}$$

$$Y = K_1 - k_2 \times X$$

where the variables X and Y are known measured functions (but not statistically independent!).

The corresponding linear 2D graph may be attractive for presentation, but is not recommended for the fitting procedure, which is much more stable in the 3D form (3). The problem is that an arterial concentration measured at the forearm is not identical to the arterial concentration at the brain, where the blood arrives faster and less dispersed. Therefore the arterial concentration needs to be corrected for delay and dispersion. The effect of this correction is dramatic at early times, especially in the 2D graph where the integral over the arterial concentration enters in the denominator: there is already tracer in the brain tissue, but the uncorrected arterial concentration from the forearm is still zero.

5.2 The Patlak-Plot

For trapping systems like the FDG-model, the IRF (1), (see section 3.1) may be regarded as a sum of pure trapping and a reversible system. One obtains:

$$C_{TIS}(t) = C_{ART}(t) \otimes K_1 \frac{k_2}{k_2 + k_3} e^{-(k_2+k_3)t} + K_1 \frac{k_3}{k_2 + k_3} \int C_{ART} \qquad (4)$$

The reversible part of C_{TIS} will at late times be proportional to C_{ART} (this condition is called pseudo-equilibrium):

$$C_{ART}(t) \otimes K_1 \frac{k_2}{k_2 + k_3} e^{-(k_2+k_3)t} \sim C_{ART}(t)$$

its factor having the units of a distribution volume V_{REV}. Therefore, if we divide (4) by $C_{ART}(t)$, we get

$$\frac{C_{TIS}(t)}{C_{ART}(t)} = V_{REV} + \frac{K_1 k_3}{k_2 + k_3} \frac{\int C_{ART}}{C_{ART}(t)}$$

which corresponds to a linear graph with

$$Y = \frac{C_{TIS}(t)}{C_{ART}(t)} \quad X = \frac{\int_0^t C_{ART}}{C_{ART}(t)} (\text{``PATLAK's funny time''})$$

the slope being the metabolic rate constant (2)

5.3 Deriving the Logan Plot

5.3.1 Logan Plot for 1-Tissue-Compartment Systems

Let's recall the equation for 1-tissue-compartment systems:

$$C_{TIS}(t) = K_1 \times \int C_{ART} - k_2 \times \int C_{TIS}$$

By reordering and dividing by k_2 and C_{TIS}, we get

$$\frac{\int\limits_0^t C_{TIS}}{C_{TIS}(t)} = \frac{K_1}{k_2} \times \frac{\int\limits_0^t C_{ART}}{C_{TIS}(t)} - \frac{1}{k_2} \qquad (5)$$

where the distribution volume K_1/k_2 appears as slope m of a linear graph $Y = mX +$ *intercept* with

$$Y = \frac{\int\limits_0^t C_{TIS}}{C_{TIS}(t)} \quad \text{and } X = \frac{\int\limits_0^t C_{ART}}{C_{TIS}(t)}$$

5.3.2 Generalization for n-Compartment System

We now consider an n-compartment system, without trapping and with final excretion. At late times, the slowest exponential of the system will govern all compartments, i.e. the system will behave like a 1-tissue-compartment system (this terminal phase is called pseudo-equilibrium). Accordingly, after some time, the Logan plot $Y = f(X)$ will become linear:

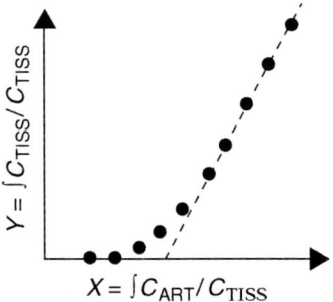

We will now show that the slope of the late linear portion of the Logan plot always represents the distribution volume of the system, regardless of the number of compartments that govern the system during the beginning. With t increasing, X and Y continuously increase to infinity, therefore the ratio Y/X approximates the slope of the Logan plot. In other words: the slope of the Logan plot corresponds to the limit of the ratio Y/X for $t \to \infty$. This limit corresponds to the distribution volume:

$$\frac{Y}{X} = \frac{\int_0^t C_{TIS}/C_{TIS}(t)}{\int_0^t C_{ART}/C_{TIS}(t)} = \frac{\int_0^t C_{TIS}}{\int_0^t C_{ART}} \to \frac{\int_0^\infty C_{TIS}}{\int_0^\infty C_{ART}} = V$$

To understand the last step in this equation, we recall that the distribution volume is defined as the ratio of concentrations during equilibrium, that the system approximates equilibrium in the presence of constant system input ("infusion"), and that the latter may be simulated by integration over time.

6 Receptor Modelling

6.1 Binding Potential

The concept of binding potential was introduced by Mintun et al.[8] As can be directly derived from mass action law, in equilibrium, the ratio of radiotracer bound to receptors over free radiotracer corresponds to B_{max}/K_D where B_{max} denotes the concentration ("density") of receptors and K_D the dissociation constant (1/affinity). A prerequisite is that the specific activity of the injected tracer is high enough to ensure that the amount of receptors occupied by the tracer is negligible compared to the total amount of receptors. This ratio is called binding potential BP:

$$BP = B_{max}/K_D$$

Measuring B_{max}/K_D requires arterial blood sampling, furthermore the free fraction of radiotracer in arterial plasma must be known. In the PET literature, the term "binding potential" is also used for two measures that are related to B_{max}/K_D, but can be derived more directly from the measured PET data. As suggested in a recent consensus paper,[9] we will call them BP_P and BP_{ND}.

To assess the binding potential in a certain target region, a reference tissue is required. In the reference tissue, it is assumed that there is only negligible specific binding and that the distribution volume of the first tissue compartment is the same as in the target region ($K_1/k_2 = K_1'/k_2'$). BP_P and BP_{ND} can be defined with respect to the distribution volumes of the target and the reference tissue. BP_P corresponds to $V_{TARGET} - -V_{REF}$, while BP_{ND} – as shown below – corresponds to $(V_{TARGET} - V_{REF})/V_{REF}$.

6.2 A Reference Region Model Accounting for Nonspecific Binding

In tissue, the radiotracer not only binds to its *specific* molecular target for which it has a high affinity, e.g. a neuroreceptor, but also to ubiquitous structures such as proteins. Even if the affinity for the latter is very low, the amount of *nonspecifically* bound tracer may be considerable due to the high concentration of (nonspecific) binding sites. Accordingly, this nonspecific binding is non-saturable and nondisplaceable.

We start by addressing volumes of distribution because of their intuitive relation to measured concentration at late times in a single injection protocol.

From Fig. 5 and from the reference tissue assumption $K_1/k_2 = K_1'/k_2'$ we get

$$V_T - V_R = \frac{K_1}{k_2} \times \left(\frac{k_3}{k_4}\right)$$

Fig. 5 Reference region
model

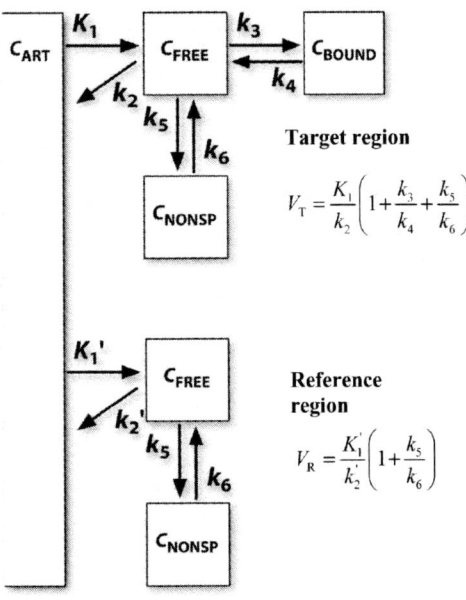

$$\frac{V_{\mathrm{T}} - V_{\mathrm{R}}}{V_{\mathrm{R}}} = \frac{\frac{k_3}{k_4}}{1 + \frac{k_5}{k_6}} = \frac{k_3}{k_4} \times f_{\mathrm{ND}}$$

where

$$f_{\mathrm{ND}} = \frac{V_{\mathrm{FREE}}}{V_{\mathrm{FREE}} + V_{\mathrm{NONSPEC}}} = \frac{\frac{K_1}{k_2}}{\frac{K_1}{k_2} \times \left(1 + \frac{k_5}{k_6}\right)} = \frac{1}{\left(1 + \frac{k_5}{k_6}\right)}$$

is the free fraction in tissue at equilibrium.

6.3 Receptor Modelling: Two Versus 3-Tissue Compartment Model

Due to the high concentration of nonspecific binding sites and a high dissociation
rate, nonspecific binding is assumed to equilibrate quickly. Due to rapid equilibration, the ratio of free over nonspecifically bound tracer is usually treated as a constant throughout the measurement interval and the related compartments are lumped
together to the compartment of *nondisplaceable* tracer. The fraction of free in relation to the total tracer in this compartment is denoted f_{ND} according to the consensus
nomenclature.[9]

Equation (4) reads in short

$$\frac{V_T - V_R}{V_R} = \frac{k_3}{k_4} \times f_{ND} \quad \text{for the 3-tissue compartment model}$$

The "lumped k_3" of the 2-tissue compartment model is the "true k_3" of the 3-tissue compartment model lowered by the factor f_{ND}, because the rate constant k_3 now refers to a higher concentration (i.e. that of the free + nonspecific ligand), which is increased by the factor f_{ND}. The product of concentration times the corresponding k_3-value (flux to receptor binding) thus remains unchanged in both models. Accordingly we get

$$\frac{V_T - V_R}{V_R} = \frac{k_3}{k_4} \quad \text{for the 2-tissue compartment model}$$

6.4 The 2 Tissue Compartment Model: Binding Potential BP_{ND} and Mass Action Law

For a single injection protocol, specific activity (i.e. the amount of injected radioactive tracer compared to the total amount of injected tracer) is usually chosen to be high enough to be in accordance with the so-called "tracer principle": the system (e.g. the synapse) is assumed to be not affected by the small amount of injected ligand. B_{avail}, the concentration of receptors available for binding, equals B_{max}, the total concentration of receptors.

The on-reaction is the product of three factors: (i) the concentration of available receptors (in the absence of cold ligand: B_{max}) (ii) the concentration of available (i.e. free) ligand: $f_{ND} \times [\text{Ligand}]$ and (iii) the rate constant k_{on}:

$$\text{on-reaction}(2 - \text{tissue-compartment model}) : k_3 \times [\text{ligand}]$$
$$\text{on-reaction}(\text{mass action law}) : k_{on} \times B_{max} \times f_{ND}[\text{ligand}]$$

Accordingly, k_3 can be interpreted as:

$$k_3 = k_{on} \quad \times \quad B_{max} \quad \times \quad f_{ND}$$

The off-reaction is the product of two factors: the concentration of ligand that is bound to the receptor, and a rate constant, denoted k_{off} in the context of mass action law, and k_4 in the context of the two-tissue-compartment model.

$$k_4 = k_{off}$$

We next interpret the ratio k_3/k_4 in terms of mass action law:

$$k_3/k_4 = k_{on} \quad \times \quad B_{max} \quad \times \quad f_{ND}/k_{off}$$

According to mass action law, k_{off}/k_{on} corresponds to the concentration of the ligand at which 50% of receptors are occupied. This is the *equilibrium dissociation constant* $K_D = k_{off}/k_{on}$. We obtain:

$$k_3/k_4 = (B_{max}/K_D) \times f_{ND}$$

Thus k_3/k_4, a common target measure in PET studies, also denoted BP_{ND}, is closely related to the binding potential $BP = B_{max}/K_D$:

$$k_3/k_4 = BP_{ND} = BP \times f_{ND}$$

6.5 Obtaining BP_{ND} Without Arterial Blood Samples

It is possible to estimate $BP_{ND} = k_3/k_4$ (k_3 and k_4 as defined in the 2-tissue-compartment model) without knowing the time course of the tracer in arterial plasma ("noninvasive quantification"). However, we need to accept the error deriving from the contribution of about 5% fractional blood volume to the total radioactivity in tissue, which can only be corrected for when the arterial concentration of radioactivity is known. This error is small for most tracers, but may be considerable for tracers with rapid metabolization, such as [11C]WAY100635.[10] An operational equation that allows for estimation of the rate constants k_2, k_3, k_4 and the ratio $R_1 = K_1/K_1$' with a nonlinear least square fit has been published by Lammertsma et al.[11] While the single rate constants k_3 and k_4 can only be estimated with poor reproducibility, estimation of the ratio k_3/k_4 is much more stable. This can be understood from the fact that the ratio k_3/k_4 is connected with the distribution volume ratio V_T/V_R ($k_3/k_4 = V_T/V_R - 1$, see above), and that this distribution volume ratio is represented in the time activity curves of the target and the reference region in a identifiable way, e.g. as the limit of the area-under-curve-ratio for $t \to \infty$. In contrast, the absolute height of k_3 and k_4 more or less reflect how fast the first and the second tissue compartment of the target region equilibrate, which is much more difficult to see from the time activity curves.

A frequently used simplified variant is known as *simplified reference tissue method* (SRTM) of Lammertsma and Hume.[11−13] In this approach, the target tissue is described by one *single* tissue compartment. Its washout constant is often denoted k_{2a} (describing the washout in relation to the total amount of radioactivity in the target tissue; index 'a' = 'apparent') to distinguish it from k_2 in the 2-tissue compartment model (describing the washout in relation to the concentration of free tracer in the first compartment). In a single compartment approach, specific binding reduces the washout k_2 by a factor of $1 + BP : k_{2a} = k_2/(1 + BP_{ND})$. Without blood data, K_1 cannot be estimated, instead, the ratio $R_1 = K_1/K_1$' (primed variables refer to the reference tissue) is calculated. The general reference tissue assumptions (see section 6.1) apply: in the reference tissue, it is assumed that there is only negligible specific binding and that the distribution volume of the first tissue compartment is the same as in the target region ($K_1/k_2 = K'_1 k'_2$). The operational equation of the SRTM then reads:

$$C_T = R_1 C_{REF} + R_1 \left\{ k'_2 - k_{2a} \right\} C_{REF} \otimes e^{-k_{2a}t} \tag{6}$$

Although this equation is frequently used, its derivation is widely unknown; here we present a simple approach that uses the Laplace transform[1], which transforms functions of t (here $C_T(t)$ and $C_{REF}(t)$) into functions of s. We start with the differential equations

$$\frac{dC_T}{dt} = K_1 C_{ART} - k_{2a} C_T \quad and \quad \frac{dC_{REF}}{dt} = K'_1 C_{ART} - k'_2 C_{REF}$$

After inserting C_{ART} from the second equation into the first one, one obtains

$$\frac{dC_T}{dt} = \frac{K_1}{K'_1} \left[\frac{dC_{REF}}{dt} + k'_2 C_{REF} \right] - k_{2a} C_T$$

Taking the Laplace transform L and replacing $R_1 = K_1 / K_1{}'$:

$$sL(C_T) = R_1 \left[sL(C_{REF}) + k'_2 L(C_{REF}) \right] - k_{2a} L(C_T)$$

resolving for $L(C_T)$ leads to

$$L(C_T) = R_1 \frac{[s + k'_2]}{[s + k_{2a}]} L(C_{REF}) = R_1 \frac{[s + k_{2a} + (k'_2 - k_{2a})]}{[s + k_{2a}]} L(C_{REF})$$

or

$$L(C_T) = R_1 L(C_{REF}) + R_1 \frac{(k'_2 - k_{2a})}{[s + k_{2a}]} L(C_{REF})$$

Now the inverse Laplace transform[2] immediately leads to equation (6).

As stated above, this "simplified reference tissue method" (SRTM) theoretically requires that the target tissue behaves like a single compartment, which is the case for some tracers with rather high kinetic rate constants k_3 and k_4. However, the SRTM is often also applied to tracers that do not fulfill this criterion, fortunately the resulting error is usually small.

---

[1] The Laplace transform is defined as follows:

$$L(C(t)) = \int_0^\infty C(t) \times e^{-st} dt$$

Because of $C(0)=0$, the derivation simplifies to a multiplication by s

$$L\left(\frac{d}{dt} C(t) \right) = s \times L(C(t)) - C(0) \rightarrow s \times L(C(t))$$

[2] To calculate the inverse Laplace transform L^{-1}, we need to know that the multiplication in the Laplace space corresponds to a convolution $\otimes$ in the time space, and we take from a standard list of inverse transforms $L^{-1} (s+k)^{-1} = e^{-kt}$.

The linearization proposed by Logan (section 5.2.) does not require a 1-tissue-compartment behavior of the target tissue starting from t = 0, instead it allows to specify a time after which the system behaves mono-exponential. It can easily be adjusted to allow for *noninvasive* calculation of the distribution volume ratio V_T/V_R by replacing the Integral of C_{ART} with a term that is derived from the differential equation of the *reference* tissue (we recall that the reference tissue is assumed to behave like a single compartment):

$$C_{REF}(t) = K_1' \times \int C_{ART} - k_2' \times \int C_{REF}$$

and after rearrangement:

$$\int C_{ART} = \frac{C_{REF}(t) + k_2' \times \int C_{REF}}{K_1'}$$

After replacing the Integral of C_{ART} in (5) we obtain

$$\frac{\int_0^t C_T}{C_T(t)} = \frac{K_1 k_2'}{K_1' k_2} \times \frac{\int C_{REF} + C_{REF}(t)/k_2'}{C_T(t)} - \frac{1}{k_2}$$

which, as (5), becomes linear, here with the slope $(K_1 k_2')/(K_1' k_2)$ being the distribution volume *ratio*.

It should be noted that Logan's method requires k_2' to be given. Fortunately, for many tracers, a rough estimate for k_2' can be used for that purpose. In order to estimate not only the distribution volume ratio, but also k_2', one might choose one of several variants of the so-called multilinear approach, described, e.g. in Ichise et al.[14]

6.6 Multiple-Injection Ligand Studies

There are various traditions of investigation and corresponding models. The co-author of the PET overview chapter, Millet, is related to the tradition of Jacques Delforge[15,16] who pioneered multiple-injection ligand studies at Orsay. As described in section 6.4, the classical approach with one injection of a high specific tracer, aims at a certain combination of binding parameters: B_{max}/K_D. In order to get both parameters separately, it is necessary to work with different states of occupancy of the receptors. In Delforge's multiple-injection ligand studies typically three injections are used within one study: a classical high specific tracer, a pure cold ligand and a coinjection of hot tracer and cold ligand corresponding to a tracer of low specificity. The curve types of multi-injection studies are easily recognized: the sites of specific binding have two up-slopes from the hot injections and one down-slope from the cold ligand injection.

7 Some Trends

In oncology PET clinical decisions are mostly based on static whole body images without kinetic modelling.

Modelling is particularly popular in brain receptor imaging e.g. to determine receptor density independent from blood flow. Not only for clinical application, but also for research projects, arterial blood sampling has been increasingly abandoned for reference tissue approaches.

References

1. Maguire, R.P., Leenders, K.L., Eds, 2007, *PET Pharmacokinetic Course Manual 2007*, Osaka, Japan.
2. Renkin E.M., 1959, Transport of potassium-42 from blood to tissue in isolated mammalian skeletal muscles. *Am. J. Physiol.*, **197**:1205–1210.
3. Crone, C., 1964, The permeability of capillaries in various organs as determined by use of the indicator diffusion method. *Acta Physiol. Scand.*, **58**:292–305.
4. James, F., Roos, M., 1976, Minuit. A System for Function Minimization and Analysis of the Parameter Errors and Correlations, CERN, Geneva.
5. Nelder, J.A., Mead, R., 1965, A simplex method for function minimization. *Comput. J.*, **7**:308–313.
6. Levenberg, K., 1944, A Method for the solution of certain problems in least squares. *Quart. Appl. Math.*, **2**:164–168.
7. Marquardt, D.W., 1963, An algorithm for least-squares estimation of nonlinear parameters. *J. Soc. Indust. Appl. Math.*, **11**:431–441.
8. Mintun, M.A., Raichle, M.E., et al., 1984, A quantitative model for the in vivo assessment of drug binding sites with positron emission tomography. *Annals of Neurology*, **15**(3):217–227.
9. Innis, B.I., Cunningham,V.C., Delforge, J., et al., 2007, Consensus nomenclature for *in vivo* imaging of reversibly binding radioligands. *J. Cereb. Blood Flow Metab.* advance online publication, 9 May 2007, **27**:1533–1539; doi:10.1038/sj.jcbfm.9600493.
10. Parsey, R.V., Slifstein, M., Hwang, et al., 2000, Validation and reproducibility of measurement of 5-HT$_{1A}$ receptor parameters with [carbonyl-^{11}C]WAY-100635 in humans: comparison of arterial and reference tissue input functions. *J. Cereb. Blood Flow Metab.*, **20**:1111–1133.
11. Lammertsma, A.A., Bench, C.J., Hume, S.P., 1996, Comparison of methods for analysis of clinical [^{11}C]raclopride studies. *J. Cereb. Blood Flow Metab.*, **16**:42–52.
12. Lammertsma, A.A., Hume, S.P., 1996, Simplified reference tissue model for PET receptor studies. *NeuroImage*, **4**:153–158.
13. Hume, S.P. et al., 1992, Quantitation of carbon-11-labeled raclopride in rat striatum using positron emission tomography. *Synapse* **12**: 47–54.
14. Ichise, M., Liow, J.-S., Lu, J.-Q., et al., 2003, Linearized reference tissue parametric imaging methods: application to [^{11}C]DASB positron emission tomography studies of the serotonin transporter in human brain.*J. Cereb. Blood Flow Metab.*, **23**: 1096–1111.
15. Delforge, J., Syrota, A., Bottlaender, M., et al., 1993, Modelling analysis of [11C]flumazenil kinetics studied by PET: application to a critical study of the equilibrium approaches. *J. Cereb. Blood Flow Metab.*, **13**(3):454–468.
16. Delforge, J. Pappata, S. Millet, P. et al., 1995, Quantification of benzodiazepine receptors in human brain using PET, [^{11}C]flumazenil, and a single-experiment protocol. *J. Cereb. Blood Flow Metab.*, **15**:284–300.

17. Laruelle, M., Slifstein, M., Huang, Y., 2002, Positron emission tomography: imaging and quantification of neurotransporter availability. *Methods*, **27**(3): 287–299.
18. Patlak, C.S., Blasberg, R.G., Fenstermacher, J.D., 1983, Graphical evaluation of blood-to-brain transfer constants from multiple-time uptake data. *J. Cereb. Blood Flow Metab.*, **3**:1–7.
19. Reimold, M., Feb 22, 2007, Binding Potential, http://en.wikipedia.org/wiki/Binding_potential

Pharmacokinetics Application in Biophysics Experiments

Philippe Millet* and Yves Lemoigne

Abstract Among the available computerised tomography devices, the Positron Emission Tomography (PET) has the advantage to be sensitive to pico-molar concentrations of radiotracers inside living matter. Devices adapted to small animal imaging are now commercially available and allow us to study the function rather than the structure of living tissues by *in vivo* analysis. PET methodology, from the physics of electron-positron annihilation to the biophysics involved in tracers, is treated by other authors in this book. The basics of coincidence detection, image reconstruction, spatial resolution and sensitivity are discussed in the paper by R. Ott. The use of compartment analysis combined with pharmacokinetics is described here to illustrate an application to neuroimaging and to show how parametric imaging can bring insight on the *in vivo* bio-distribution of a radioactive tracer with small animal PET scanners. After reporting on the use of an intracerebral β^+ radiosensitive probe (βP), we describe a small animal PET experiment used to measure the density of 5HT 1_A receptors in rat brain.

1 Why a Small Animal PET Device?

In modern medical imaging there are several ways to study *in vivo* pharmaceutical drugs injected at lower concentration in living beings, human patients as well as small animals. The most widespread are magnetic resonance imaging (MRI) and positron emission tomography (PET), but PET appears to be more sensitive to pico-molar concentrations at a level lower than pharmaceutical effects.[1]

P. Millet*
Unité de Neurophysiologie Clinique et Neuroimagerie, Geneva University Hospitals (HUG), Geneva, Switzerland
e-mail: Philippe.Millet@sim.hcuge.ch

Y. Lemoigne
Laboratoire ABC, European Scientific Institute, Archamps, France

Y. Lemoigne, A. Caner (eds.) *Molecular Imaging: Computer Reconstruction and Practice,* 211
and Experiments,

Despite the small number of available (and easy to use) positron emitting isotopes (^{18}F,^{11}C,^{13}N,^{15}O...), radiochemists are able to label many compounds. Such radiolabelled tracers make PET a powerful tool to study therapeutic agents and help in the development of new drugs.[2] This paper will give examples of the application of small animal PET in biophysics and medical research.

2 Physics Overview

Computed tomography (CT) measures attenuation of X-rays passing through an object and has been called "transmission method of imaging" because both radiation sources and detectors are external to the object investigated. Conversely, nuclear medicine can exploit "emission methods", where the source of photons or gamma rays is inside the body, i.e. emitted by tissues containing radiopharmaceuticals labeled with radioactive nuclei and injected into patients. Transmission tomography gives a representation of the spatial distribution of the object properties. Emission Tomography has the property to reflect the function rather than the structure of living tissues. Depending on the number of photons detected, the emission CT is called SPECT (Photon Emission Computerised Tomography) when just one photon is required for triggering, whereas it is called PET (Positron Emission Tomography) when two photons from $e + e^-$ annihilation, of predefined energy, time and directions are required (Fig. 1). The more sever trigger requirements used in PET result in enhanced ability to suppress background and, consequently, in improved sensitivity with respect to SPECT.

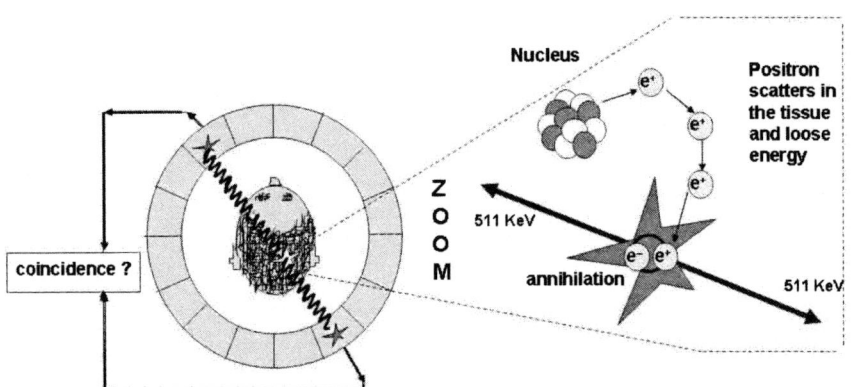

Fig. 1 Electron-Positron annihilation resulting from a β^+ decay. The positron (e^+) emitted at high velocity looses its energy by collisions until approximately at rest. Subsequently it forms an unstable particle (100 ps half-life) with an atomic electron (e^-) which disintegrate into two 511 keV back-to-back photons. For this reason PET cameras require signal coincidence of two back-to-back detector units

Images are reconstructed from the lines-of-response (LOR) defined by the coincidence channels and are used to reconstruct the three-dimensional (3D) distribution of the tracer concentration. Modern tomographs acquire data in 2D (with septa) or in 3D (no septa) mode. LORS's are characterized by their distance with respect to the field of view (FOV) center, an azimuthally angle and a polar angle. In 2D mode the polar angle is close to zero and projections (Random transform) are often stored as word sinograms for direct and first order cross planes. This scheme is ideal for the reconstruction of uncorrelated planes, using backprojection of filtered projections (see paper from C. Comtat in this book). In 3D mode, LOR's are more often stored as planar views and images are reconstructed using the reprojection algorithm.[3] 3D reconstruction allows improving signal-to-noise ratio in reconstructed images by using oblique sinograms, often followed by rebinning of the oblique sinograms into a stack of direct sinograms.[4]

3 Biophysics Overview

3.1 PET Tracers

There is a huge quantity of PET tracers which have been developed during the last decades, and this number is constantly growing. These labeled molecules have been developed to explore a large number of biological systems such as cerebral blood flow and volume, oxygen consumption, glucose metabolism and many neurotransmitter systems.

Radiotracers can be divided into two categories, *non-specific radiotracers* and *specific radioligands*. The first category is used to measure a tissue extraction or metabolism by means of the biochemical pathway of the molecule. For example, ^{15}O-water (H_2 ^{15}O) used to measure cerebral blood flow, or ^{18}FDG allowing tissue glucose metabolism measurement. The second category includes radioligands involved in interactions with receptor sites.

The [^{18}F]Fluorodeoxyglucose (FDG) is incontestably the most used tracer in clinical routine, notably in oncology where it is used to investigate the metabolic activity of human brain tumors, at diagnosis level and during follow-up. However, neurotransmission studies have been increasing in the last few years, due to the enhanced availability of radioligands.

As an example, the Gabaergic system has been widely investigated using the benzodiazepine antagonist [^{11}C]flumazenil. Dopamine D2 receptors can be studied using [^{18}F]dopa and [^{11}C]raclopride and 5HT1a receptors can be studied using [^{18}F]MPPF. The [^{11}C]arachidonic acid is used to explore the brain phospholipids turnover.

3.2 Compartment Models

3.2.1 Introduction

Kinetic analysis is an essential aspect of the quantification of brain functions with PET. After injection of a labeled molecule, PET enables to follow in vivo spatial and temporal distribution of a molecule. The main interest of this technique is the quantification, namely the capability to measure the radiotracer concentration anywhere in an organ like the brain. Time-concentration curves of a radiotracer can be obtained for a region of interest, from which it is possible to derive biological parameters such as cerebral blood flow, glucose consumption or receptor concentration.

Kinetic PET models are usually based on a compartmental analysis, where the studied organ is represented by a number of compartments corresponding to the different states of the tracer. In the following sections, we discuss several examples to illustrate the interest of compartment models in PET. We explain how it is possible to analyse cerebral blood flow with a two compartment model or the glucose consumption with a three compartment model. The last example on ligand receptor interactions is particularly interesting, showing the capability of compartment models to extract all binding parameters when a multi-injection protocol is associated.

3.2.2 Cerebral Blood Flow Estimate from a Two Compartment Model

After the pioneering work of Kety et al.,[5-7] several techniques have been developed to measure regional cerebral blood flow (rCBF). The most commonly used is based on a single PET scan of about 60 s, performed following an intravenous injection of ^{15}O-labeled water ($H_2^{15}O$). The rCBF is estimated using a simple model (Fig. 2) that includes only two compartments representing the plasma and the tissue $H_2^{15}O$ concentration (C_P and C_T, respectively), and two kinetic parameters, K_1 and k_2, representing forward and reverse exchanges through blood brain barrier. The differential equation relating the concentration of radioactivity in the tissue at time t, $C_T(t)$, and the arterial input function, $C_P(t)$, is expressed as:

$$\frac{dC_T(t)}{dt} = K_1 C_P(t) - k_2 C_T(t) = E \times rCBF \times C_P(t) - \frac{rCBF}{V_d}C_T(t), \quad (1)$$

where K_1 is related to the cerebral blood flow and the extraction fraction of the tracer into the brain, $E, (K_1 = E \times rCBF)$. The extraction fraction related to the regional capillary permeability-surface (PS) area product is considered close to unity for ^{15}O-labeled water. The variable V_d is the distribution volume of the tracer in

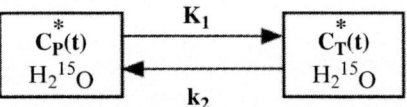

Fig. 2 Cerebral blood flow model

the tissue. Therefore, solving (1), rCBF values are directly deduced from the K_1 parameter, where the asterisk denotes the convolution integral:

$$C_T(t) = rCBF \int_0^t C_P(s)\, e^{-\frac{rCBF}{V_d}(t-s)} ds = C_P(t) * rCBF\, e^{-\frac{rCBF}{V_d}t}. \qquad (2)$$

3.2.3 Glucose Consumption Estimates Using a Three Compartment Model

To measure the regional metabolism of glucose, it is common to use the molecule of deoxy-glucose labeled with [18F]. The deoxy-glucose, like the glucose, crosses the blood brain barrier and is phosphorylated by the hexokinase to deoxy-glucose-6-phospate. However the deoxy-glucose-6-phosphate cannot be isomerized to fructose-6-phosphate and thus it accumulates in the tissues.[8] A small amount of deoxy-glucose-6-phosphate is transformed to deoxy-glucose by the glucose-6-phosphatase. To take into account all of these factors the deoxy-glucose model was developed by Sokoloff et al.[9] for autoradiography and adapted by Phelps et al.[10] for PET studies. The model described by Phelps, assumes that deoxy-glucose-6-phospate can be transformed back to deoxy-glucose.

The deoxy-glucose model is presented in Fig. 3, based on Phelps et al.[10] Only dynamic aspects of the model are considered here. The model has four parameters. Parameters K_1 and k_2 have already been explained for the CBF model (Fig. 2). The rate constant k_3 represents the fraction of deoxy-glucose transformed to deoxy-glucose-6-phosphate per unit of time, and the rate constant k_4 is the fraction of deoxy-glucose-6-phosphate transformed back to deoxy-glucose per unit of time. For any given tissue, [18F]-deoxy-glucose activity measured with PET and plasma activity known, it is possible to calculate the rate constants for the deoxy-glucose. Since glucose follows the same kinetics as the deoxy-glucose, it is possible to calculate the regional cerebral glucose utilization, also called the regional cerebral metabolic rates for glucose (rCMRglu), instead of the deoxy-glucose utilization. For this purpose, we have to take into account the glucose concentration in plasma, and calibrate the differences in transport, phosphorylation, and hydrolysis rates between deoxy-glucose and glucose. This calibration is accomplished by using the lumped constant term.[9] The rCMRGlu values are calculated with the following equation:

$$rCMRGlu = \frac{C_p}{LC} \frac{K_1 k_3}{(k_2 + k_3)}. \qquad (3)$$

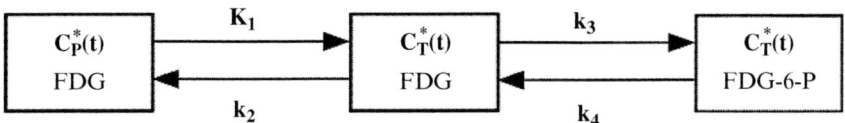

Fig. 3 FDG model

3.2.4 Ligand Receptor Interactions

Ligand-receptor interaction is particularly interesting in PET since it is the only method to quantify in vivo the receptor concentration. Indeed, it is possible to extract binding parameters from time-concentration curves. However, the receptor concentration, B'_{max}, is not directly obtained from the PET measurements, since they include not only the ligand bound to receptors, but also the different steps of the labeled molecule in a tissue volume such as free ligand, non specifically bound ligand and also radioligand in blood vessels present in the tissue. The quantification of binding parameters, such as receptor concentration and ligand affinity, can therefore be performed using a compartmental model.

Many kinetic approaches have been described for modeling ligand receptor interactions. Some are based on an equilibrium state between bound and free plus non-specifically bound ligand, which must be obtained for different free ligand concentration.[11,12] This approach is derived from the well known Scatchard graphical method[13] used for in vitro experiments. Others are dynamic approaches using a single injection with a high specific activity or two injections at different specific activities values. In this case, time-concentration curves of both tissue and arterial plasma are used to estimate model parameters. All these methods are based on a compartment model.

The Ligand Receptor Model

The general model is more complex than those described previously, since it includes exchanges of the radio-ligand through blood brain barrier and its interactions with receptors, namely association and dissociation kinetics for specific and non specific binding, as shown in Fig. 4. Parameters K_1 and k_2 are associated with the exchanges between the plasma and the free ligand compartment; B'_{max} represents the concentration of receptors available for binding; k_{on} and k_{off} are the association

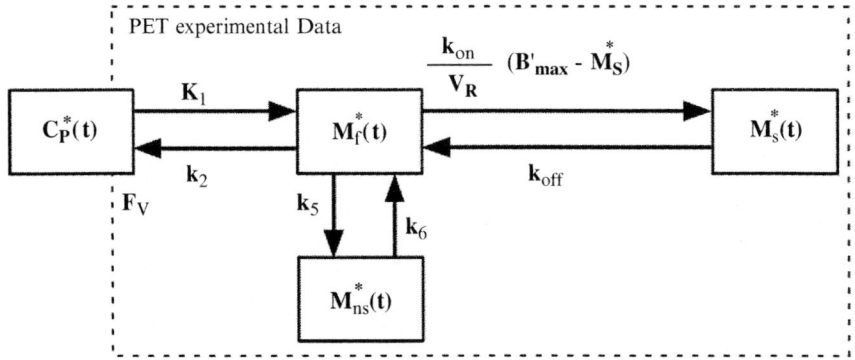

Fig. 4 Ligand-receptor model

and dissociation rate constants for specific binding, k_5 and k_6 are the association and dissociation rate constants for non-specific binding, respectively; V_R is the volume of reaction, which accounts for tissue inhomogeneities. All ligand transfer probabilities between compartments are linear, except for the specific binding probability, which depends on the bimolecular association rate constant (k_{on}), the local free ligand concentration ($M_f(t)/V_R$), and the local concentration of free receptor sites $B'_{max}-M^*_b(t)$. The PET experimental data correspond to the sum of the labeled ligand in the free and bound ligand compartments and of a fraction, F_V, of the blood compartment.

Identification of all model parameters is not possible with a single injection protocol because of the poor experimental design which leads to inaccurate parameter estimates. Many studies have been dedicated to propose a solution for PET modeling overcoming this problem. Most of them have proposed a simplification of the ligand receptor model by including for example, the free and the non-specific bound ligand in the same compartment.[14,15] In this case, the association and dissociation parameter values for the non-specific binding are larger than the other parameters which lead to a rapid equilibrium between compartments. In this way, it is possible to improve parameter estimates without increasing the complexity of the experimental protocol.

However, according to which modeling approach is adopted, time-concentration curves are used either to estimate directly the receptor concentration and ligand affinity or to calculate an index assumed to be correlated with the receptor concentration, for example, the distribution volume approach proposed by Koeppe et al.[16]

Recently, a new approach has been proposed[17–19] whose objective was to access input function without arterial catheter. This method, named "reference tissue method", uses a region without specific binding of the ligand as an indirect input function, from which it is possible to express a time-concentration curve in the tissue of interest using a compartment model. In this simplified method only the binding potential, an indirect measure of receptor density, can be obtained using a single injection protocol.

The Multi-Injection Approach

The main limitation of PET quantification is the experimental protocol, which must be adapted to obtain correct parameter estimates during the identification process. However, a complex experimental protocol is applicable with difficulty for humans, due to ethical problems: experiment time duration or pharmacological effects produced by cold ligand injection. Several studies have therefore been dedicated to identify receptor concentration information by using a simple model, including few parameters which can be estimated with a single injection protocol. However, these simple models have to be validated with a reference approach.

In this way, Delforge et al.have proposed a multi-injection approach[20,21] whose goal was to quantify exactly all model parameters. Although this method has been used in human studies,[21,22] it is more adapted to animal studies, where it is possible

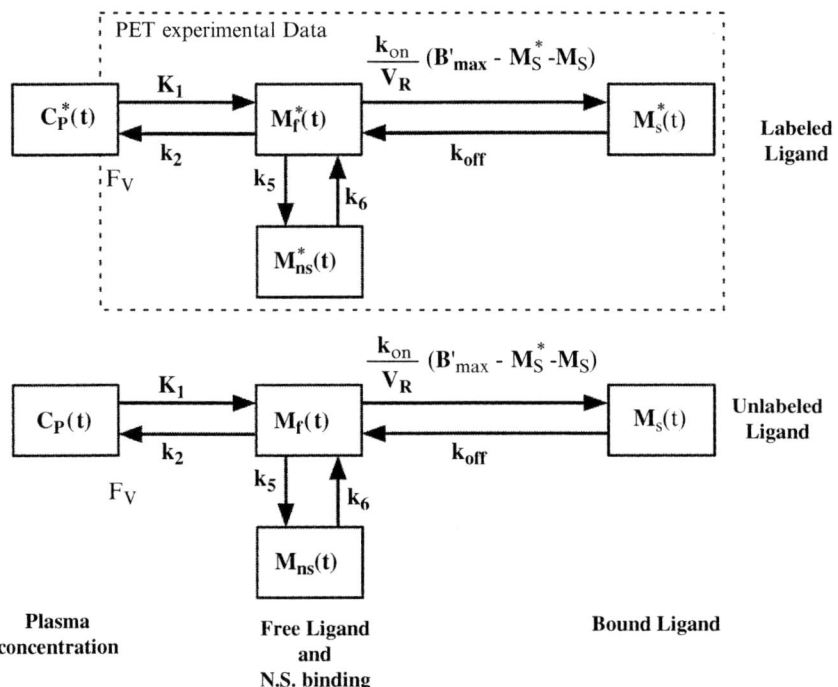

Fig. 5 Ligand-receptor model used with a multi-injection protocol

to apply complex experimental protocols allowing to estimate exactly biological parameters. In this case, a full model can be used as shown in Fig. 5.

A four-compartment model (non metabolized radioligand in plasma, free and non specific ligand, ligand bound to receptor sites) is used to estimate seven parameters $(B'_{ma}, k_1, k'_2, k_{on}/V_R, k_{off}, k_5, k_6)$ with the multi-injection approach.[20] The parameters k_1 and k'_2 are associated with the exchanges between the plasma and free ligand compartment. The parameters k_5 and k_6 correspond to the rate constants describing exchanges between non specific and free compartments. B'_{max} represents the concentration of receptors available for binding (effects of endogenous ligands are not included in model); k_{on} and k_{off} are the association and dissociation rate constants, respectively, and V_R is the volume of reaction, which accounts for tissue inhomogeneity.[23] The k_{on} and V_R parameters cannot be estimated separately and only the k_{on}/V_R ratio is identifiable. Consequently, only the apparent equilibrium dissociation constant, $K_d V_R$ can be estimated. K_d is the equilibrium dissociation constant, defined as the ratio of k_{off} to k_{on}. The parameter F_V represents the fraction of blood present in the tissue volume.

The multi-injection protocols include injection of unlabelled ligand (with or without simultaneous labeled ligand injection). The kinetics of the unlabelled ligand affects the local concentration of free receptor sites and must therefore be taken into account. The unlabelled and labeled ligand kinetics is assumed to be similar. Thus, the model contains two components with the same structure and parameters. The plasma concentration of the non metabolised unlabelled ligand was simulated using the curve corresponding to the labeled ligand.

3.3 Application to Neuroimaging

3.3.1 Introduction

As previously described, quantitative approaches are frequently simplified because of their difficulties to be applied to humans. In many cases, new radioligands are evaluated using a single injection protocol and, therefore, few binding parameters are estimated from a one-tissue compartment model. Multiple injections protocols involving displacements with a large dose of cold ligand or other different specific activities are impossible to perform in human studies due to pharmacological effects.

Recently, a new small animal tool has been developed for in vivo investigations, in which an intracerebral β^+ radiosensitive probe (βP)[24] implanted in the brain of anesthetized or conscious animals enables researchers to measure the local kinetics of β^+ labeled molecules in a stereotactically defined brain region. Several studies have shown that this tool is very promising to study time-concentration curves of a PET tracer under different experimental conditions. This probe has been shown to acquire time-concentration curves with high temporal and spatial resolution, and it is being applied to metabolic and neurotransmission studies. In Fig. 6, we show an example of time-concentration curves obtained with $[^{18}F]$MPPF, an antagonist of $5HT1_A$ receptors. In this case, multiple injections of radioligand with different specific activities have been performed in order to define correctly binding properties during the parameter identification procedure.

3.3.2 Quantification of Benzodiazepine Receptors in Human

The multi-injection approach has been applied successfully in many studies.[21,25–27] We present here a quantitative study applied to benzodiazepine receptors. Benzodiazepine receptors have been widely studied in PET by using $[^{11}C]$flumazenil $([^{11}C]FMZ)$, an antagonist ligand with high affinity and selectivity for the central benzodiazepine receptors.[28–30] As described previously, such quantification allows estimating all binding parameters, including receptor concentration and ligand affinity.

The ligand-receptor model is derived from Fig. 5. In the context of FMZ studies, a simplified configuration of the model is commonly used that combines the free and non specific compartments in a single compartment.[27–31] Consequently, the k_2 parameter is modified and can be expressed by the following expression: $k_2' = k_2/(1 + k_5/k_6)$, where k_5 and k_6 correspond to the rate constants describing exchanges between non specific and free compartments. A three injection protocol

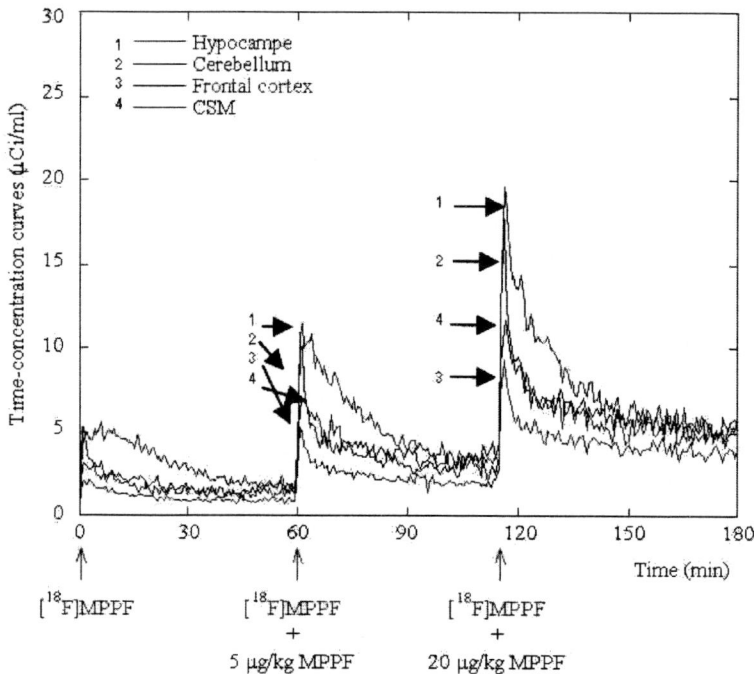

Fig. 6 Time concentration curves obtained with βP after injection of [^{18}F]MPPF, an antagonist of 5HT1$_A$ receptors. A three injection protocol with different specific activities has been used to quantify binding parameters

is required to estimate correctly all model parameters $(B'_{max}, K_1, k_2, k_{on}/V_R, k_{off})$. Figure 7 shows an example of time-concentration curve in a brain region. The experimental protocol consists of an injection of labeled ligand ([^{11}C]FMZ) at the beginning of the experiment, a displacement procedure with a cold ligand (FMZ) after 30 min and, finally, a co-injection of [^{11}C]FMZ and FMZ after 60 min.

The multi-injection approach allowed to obtain absolute values of all binding parameters for all brain regions.[21–23]

3.3.3 Parametric Imaging

Parametric imaging is becoming an important tool in PET. Several studies have shown that it is possible to obtain quantitative information for the whole brain.

Indeed, the usual approach consists in performing a kinetic PET experiment, in defining some regions of interest on brain activity maps and in deducing the corresponding time-concentration curves. Depending on which modeling approach is implemented, these curves are used either to calculate an index assumed to be correlated with the receptor concentration or to estimate directly the receptor concentration. However, whereas PET data correspond to ligand concentration images,

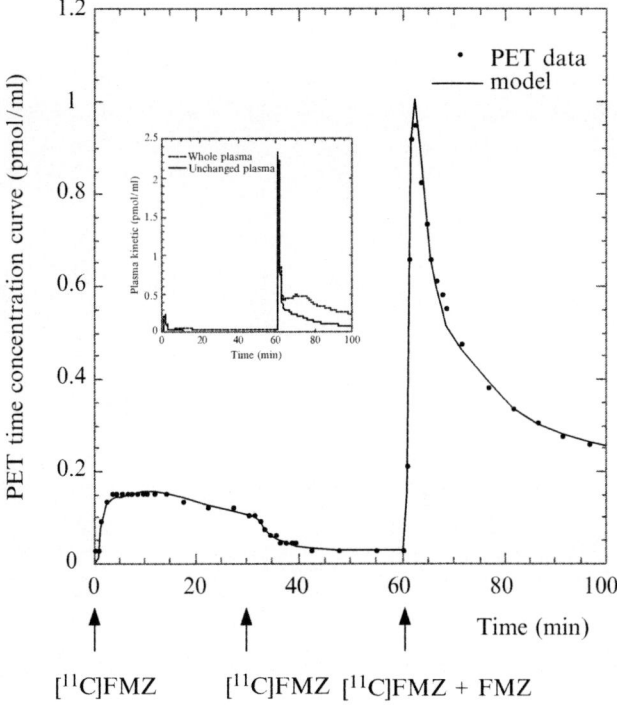

Fig. 7 Time concentration curves in a brain region and in plasma

estimates of the model parameters are usually obtained only for a small number of regions of interest. Methods have been developed to obtain parametric images of the receptor concentration and of some kinetic parameters. The advantage of such images is to allow visual screening of ligand transport and receptor site concentration in the whole brain.

Figure 8 shows an example of parametric images obtained with the multi-injection approach and [^{11}C]FMZ. In this case, the use of a wavelet-based filter on dynamic PET data provided a definite improvement for parametric imaging.

4 Quantification with PET Scanner

4.1 Principles

PET scanner technology can be considered as a daughter of experimental particle physics (positron-electron annihilation) and thus it benefits of new detector development in that field, fostering progress in the fight for higher spatial resolution and higher signal-to-noise ratio, a common challenge for particle physics and medical

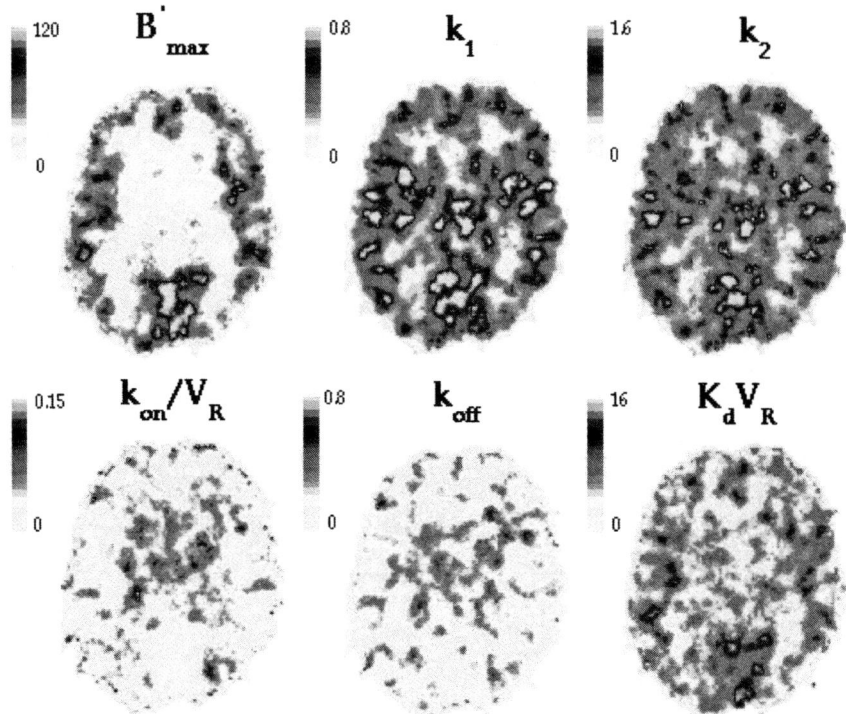

Fig. 8 Parametric images of benzodiazepine receptor – FMZ interactions

imaging. Better measurement of tracer activity distribution, needed for pharmacokinetic pathway optimisation and its molecular target, became possible thanks to the development of various other techniques, ranging from radiochemical synthesis of tracers to the use of sophisticated mathematics in the analysis of recorded data. After this achievement, the next advancement in the comprehension of biological processes consists of modeling with compartment analysis, as discussed earlier in this paper.

A few years ago PET was reserved to human studies, characterised by a modest space resolution, of the order of half a centimeter. Recently, dedicated small animal PET were made available achieving 1 mm spatial resolution, whereas high temporal resolution β-microprobes begun to work. As a consequence, more complex experimental protocols for *in vivo* new molecular issues can be implemented in small animal studies.

Due to the modest spatial resolution of PET, the trend is now oriented to multimodal imaging, which allows simultaneous recording of anatomical and functional devices. MRI and PET seems to be the most promising association for human devices as well as for small animal ones.[32,33]

4.2 Small Animal PET Scanners

The first small animal PET was built at Hammersmith Hospital to study rodents who were already well known species and useful tool in neuroscience.[34] Whereas the first small animal devices were just miniaturized version of clinical PET systems, new and specific small animal PET scanners were developed in universities and/or research centres and later on commercialised by medical equipment companies.

Table 1 shows the parameters and performance of some currently available small-animal PET systems. The scanner performance can be characterised by space resolution (given here as axial resolution and volumic resolution) and by sensitivity (i.e. ability to measure accurately a small signal above background) or efficiency.

In this table, sensors are MultiChannel PhotoMultiplier Tubes (MC-PMT), Position-Sensitive photomultipliers (PSPMT), Avalanche PhotoDiodes (APD) or MultiWire Proportional Chambers (MWPC). Scintillators used are Lutetium-oxyorthosilicate (LSO), Lutetium-yttrium-oxyorthosilicate (LYSO), Gadolinium-orthosilicate (GSO), Lutetium orthoaluminate (LuYAP) or Yttrium-aluminium-perovskite (YAP).

The YAPPET presents performances comparable to the other devices. This machine is an instrument for Small Animal Emission Tomography matching requirements for an accurate, easy-to-use, versatile and cost effective instrument. The YAPPET has a sensitivity of 19 cps/kBq at the centre of the FOV and a FWHM spatial resolution of 1.5 mm at the centre of the scanner. The field of view is 4 cm axially and 4 cm in diameter. The scanner is made of four detector modules that can rotate up to 360°. Each detector is composed of a YAP:Ce matrix (20×20 match-like crystals of dimension $2 \times 2 \times 30 \, mm^3$) coupled to a position-sensitive photomultiplier (Hamamatsu R2486-06). This device can be used in PET (with coincidence between opposite blocks) or SPECT (blocks triggered according to an OR logic with collimators ahead). The prominent feature of YAP-(S)PET scanner is the ability to perform both PET and SPECT imaging studies, due to the use of the YAP:Ce crystals, which are effective in detecting 511 KeV radiation for PET and, at the same time, generate enough light to allow detection of low-energy gammas in SPECT applications. For this reason this device is also called "YAP-(S)PET". In dedicated studies, e.g. the YAP-(S)PET being fitted with four detector heads on a rotating gantry, the scanner may be operated with combined functionalities, e.g. with two heads in coincidence for PET and two heads equipped with collimators and in anticoincidence for SPECT. The very fast response of the YAP:Ce scintillator, coupled to the very fast coincidence gating for PET studies, yields a measured FWHM of less than 5 ns.

The spatial resolution (after 2D-EM reconstruction of a source of 1 mm diameter) is practically constant over the entire field view (Fig. 9a). Due to the high efficiency YAP:Ce scintillators and to the possibility to use the whole energy spectrum (50–850 KeV), the YAP-(S)PET scanner can reach a maximum sensitivity of 1.7% at the centre of field-of-view (Fig. 9b).[42–44]

Table 1 Some currently available small-animal PET systems

Device	Units	MicroPET	Focus 220	Quad HIDAC	LabPET	ClearPET	YAP(S)PET
Solder		CTI-Concorde	CTI-Concorde	Oxford Instr.	GammaMedica	Raytest	ISE
Detector type		LSO	LSO	Lead+Mwpc	LSO+LGSO	LYSO+LuYAP	YAP:Ce
Crystal size	mm	1 × 1 × 12.5	1.5 × 1.5 × 10		2 × 2 × 3	2.2 × 2.2 × 10	2 × 2 × 30
Crystal number		10.572	13.824		3.072	10.240	1.600
Sensor type		MC-PMT	PSPMT	MWPC	APD	PSPMT	PSPMT
Useful diameter	mm	170	250	170	156	120	150
FOV (φ,l)	mm	160,49	190,80	165,250	110,75		40,40
Axial resolution	mm	1.3	1.3	1.04	0.9	1.5	1.8
Volume resolution	µl	2.35	2.5	1.1	2.4	3.4	5.8
Efficiency		2.26%	3.4%	1.8%	2.6%	3.8%	1.8%
Sensitivity	Cps/µ Ci	810		660	960	630	640
Energy window	KeV	250–650	250–750		250–650	250–750	
References		[35]	[36]	[37]	[38,39]	[40]	[41]

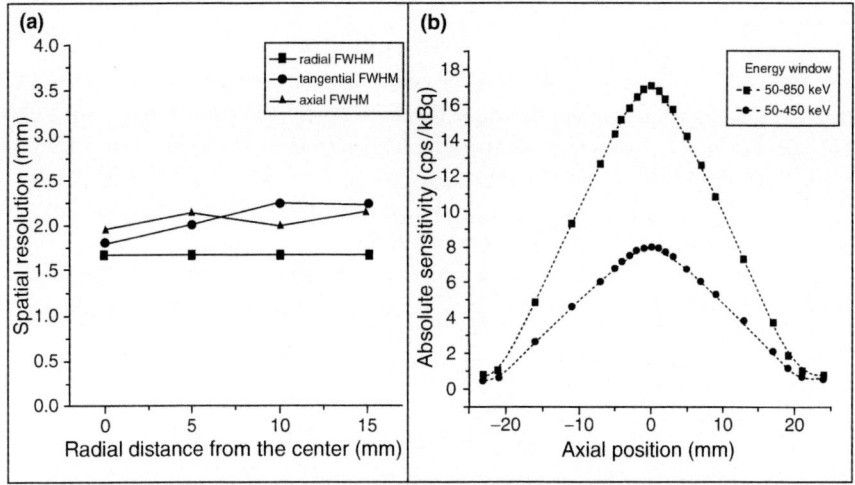

Fig. 9 (**a**) Spatial resolution and (**b**) sensitivity of the YAP-(S)PET scanner

4.3 In Vivo Quantification of 5-HT$_{1A}$-$[^{18}F]$MPPF Interactions in Rats

Several studies have provided evidence that PET can be used to measure fluctuations in synaptic concentration of neurotransmitters. In this context, $[^{18}F]$MPPF, a 5-HT$_{1A}$ radioligand with an affinity close to that of serotonin, has been proposed as a tool to evaluate 5-HT levels. However, discordant results have been reported regarding the sensitivity of this radioligand to endogenous 5-HT. In the study of Millet et al. (2008),[45] a full quantitative analysis of the pharmacokinetic properties of MPPF binding was performed, using the β-microprobe (βP) (1) and the YAP-(S)PET (YAP) (2) to improve the understanding of $5 - HT_{1A}$-$[^{18}F]$MPPF interactions.

4.3.1 Methods

Sixteen Wistar rats were used for βP(n = 5) and YAP-(S)PET (n = 5) acquisition and metabolite studies (n = 6). Time–concentration curves were obtained in hippocampus, raphe dorsalis, frontal cortex and cerebellum, using three injections of $[^{18}F]$MPPF at different specific activities. B$'_{max}$ values were estimated from a two (2T-5k) and three (3T-7k) tissue-compartment model, with βP and YAP-(S)PET time–concentration curves. The simplified reference tissue model (SRTM) was used to estimate binding potential (BP$_{SRTM}$) values from data obtained with the first injection and the cerebellum as the reference region.

4.3.2 Results

Overall, the 3T-7k model provided a better fit than the 2T-5k model, as evaluated from AIC criteria in all experiments. The rank order of receptor density (B'_{max}) values was as follows: hippocampus > raphe $\approx$ frontal cortex > cerebellum. Non-negligible specific binding was observed in the cerebellum ($B'_{max}(\beta P) = 1.5 \pm 0.9\,pmol/ml$). Significant correlations ($p < 0.001$) between B'_{max} and BP_{SRTM} values were observed with both $\beta P(r = 0.895)$ and YAP-(S)PET ($r = 0.695$). The YAP-(S)PET system underestimated the [^{18}F]MPPF binding levels in brain due to limited resolution (i.e. partial volume), leading nevertheless to similar conclusions.

4.4 Conclusion

A multi-injection protocol using both the βP and the YAP-(S)PET systems was employed to determine all the binding parameters of interactions between [^{18}F]MPPF and 5-HT$_{1A}$ receptors. Both modalities reveal a large quantity of radiotracer specifically bound to 5-HT$_{1A}$ receptors. However, modeling results were significantly different, particularly for main parameters, such as receptor density and BP_{SRTM}. Underestimated values were obtained with YAP-(S)PET, since no corrections for partial volume, scattering or attenuation were included. Further studies are necessary to provide corrected images for improved quantitative results.

However, beyond discrepancies in absolute modeling parameter values, the consistent findings obtained with both methods lead to reasonably expect that they would be valid in a more realistic and feasible quantitative PET scanning.

Nevertheless, the YAP-(S)PET studies provide promising results and support the feasibility of performing full modeling studies in animals. Indeed, YAP-(S)PET is better suited than βP for *in vivo* studies, as it is less invasive and it allows the analysis of the whole brain activity during the chronic treatment of the animal.

Finally, based on modeling data, we have shown that the BP_{SRTM} parameter can be used as an index of receptor density, but that the cerebellum used as a reference region can attenuate BP sensitivity during challenge studies.

References

1. Chatziioannou AF, "Molecular imaging of small animals with dedicated PET tomographs," *Eur. J. Nucl. Med.* **29** (2002) 98–114.
2. Jones T and Lemoigne Y, "PET - clinical and pharmacological potential," in *Proc. of the ESI International Seminar on Medical Imaging and New Types of Detectors*, edited by Y. Lemoigne, Physica Medica, Vol. XII, Supp. 1, Archamps, France, 1996, pp. 4–7.
3. Kinahan PE et al., "Three dimensional reconstruction in object space," *IEEE Trans. Nucl. Sci.* **NS-35** (1998) 635–638.

4. Defrise M, Kinaha P et al., "Exact and approximate rebinning algorithms for 3D PET Data," *IEEE Trans. Nucl. Sci.* **16** (1997) 145–158.
5. Kety S S, "Basic principles for the quantitative estimation of regional cerebral blood flow," *Res. Publ. Assoc. Res. Nerv. Ment. Dis.* **63** (1985) 1–7.
6. Kety S S, "Relationship between energy metabolism of the brain and functional activity," *Res. Publ. Assoc. Res. Nerv. Ment. Dis.* **45** (1967) 39–47.
7. Kety S S, "Recent approaches to the measurement of cerebral blood flow and their underlying principles," *Res. Publ. Assoc. Res. Nerv. Ment. Dis.* **41** (1966) 226–236.
8. Clarke D D and Sokoloff L, "Basic neurochemistry: Molecular cellular and medical aspects," Lippincott-Raven, Philadelphia PA, 1999.
9. Sokoloff L et al., "The [14C]deoxyglucose method for the measurement of local cerebral glucose utilization: theory, procedure, and normal values in the conscious and anesthetized albino rat," *J. Neurochem.* **28** (1977) 897–916.
10. Phelps M E et al., "Tomographic measurement of local cerebral glucose metabolic rate in humans with (F-18)2-fluoro-2-deoxy-D-glucose: validation of method," *Ann. Neurol.* **6** (1979) 371–388.
11. Persson A et al., "Saturation analysis of specific (11)C Ro 15-1788 binding to the human neocortex using positron emission tomography," *Hum. Psychopharmacol.* **4** (1989) 21–31.
12. Pappata S et al., "Regional specific binding of [(11)C]RO 15 1788 to central type benzodiazepine receptors in human brain: Quantitative evaluation by PET," *J. Cereb. Blood Flow Metab.* **8** (1988) 304–313.
13. Scatchard G, "The attractions of proteins for small molecules and ions," *Ann. N. Y. Acad. Sci.* **51** (1949) 660–672.
14. Mintun M A, Raichle M E, Kilbourn M R, Wooten G F and Welch M J, "A quantitative model for the in vivo assessment of drug binding sites with positron emission tomography," *Ann. Neurol.* **15** (1984) 217–227.
15. Perlmutter J S et al., "Strategies for in vivo measurement of receptor binding using positron emission tomography," *J. Cereb. Blood Flow Metab.* **6** (1986) 154–169.
16. Koeppe R A, Holthoff V A, Frey K A, Kilbourn M R and Kuhl D E, "Compartmental analysis of [11C]flumazenil kinetics for the estimation of ligand transport rate and receptor distribution using positron emission tomography," *J. Cereb. Blood Flow Metab.* **11** (1991) 735–744.
17. Lammertsma A A and Hume S P, "Simplified reference tissue model for PET receptor studies," *Neuroimage* **4** (1996) 153–158.
18. Lammertsma A A et al., "Comparison of methods for analysis of clinical [11C]raclopride studies," *J. Cereb. Blood Flow Metab.* **16** (1996) 42–52.
19. Hume S P et al., "Quantitation of carbon-11-labeled raclopride in rat striatum using positron emission tomography," *Synapse* **12** (1992) 47–54.
20. Delforge J et al., "Modeling analysis of [11C]flumazenil kinetics studied by PET: application to a critical study of the equilibrium approaches," *J. Cereb. Blood Flow Metab.* **13** (1993) 454–468.
21. Delforge J et al., "Quantification of benzodiazepine receptors in human brain using PET, [11C]flumazenil, and a single-experiment protocol," *J. Cereb. Blood Flow Metab.* **15** (1995) 284–300.
22. Millet P et al., "Parameter and index images of benzodiazepine receptor concentration in the brain," *J. Nucl. Med.* **36** (1995) 1462–1471.
23. Delforge J, Syrota A and Bendriem B, "Concept of reaction volume in the in vivo ligand-receptor model," *J. Nucl. Med.* **37** (1996) 118–125.
24. Pain F et al., "SIC, an intracerebral radiosensitive probe for in vivo neuropharmacology investigations in small laboratory animals: theoretical considerations and physical characteristics," *IEEE Trans. Nucl. Sci.* **47** (2000) 25–32.
   F. Pain et al., *PNAS* **99**(16) (2002) 10807.
25. Millet P, Graf C, Buck A, Walder B and Ibanez V, "Evaluation of the reference tissue models for PET and SPECT benzodiazepine binding parameters," *Neuroimage* **17** (2002) 928–942.

26. Costes N et al., "Modeling [18 F]MPPF positron emission tomography kinetics for the deter-
    mination of 5-hydroxytryptamine(1A) receptor concentration with multiinjection," *J. Cereb.
    Blood Flow Metab.* **22** (2002) 753–765.
27. Delforge J et al., "Quantitation of extrastriatal D2 receptors using a very high-affinity lig-
    and (FLB 457) and the multi-injection approach," *J. Cereb. Blood Flow Metab.* **19** (1999)
    533–546.
28. Maziere M et al., "11C-Ro15-1788 and 11C-flunitrazepam, two coordinates for the study by
    positron emission tomography of benzodiazepine binding sites," *C. R. Séances Acad. Sci. III*
    296 (1983) 871–876.
29. Persson A et al., "Imaging of [11C]-labelled Ro 15-1788 binding to benzodiazepine receptors
    in the human brain by positron emission tomography," *J. Psychiatr. Res.* **19** (1985) 609–622.
30. Samson Y et al., "Kinetics and displacement of [11C]RO 15-1788, a benzodiazepine antago-
    nist, studied in human brain in vivo by positron tomography," *Eur. J. Pharmacol.* **110** (1985)
    247–251.
31. Goeders N E and Kuhar M J, "Benzodiazepine receptor binding in vivo with [3H]-Ro
    15-1788," *Life Sci.* **37** (1985) 345–355.
32. Shao Y et al., "Development of a PET detector system compatible with MRI/NMR systems,"
    IEEE Trans. Nucl. Sci. **44** (1997) 1167–1171.
33. Garlick PB, Southworth R, Medina R and Marsden PK, "Design and application of the
    "PANDA" system: simultaneous acquisition of PET and NMR data from a dual-perfused iso-
    lated rat heart," in *Conf. Rec. of the Conf. on High Resolution Imaging in Small Animals with
    PET, MR and Other Modalities*, Amsterdam, Sept. 1999.
34. Bloomfield PM Rajeswaran S, Spinks TJ, et al., "The design and physical characteristics of a
    small animal positron emission tomograph," *Phys. Med. Biol.* **40** (1995) 1105–1126.
35. Tai Y C et al., "MicroPET II: design, development and initial performance of an improved
    microPET scanner for small-animal imaging," *Phys. Med. Biol.* **48**(11) (2003 Jun 7) 1519–37.
36. Tai Y C et al., "Performance evaluation of the microPET focus: a third-generation microPET
    scanner dedicated to animal imaging," *J. Nucl. Med.* **46**(3) (2005 Mar) 455–463.
37. Schäfers K P, Reader A J, Kriens M, Knoess C, Schober O, Schäfers M, "Performance eval-
    uation of the 32-module quadHIDAC small-animal PET scanner," *J. Nucl. Med.* **46**(6) (2005
    Jun) 996–1004.
38. Dumouchel T., Bergeron M., Cadorette J., Lepage M., Selivanov V., Lapointe D., DaSilva J.,
    Lecomte R. and deKemp R., "Initial performance assessment of the LabPETTM APD-based
    digital PET scanner," *J. Nucl. Med.* **48** (Supplement 2) (2007) 39P.
39. Lecomte R. et al, *IEEE NSS-MIC* (2006).
40. http://crystalclear.web.cern.ch/crystalclear/.
41. Del Guerra A. et al. High spatial resolution small animal YAPPET," *Nucl. Instr. Meth. Phys.
    Rev.* A409 (1998) 537–541.
    Del Guerra A., et al., *IEEE Trans. on Nucl. Sci.* **53**(3), 1078 (2006).
42. Di Domenico G. et al. "YAP-(S)PET small animal scanner: quantitative results," *IEE Trans.
    Nucl. Sci*, **50**(5) (2003), 1351–1356.
43. Del Guerra A. et al., " *Calibration and performance of the fully engineering YAP-(S)PET
    scanner for small rodents*". Presented at the first "*Workshop on Small Animal SPECT*",
    Tucson, AZ January 14–16, 2003.
44. Cesca N. et al., "*A Triple Modality Device for Simultaneous Small Animal CT and PET-
    SPECT Imaging*" Presented at the "*IEEE Nuclear Science Symposium & Medical Imaging
    Conference (NSS-MIC)*", Rome, Italy, October 16–22, 2004.
45. Millet P, Moulin M , Bartoli A, Del Guerra A, Ginovart N, Lemoucheux L, Buono S, Charnay
    Y, Ibáñez V, "In-vivo quantification of 5-HT1A-[18F]MPPF interactions in rats using the
    YAP-(S)PET and the β-microprobe," In press in *Neuroimage* 2008.

The DICOM Standard: A Brief Overview

Bernard Gibaud

Abstract The DICOM standard has now become the uncontested standard for the exchange and management of biomedical images. Everyone acknowledges its prominent role in the emergence of multi-vendor Picture Archiving and Communication Systems (PACS), and their successful integration with Hospital Information Systems and Radiology Information Systems, thanks to the Integrating the Healthcare Enterprise (IHE) initiative. We introduce here the basic concepts retained for the definition of objects and services in DICOM, with the hope that it will help the reader to find his or her way in the vast DICOM documentation available on the web.

Keywords: Medical imaging · PACS · DICOM

1 Introduction

Exchanging images between various kinds of equipment has been an issue since the very beginning of digital medical imaging. The ACR/NEMA standard, issued in 1985 and 1988, did not provide a satisfactory solution and the medical imaging community had to wait until 1993 to have a real usable standard available. The publication of this standard was followed by several demonstrations in key international congresses, especially the RSNA in 1993, which convinced the whole community that the standard actually led to successful implementations, and that the manufacturers were willing to use it, and turn the page of proprietary formats and solutions. Since that time, the position of DICOM was significantly reinforced, with the creation of the DICOM Committee, and the collaboration with other standards

B. Gibaud
Unit/Project VISAGES, INSERM/INRIA/CNRS/University of Rennes 1.
2, Avenue du Pr Léon Bernard, F-35043 Rennes, France
e-mail: bernard.gibaud@irisa.fr

Y. Lemoigne, A. Caner (eds.) *Molecular Imaging: Computer Reconstruction and Practice,*   229
and Experiments,
© Springer Science+Business Media B.V., 2008.

development organizations such as the Comité Européen de Normalisation (CEN) in the mid-nineties, and ISO TC 215 "Health Informatics" in 1999.

DICOM has now become the international standard in the field of biomedical imaging. Its influence was critical in the emergence of multi-vendor technical solutions for Picture Archiving and Communication Systems (PACS), and in providing appropriate solutions for the integration with the other information systems involved, especially the Hospital Information Systems and the Radiology Information Systems. These integration issues were addressed by the Integrating the Healthcare Enterprise Initiative (IHE) – initially in the USA in 1998, under the auspices of the RSNA and HIMSS, then world-wide, with the launching of IHE-Europe in 2001, and similar initiatives in Asia more recently.

2 The DICOM Committee

The DICOM committee gathers various kinds of contributors. All the big manufacturers of the field of biomedical imaging are members of the DICOM Committee. There are also many professional societies involved, in radiology, but also from other fields such as cardiology, dentistry, pathology, ophthalmology, as well as stakeholder institutions such as the American National Cancer Institute.

The scope is quite broad with currently 26 working groups addressing the various aspects involved (Table 1). Among those working groups, one has to distinguish Working Group 6 ("Base standard"), whose function is to guarantee the overall technical consistency of the standard, and Working Group 10 ("Strategic advisory"), in charge of advising the DICOM Standard Committee on long term orientations that should be followed.

Table 1 The various working groups of the DICOM Committee

DICOM Committee working groups	
WG-01: Cardiac and vascular information	WG-14: Security
WG-02: Projection radiography and angiography	WG-15: Digital mammography and CAD
WG-03: Nuclear medicine	WG-16: Magnetic resonance
WG-04: Compression	WG-17: 3D
WG-05: Exchange media	WG-18: Clinical trials and education
WG-06: Base standard	WG-19: Dermatologic standards
WG-07: Radiotherapy	WG-20: Integration of imaging and information systems
WG-08: Structured reporting	WG-21: Computed tomography
WG-09: Ophthalmology	WG-22: Dentistry
WG-10: Strategic advisory	WG-23: Application hosting
WG-11: Display function standard	WG-24: Surgery
WG-12: Ultrasound	WG-25: Veterinary medicine
WG-13: Visible light	WG-26: Pathology

3 Organization of the Standard

3.1 A Multi-Part Document

The DICOM standard is organized as a multi-part document, made of 18 independent parts (Fig. 1).[1]

The most important ones are Parts 3, 4, 16, 5 and 2. Part 3 "Information object definitions" provides the specification of the information objects to be exchanged (more than 1,000 pages of text), as well as the definition of the semantics of each data element. The main reason why this part is so long and complex is related to the many existing imaging modalities (such as computed tomography, ultrasound, magnetic resonance, positron emission tomography etc.), that require many technical parameters. Part 4 "Service Class specifications" defines the services for exchanging information, either images or information that is useful to manage images. Part 16 "Content Mapping Resource" addresses the question of terminology, i.e. on the one hand, it defines how existing terminological resources can be used in DICOM (e.g. SNOMED, LOINC, UCUM), and on the other hand, how content items can be grouped together and re-used in DICOM Structured Reporting documents (notion of Templates). Part 5 "Data structure and encoding" specifies how the information objects specified in Part 3 can be organized into a linear bit stream, in order to be sent over a network connection or stored in a file. All aspects related to image compression are addressed in Part 5. Finally, Part 2 "Conformance" specifies how a manufacturer can claim conformance to the DICOM standard for a particular product or implementation, by writing a document called a "conformance statement". Part 2 explains in detail how this document must be written and the information it must contain.

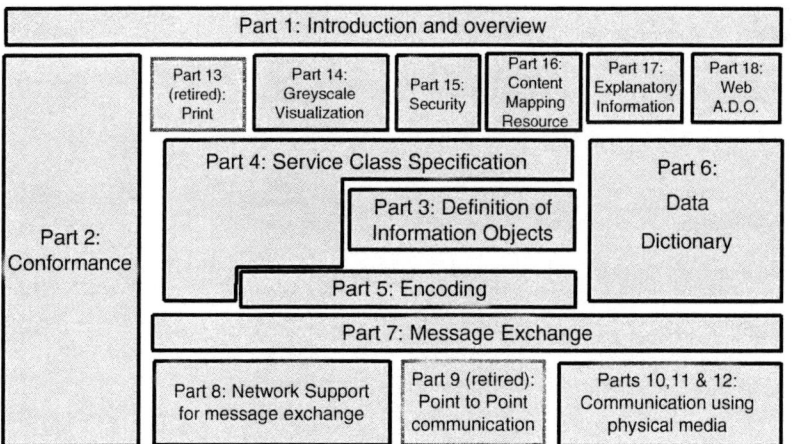

Fig. 1 The different Parts of the DICOM standard (numbered 1 to 18)

3.2 Extensions and Maintenance

The standard is being continuously updated, either to address new fields of biomedical imaging (for example, surgery applications), or to revisit existing fields (e.g. Magnetic resonance or PET images), in order to face new needs resulting from the evolution of imaging techniques.

Such extensions are made through the publication of "Supplements" to the standard. Several such Supplements are issued every year. Each Supplement includes text modifying the existing parts, e.g. to add a new information object in Part 3, or a new service class in Part 4, or new codes in Part 16, or new data elements in Part 6 "Dictionary" etc. All those changes are put together once a year to deliver a new issue of the standard, e.g. "DICOM 2008". However, this yearly issue should not be seen as a new "version" of the standard. There is only one single version of the standard, since all changes that are made are supposed to be "backward compatible": it means that new specifications bring new functionalities, but they never change the way previous services work. The only exception to this rule is the case of retirement, i.e. when service classes are retired from the standard (this happens only to services that are almost never used). The general motivation for this kind of approach is that imaging equipment can be in operation during quite long periods of time (6, 7, up to 10 years) and that it would be non-acceptable to change the standard in ways that would "break" existing implementations, and cause troubles to installed and working equipment.

A second way of updating the standard exists, through the mechanism of Correction Proposals (CPs). CPs provide the means to raise and fix minor issues and imprecision that cause (or may cause) interoperability problems. CPs are submitted, accepted, worked out and balloted as any other parts of the standard.

4 Basic Concepts

4.1 Data Syntax

The syntax used in the DICOM standard is a binary one, based on the transfer of (Tag-Length-Value) triplets. It means that any elementary Data Element is given a binary Tag, e.g. Data element *(0028,0010) Rows* represents the number of rows in an image. This Tag, composed of 4 bytes is sent first, followed by the corresponding data type, represented in this particular case by ASCII characters 'U' and 'S' (for Unsigned Short), followed by the length of the value field, in this case "2" because the Data element *(0028,0010) Rows* is represented by an Unsigned Short, represented by 2 bytes, and finally comes the value itself, e.g. "256", represented in binary ('00' in the least significant byte, and '01' in the most significant byte). So the binary bit stream corresponding to this Data element would be in this example:

... '28' '00' '10' '00' '55' '53' '02' '00' '00' '01' ...

As can be seen from this example the values are binary ones, which explains that images in DICOM cannot be edited like, e.g. XML documents. One must use a parser that decodes the TLV triplets in order to put the information in human-readable form.

4.2 Information Objects

The specification of an information object (called in DICOM an Information Object Definition or IOD) consists in defining the set of Data elements that shall or may be transmitted. They are organized in groups called 'modules'. For example, the CT image IOD is depicted by a number of modules focusing on: (1) the general context of image acquisition (essential information on the patient, the study, the series), (2) the acquisition procedures (particularly the physical acquisition methods, the reconstruction algorithm etc.), (3) the image's characteristics (size, resolution etc.), and (4) the pixel data themselves. The 'module' concept, by gathering together data elements relating to one same information entity (for example 'Patient Module', 'General Study Module', 'General Image Module', 'Image Plane Module', 'Image Pixel Module'), facilitates their reuse in different IODs. These information entities are defined using information models, following the "entity-relationship" formalism. The basic DICOM model is a hierarchy in which a Patient can have one or more Studies, each containing one or more Series, each containing one or more Composite Objects (such as Images, Presentation States, etc.)

A distinction is introduced between Composite objects and Normalized objects. Composite objects gather data elements describing several entities of the real world, whereas Normalized objects focus on one single real world entity.

All sorts of images, whatever the modality (Computed Tomography, Magnetic Resonance, Ultrasound, XRay angio, etc.) are represented by Composite objects, as well as Presentation States, Fiducials, Registration objects, Structured Reports etc.

At the beginning of the standard, image objects were essentially 2D images. Multi-frame objects were introduced to represent in a single Dataset several related images, e.g. in ultrasound (Multi-Frame Ultrasound IOD) and nuclear medicine (Nuclear Medicine IOD). However, the new image objects developed recently for CT, MR, PET etc. (Enhanced CT, Enhanced MR, Enhanced PET, respectively) make use of a new extended multi-frame mechanism. This leads to a better and more efficient organization of common attributes (using so-called functional groups), as well as to faster transmission times.

4.3 Services

Two sets of services exist. A first set of services (called Composite services), applies to Composite objects, whereas a second set (called Normalized services) concerns Normalized objects. The first provide facilities to 'push' an object (service

C-STORE), or to query a set of objects based on some search criteria (services C-FIND, C-GET and C-MOVE). A last service exist (C-ECHO) to simply test connectivity between two peer application entities.

Normalized services provide facilities to act upon a single object managed by a remote application entity, to create an instance (N-CREATE), delete it (N-DELETE), get information about it (N-GET), modify it (N-SET), launch an action (N-ACTION), report an event (N-EVENT-REPORT).

4.4 Service-Object Pair

Objects, i.e. IODs, can be associated with services to form what is called a 'Service Object Pair'. For example, a CT image IOD is associated with the Storage service to form the CT Image Storage SOP Class. This abstraction defines a service that can be provided by an application entity (this entity is called the SCP, for Service Class Provider), or used by another application entity (this entity is called the SCU, for Service Class User). These notions of SOP, SCU and SCP are extensively used in DICOM Part 2 to claim the conformance to DICOM of an implementation. A SOP Instance corresponds to the instantiation of the use of a service, e.g. for a particular image. SOP Classes and SOP instances have unique identifiers based upon the OSI Object Identification (numeric form) as defined by the ISO 8824 standard. The SOP instance UIDs provide identifiers for persistent objects such as images.

5 DICOM Major Achievements

5.1 Storage and Query/ Retrieve of Composites Objects

Storage and Query & Retrieve are the services classes used to communicate any sorts of Composite objects. In fact the exchange and query mechanisms are totally generic with respect to the objects to be exchanged. Requests are based on the use of unique identifiers for Studies, Series and Composite objects (using the SOP Class and SOP instance UIDs).

Composites objects may be exchanged in two ways, either in connected mode, or using removable media. In the first case, composite objects are transmitted as one message per Dataset (a Dataset corresponds to a Composite object instance), after a preliminary stage called Association establishment, during which the two peer application entities negotiate the SOP Classes and Transfer Syntaxes they want to use in the exchange. In the second case, the composite objects are copied on removable media, together with a file called DICOMDIR, providing a structured description of the Patient/Study/Series/Composite Objects relationships, with links to corresponding files on the media. DICOM does not specify any particular convention for the naming of files. Many kinds of media are supported like Cederoms, DVD, Flash memory etc.

5.2 Image Management Services

After providing the basic services for exchanging images, DICOM developed services for image management, in close collaboration with CEN TC251/WG4 "Medical Imaging", a standardization body that was quite active at that time in the area of image management.[2] Such services were acutely needed, both in the US and in Europe, for the development of multi-vendor PACS systems.

Modality Worklist was one of those services whose impact was prominent for successful PACS deployment. Modality Worklist provides the means for a modality to query an information system (e.g. a RIS) about the tasks that have been scheduled for this particular piece of equipment for a certain period of time. More precisely, the modality queries the Modality Scheduled Procedure Steps that were registered in the information system, and this mechanism provides the modality with essential identifying information about the Patient and Study concerned. This is really essential because it is the key for having consistent identifiers in the RIS and the PACS.

A second important service was developed at the same time, it is called "Storage Commitment". The basic idea is for the modality (or any kind of equipment producing images) to get an explicit confirmation that all the images that have been sent to the image manager have been well received and stored in a safe place, so that the modality can delete those images and free the corresponding storage spaces.

A third image management service is "Performed Procedure Step", allowing to "close the loop" with the information system. This service allows the information system to be notified about events such as the beginning and completion of the examination and the creation and updating of Performed Procedure Step objects, documenting what was precisely done during a procedure step (which could eventually differ from what was scheduled).

Additional services were added, such as "General Purpose Worklist", to support the organization of any sort of task (e.g., quality control, image processing, Computer Assisted Detection or CAD, reporting, transcription), based on a local definition of these tasks.

All these management services have been extensively used in the IHE Integration Profiles specified for Radiology applications, especially those addressing the issue of workflow (for example Scheduled Workflow, Post-Processing Workflow, Reporting Workflow).

5.3 Structured Reporting

DICOM's major focus is on images. However, it became obvious that standards were needed to support the exchange of observations made on images, whatever the origin of such observations, i.e. the analysis of images by a human observer, or through the use of an image analysis algorithm.

DICOM tried (and succeeded!) in addressing such issues in a generic way. The solution was provided by the Structured Reporting concept (SR), introduced in DI-COM with Supplement 23, published in 1999.

The basic idea is to represent the observations related to a Structured Reporting document as a tree, composed of content nodes (called content items) connected by relationships.[3] Such observations may represent measurements made on images, refer to regions of interest in specific images, or document the context of observation. DICOM introduced 14 content item types, listed in Table 2. What is important to note is that each node has a concept name, that is necessarily a coded entity, and a value, which may be coded or not, depending on the kind of node.

Similarly, DICOM introduced a limited number of relationships, with relatively precise semantics such as 'CONTAINS', 'HAS OBSERVATION CONTEXT', 'HAS ACQUISITION CONTEXT', 'HAS PROPERTIES', 'INFERRED FROM', 'SELECTED FROM' and 'HAS CONCEPT MODIFIER'.

In terms of new DICOM IODs and SOP Classes, Supplement 23 introduced three new classes, addressing various levels of complexity in the use of available content items. The first, called Basic Text SR IOD, addresses the needs of applications exchanging very simple documents, limited to text referring to images. The second, called Enhanced SR IOD, provides the same features plus the possibility to represent measurements and to refer to spatial and temporal coordinates. Finally, the third, called Comprehensive SR IOD may involve any kind of content nodes and authorizes the use of 'by reference' relationships (ability to refer to a node anywhere in the SR tree).

Table 2 The various types of content items of SR trees

Type of content item	Comment
CONTAINER	A such node has no proper content: it contains the nodes that are connected to it using the CONTAINS relationship
TEXT	The content is free text, coded in ASCII, without any presentation features
PNAME	Contains a Person Name; often used to define the observation context
DATETIME	Contains a Date and Time, also often used to define the observation context
TIME	Contains a Time, also often used to define the observation context
DATE	Contains a Date, also used to define the observation context
NUMERIC	Contains a measurement made on an image
IMAGE	Contains a reference to a DICOM image (only DICOM images are allowed)
WAVEFORM	Contains a reference to a DICOM waveform
COMPOSITE	Contains a reference to another DICOM composite object, for example another DICOM SR document
UIDREF	Contains a UID of a DICOM entity
SCOORD	Contains a spatial coordinate ROI; can be a single point, or a polyline etc.
TCOORD	Contains a temporal coordinate, e.g. in a time series of images
CODE	Contains the value of a code, taken from the list of codes available in DICOM Part 16 "Content Mapping Resource"

Another important feature of DICOM SR was the introduction of Templates. A Template is the specification of a sub-tree that can be instantiated in an SR document. It groups and specifies content item nodes in roughly the same way as modules do, but in a way that is more formalized and dedicated to tree-like structures. Basically, a Template is a table of all content item nodes of a sub-tree, specifying things like requirement type (mandatory or optional), multiplicity, code values allowed for the Concept name attached to each content item node, and eventually, code values allowed for the value of content item node (when represented by a coded entry).

Interested readers can refer to the book written by David Clunie[4] for (far) more details about DICOM SR.

Since the publication of Supplement 23, a wide range of SR documents templates have been standardized to address specific needs in the domains of CAD (e.g. Chest CAD SR, Mammo CAD SR), ultrasound procedure reports (e.g. Ultrasound OB-GYN Procedure report, Vascular Ultrasound Procedure Report, Echocardiographic Procedure Report, Intravascular Ultrasound Procedure Report) and CT/MR reports in cardiology (CT/MR Cardiovascular Analysis Report), among others.

5.4 Web Access to DICOM Objects

Web access to DICOM persistent objects (WADO) was introduced in the standard to facilitate the referencing and retrieval – via the internet protocols http and https – of DICOM persistent objects (images, structured reports, etc.), using URL/URI (Uniform Resource Locator/Identifier). This extension was made in 2003 with Supplement 85, in the form of Part 18 of the standard. It was also recognised by the ISO TC 215 (ISO 17432).[5]

6 Conclusion

The development of the DICOM standard allowed to turn the page of proprietary standards for representing biomedical images and associated metadata. Initial efforts concerned primarily the exchange and management of images acquired on imaging equipment. Image processing is now being considered as well, with extensions of the standard dedicated to radiation therapy, CAD structured reports, spatial registration and fiducials, segmentation objects and even more recently segmentation surfaces such as meshes.

Besides, DICOM is open to the web technology to communicate and refer to images (or any other composite objects), as demonstrated with the WADO extension. However, DICOM remains based on its original binary syntax, in spite of growing pressure to move to some sort of XML encoding. For example efforts are being devoted in Working Group 20 to facilitate the translation of DICOM SR documents into HL7 Clinical Document Architecture documents, based on XML.

Finally, DICOM has started studying how processing tools, such as plugins or web services, could be more easily shared in the future, based on standard Application Programming Interfaces (API). This initiative, led by Working Group 23 "Application Hosting" received a strong support from the CaBIG initiative (Cancer Biomedical Informatics Grid) and the NCI, because of the perspectives it opens for in vivo imaging in cancer research, especially in molecular imaging, and CAD.

Acknowledgements Many thanks to Joël Chabriais, main representative of the French Society of Radiology in the DICOM Standards Committee, and currently radiologist in the hospital of Aurillac (France), for his continuous and enthusiastic participation to the DICOM standard's evolution and promotion.

References

1. Digital Imaging and Communications in Medicine (DICOM) – National Electrical Manufacturers Association, Parts 1 to 18, 2008.
2. GIBAUD B, GARFAGNI H, AUBRY F, TODD POKROPEK A, CHAMEROY V, BIZAIS Y, DI PAOLA R, Standardisation in the field of medical image management : the contribution of the MIMOSA model, *IEEE Trans Med Imaging*, 17(1), 62–73, 1998.
3. BIDGOOD WD, Jr., Clinical importance of the DICOM structured reporting standard. *Int J Card Imaging*, 14, 307–315, 1998.
4. CLUNIE D, DICOM Structured Reporting. Bangor: PixelMed Publishing, 2000.
5. ISO 17432:2004 – Health informatics – messages and communication – web access to DICOM persistent objects, TC 215, 18, 2004.

PACS: Concepts and Trends

David Bandon*, Osman Ratib, Christian Lovis, and Antoine Geissbuhler

Abstract Picture Archiving and Communication Systems (PACS) are becoming part of the core components of today's infrastructure for digital radiology. In modern setting they also extend to the whole medical enterprise and integrate into the patient medical record. While local architectures may vary significantly, the major features such as image archiving and retrieval, workflow management, workstations and image acquisition and quality control systems constitute the key elements of such systems. PACS has also vocation to extend beyond the radiological images to become the common platform for hospital wide management of all images. The wireless access can contribute to expand the scope of image distribution.

Keywords: Imaging informatics · PACS · DICOM · electronic patient record

1 Introduction

Picture archiving and Communication Systems (PACS) are today considered the gold standard by the radiologist community. Hence PACS are widely implemented and sometimes at very large scales.[1,2] The PACS success is largely supported by a mature industry offering archiving solutions and reading stations that fulfill the needs of the users in radiology. PACS also interacts with the other IT systems which support clinical activities like the electronic patient record (EPR). The Integrating Healthcare Enterprise (IHE) initiative provides a solid framework to ensure a good integration between the PACS and the various components of a hospital information system (HIS), such as the clinical information system (CIS) and its subcomponents: radiology information system (RIS), computerized provider order entry (CPOE),

D. Bandon,* O. Ratib, C. Lovis, and A. Geissbuhler
Dept. of Medical Imaging and Information Science. University Hospitals of Geneva,
24 Rue Micheli-du-Crest. CH-1211 Geneva 14, Switzerland
e-mail: david.bandon@sim.hcuge.ch

Y. Lemoigne, A. Caner (eds.) *Molecular Imaging: Computer Reconstruction and Practice,*   239
and Experiments,
© Springer Science+Business Media B.V., 2008.

amongst others, the admission-discharge transfer (ADT) component, billing components, scheduling, etc. This paper reviews some of the trends and innovations that are emerging within the PACS field.

2 PACS Basic Concepts

2.1 Interpretation

A key element to ensure the PACS acceptation by the users is a good reading software as it is the interface that radiologists use for image interpretation. It is a multimodality viewer allowing to visualize all kinds of images. It also increases productivity through:

- Customized display protocols which automatically load and present studies in a predefined order.
- A transparent access to all information and tools needed for the interpretation activity[3]: seamless access to clinical indication, access to previous reports and dictation tools. This seamless access requires advanced synchronization among the different software modules (RIS, PACS, EPR, ..).

We are seeing more and more reading software that also include multidimensional image visualization capabilities (3D, 4D,..)[4] (see Fig. 1) along with computed-aided diagnosis tools (CAD[†]) facilitating the interpretation activity.

2.2 Archive

The cost reduction and the rapid technology evolution have introduced a significant paradigm shift in the design of storage architecture. The old hierarchical storage model in which images were migrated from high-speed high-cost storage to long-term affordable storage devices is now outdated: images are now archived on magnetic disks. It enables a "pay as you grow model", i.e. capacity can be expanded continuously following the need.

2.3 Distribution Through the Electronic Patient Record

A convenient clinical access to image data may improve clinical decision making and patient management. The patient-centric EPR is the ideal platform to access the

---

[†] CAD: software programs to help the radiologist to detect and characterize abnormal tissues or tumors using automated image analysis and classifications. Applications: screening or diagnosis of breast cancer, lung cancer, colon cancer.

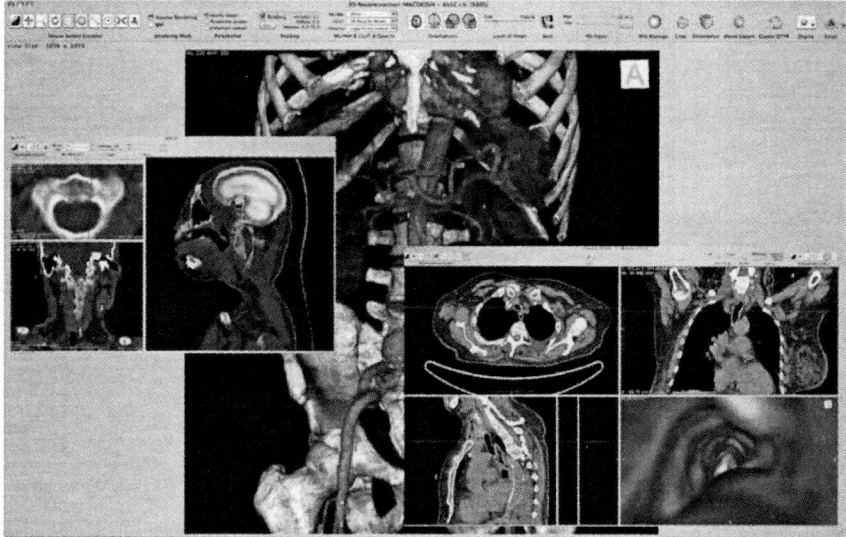

Fig. 1 Example of reading software with multidimensional image visualization capabilities (OsiriX software[5]): volume rendering, virtual endoscopy, multiplanar reconstruction. See Appendix for color version of the picture

medical records in an exhaustive way and allows a multidisciplinary image review. It provides a unified access to all medical results as well as an order entry system allowing ordering of drugs, care, and radiological procedures. Therefore image distribution through the electronic patient record is the preferred model.[6,7]

In most of the EPR implementations, clinicians can display images from a viewer directly launched from the PACS and fully embedded in the EPR user interface (see Fig. 2). The content of the viewer is driven by a PACS portal, insuring the correct correspondence between images and other data.

2.4 DICOM and IHE

Implementation of large PACS systems relies heavily on existing standards such as DICOM[8] and HL7 to ensure a smooth integration of the imaging acquisition devices with PACS and other clinical information systems. DICOM (Digital Imaging and Communications in Medicine) can manage all kind of medical images. It includes a file format definition and communication services to exchange images and associated information of any kind. It particularly specifies how to implement key images and provides templates for structured and multimedia reports. HL7 (Health Level 7) is devoted to administrative (patient registration), financial and medical activities (report, lab, order).

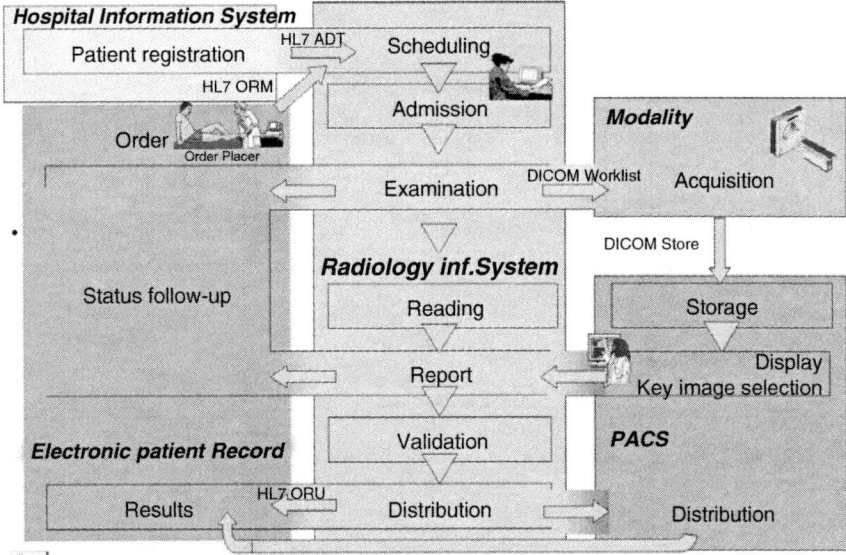

Fig. 2 Integrated workflow among the various information systems and implemented thanks to IHE recommendations: DICOM and HL7 are used to exchange information

A new step in a higher level of standardization is underway with IHE (Integrated Healthcare Enterprise)[9] initiative that focuses on clinical workflow and practical scenarios to define the basic guidelines for an integrated clinical workflow. This initiative provides a solid framework to ensure a seamless integration of the various systems (see Fig. 2). It promotes a coordinated use of established standards – such as DICOM and HL7 – to address specific clinical needs in support of optimal patient care. It defines integration profiles covering the phases of the workflow such as patient admission, scheduling, order entry and image acquisition.

3 Trend: Enterprise PACS

PACS are now widely used as routine within hospitals. However they still continue to evolve:

- They address some emerging needs such as managing images beyond radiology to become an enterprise PACS.
- They take benefit from new technologies such as wireless distribution to facilitate the image distribution in bedsite environment or emergency case.

3.1 Clinical Motivations

The complexity generated from the medical care of an aging population explains the increasing number of health specialties in charge of a patient. As a result, the quantity and variety of information generated are exploding. The distribution of consolidated patient information between departments is therefore a critical point and sets a new challenge for the information systems. PACS has definitely vocation to extend beyond the radiological images to become the common platform for hospital wide management of all images from biomolecular images up to whole body pictures: cardiology, dermatology, ophthalmology, surgery, ear nose throat, hematology pathology, gastroenterology, obstetrics, gynecology and surgery.

An enterprise PACS offers all the advantages of a traditional radiological PACS (large availability and ubiquity of the images, reliable retrieval of old exams, improved navigation inside the image series) as well as new opportunities:

- It allows the inter modalities comparison by simplifying the images registration. For instance, the recent development of new imaging techniques (PET-CT, functional MRI) will offer multidisciplinary collaboration between radiologists and cardiologists.[10] In that respect, PACS can be a fostering force for team working.
- It reinforces the diagnostic power of a traditional PACS by offering to the clinicians a complete view of the patient multimedia record. Finally, it may increase the strength of content-based image retrieval systems by creating extended image databases.[10]

3.2 Architecture

So far, non-radiological specialties had only access to very specific solutions. Those solutions usually managed images locally without any real perspective for inter-department communication. For example, in the pathology specialty, a commercial solution covering the following tasks could be found: image acquisition from a network of microscopes, reading and reporting. Images however could not be exported or retrieved from external sources.

In place of the currently large number of information systems addressing the specific needs of each specialty we envisage an enterprise PACS storing all diagnostic images acquired within the hospital with a unique distribution point via the EPR platform. Some institutions have also adopted a similar approach such as the US Department of Veteran Affairs[12,13] or the University Hospitals of Geneva (Switzerland)[14] (see Fig. 3). The challenge is to develop a common system that is flexible enough to address the specificities of each specialty. Two alternatives exists to implement such an enterprise PACS:

- A hierarchical model coupling multiple mini PACS to an enterprise PACS: each mini PACS fulfils the needs of its medical specialty and stores the most recent images while the enterprise PACS archives for the long-term. The rationale for this

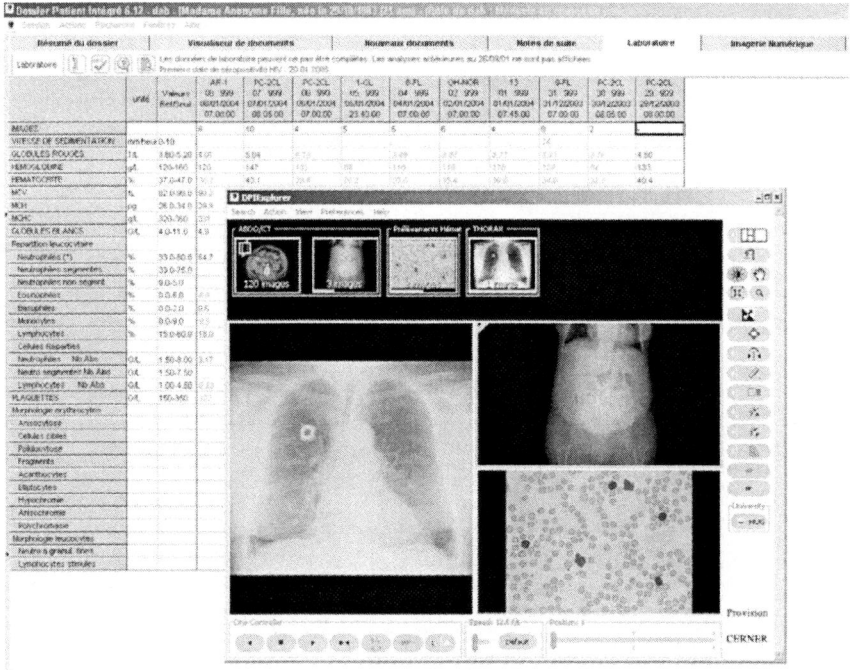

Fig. 3 The electronic patient record and its graphical user interface used at the University Hospitals of Geneva (Switzerland). This home-made EPR [11] offers an integrated access to medical records and allows access to all kinds of images through the same PACS image viewer (PACS supplier: Cerner). This image viewer is launched within a legacy container which encapsulates a web browser

mini PACS is to address the specific needs in terms of processing and quantification tools (for instance in cardiology) and to provide high performance. According to the imaging speciality, this mini PACS is either a single acquisition station or a network of stations connected to a temporary storage server. This approach is well suited to medical specialties needing some sophisticated processing tools and advanced structured reports combining text, customized measurements and images.

- A direct storage to the PACS: the imaging devices directly pushes data to the enterprise PACS. However, most commercial PACS have been developed for radiology and this direct storage requires some structural evolution to transform the legacy PACS into an enterprise PACS.[14] Such evolutions are needed to support video capabilities in surgery,[15] various reporting practices[16] in gynecology and obstetrics and specialized quantification tools in cardiology. Another strong requisite is the integration to the other information systems to deliver an integrated workflow with for example patient demographics injected in the imaging devices. For many departmental information systems not supporting DICOM, some gateways are required to transform the data from the HIS like patient demographics into DICOM compliant format.

4 Trend: Wireless Access to Images

Recent emergence of disruptive technologies such as wireless network, web-based applications, remote and mobile services have introduced significant paradigm shift in concepts and designs of medical imaging networks. Mobile computing has become prevalent in consumer market while it is still slowly emerging in medical informatics. While access to medical data and patient records through handheld and portable computers is becoming more widely available, accessing images still represents a technical and logistic challenge. Concurrently, the demand for accessing images and image data outside radiology and imaging departments is increasing dramatically. Physician and care provider rely more and more on imaging results for patient management and patient care. The number of instances where images are being used for treatment planning and for specialized procedures is increasing. Solutions using wireless technology and improved performance of portable computers allow more realistic implementation of mobile alternative to conventional imaging workstations. These innovative solutions are more suitable for clinical wards, bedside patient care and on-call radiologists. Many technical solutions are emerging as alternatives for remote access to image data[17]:

- Image access on portable tablet computers: The development of higher performance laptop and tablet computers allows today for a better and more efficient management of large volume of data with sufficient processing power for image manipulation and processing. Several prototypes of such implementations have already been reported using web-based image viewing applications.
- Hand-held devices and PDAs: Numerous systems were developed to allow extracts of electronic patient records to be downloaded and stored in handheld devices to be readily accessible when needed.[18] There have been recent reports of a variety of clinical applications of electronic medical record on hand-held devices that include medical images displayable on the low-resolution screens of these devices.[19,20] However, experience with handheld devices for displaying medical images shows that today's handheld devices suffer from three major limitations: low display resolution, low storage capacity and relatively limited communication speed. These technical limitations restrict the usage of these devices for accessing medical images in clinical routine.

A current upgrade to new 802.11 technology – 54 Mbit/sec – combined with the use of image quality on demand improve performance. Images are retrieved lossy compressed using JPEG 2000 technique. Once displayed, the user can refine the image quality factor up to the diagnostic standard. The image loading time is similar to the one under wired distribution.

5 Conclusion

PACS left the research arena and are on their way toward commoditization as current versions successfully fulfill functional and technical needs. Those systems can nowadays be fully integrated with other clinical information systems to facilitate the patient follow-up. Images can be distributed within the hospital and also outside the hospital. However there are still some improvements to be made for specific environments such as operating rooms where a better is needed to allow surgeons to easily interact with the viewer while operating.

References

1. Huang H.K., Enterprise PACS and image distribution, Computerized Medical Imaging and Graphics, 2003, 27:241–253.
2. Ratib O., Swiernik M., McCoy J.M., From PACS to integrated EMR, Computerized Medical Imaging and Graphics, 2003, 27:207–215.
3. Geis R.J., Medical imaging informatics: how it improves radiology practice today, Journal of Digital Imaging, 2007, 20(2):99–104.
4. Rosset A., Spadola L., Pysher L., Ratib O. Informatics in radiology (infoRAD): navigating the fifth dimension: innovative interface for multidimensional multimodality image navigation, Radiographics, 2006, 26(1):299–308.
5. Rosset A., Spadola L., Ratib O. OsiriX: an open-source software for navigating in multidimensional DICOM images, J Digit Imaging, 2004 Sep;17(3):205–216.
6. Ratib O., Ligier Y., Bandon D., Valentino D. Update on digital image management and PACS, Abdom Imaging, 2000, 25:333–340.
7. Münch H., Engelmann U., Schroeter A., Meinzer H.P. The integration of Medical Images with the Electronic Patient Record and their Web-Based Distribution, Academic Radiologic, 2004, 11:661–668.
8. Mildenberrger P., Eichelberg M., Martin E. Introduction to the DICOM Standard, European Journal of Radiology, 2002, 12:920–927.
9. Integrating the Healthcare Enterprise (IHE). *http://www.ihe.net* . Accessed June 2007.
10. Muller H., Michoux N., Bandon D., Geissbuhler A. A review of content-based image retrieval systems in medical applications—clinical benefits and future directions. Int J of Medical Inf 2004, 73(1):1–23.
11. Lovis C., Baud R.H., Rassinoux A.M., Scherrer J.R. Value to add to the patient record when making the EPR. Iwg, Editor, Eprimp, Rotterdam, 1998, 17:225–227.
12. Siegel E., Reiner B. Filmless radiology at the Baltimore VA Medical Center: a 9 year retrospective, Computerized Medical Imaging and Graphics, 2003, 27:101–109.
13. Kuzmak P., Dayhoff R. Integrating Non-Radiology DICOM Images into the Electronic Medical Record at the Department of Veterans Affairs, Medical Imaging 2001: PACS and integrated Medical Information Systems, E. Siegel, H.K. Hunag, Editors, Proceedings of the SPIE, (2001) 4323: 216–221.
14. Bandon D. et al. Enterprise-wide PACS: beyond radiology, an architecture to manage all medical images, Academic Radiology, 2005, 12(8):1000–1009.
15. Dubuisson J.B., Chapron C. The endoscopic operating room OR 1, Gynécologie Obstétrique & Fertilité, 2003 31:382–387.
16. Siegel E., Reiner B., Clinical challenges associated with incorporation of nonradiology images into the electronic medical record, Medical Imaging 2001: PACS and integrated Medical Information Systems, E. Siegel, H.K. Hunag, Editors, Proceedings of the SPIE, (2001) 4323: 287–291.

17. Bandon D., Ratib O., Rosset A. Wireless access to patient records in a Clinical Environment, Imaging Management, 2006, 6(5):21–24.
18. Tschopp M. et al. Understanding usage patterns of handheld computers in clinical practice, Proceedings of AMIA Symposium, 806–809.
19. Servadei F. et al. Integration of image transmission into a protocol for head injury management: a preliminary report Journal of Neurosurg., 2002, 16(1):36–42.
20. Yoo, S.K., Park J.J., Kim S.H., PDA-phone-based instant transmission of radiological images over a CDMA network by combining the PACS screen with a bluetooth-interfaced local wireless link, Journal of Digital Imaging, 2007, 20(2):131–139.

Biomedical Multimodality Imaging for Clinical and Research Applications: Principles, Techniques and Validation

Biomedical Multimodality Imaging

Luc Bidaut[*] and Pierre Jannin

Abstract As more imaging modalities became available to clinical and research applications, multimodality imaging's importance and relevance to the biomedical scene have increased regularly over the years. Most recently, the advent of hybrid modalities has further eased clinical applications and moved this complex discipline to the forefront of biomedical imaging. This chapter summarizes the context of multimodality imaging from a historical viewpoint as well as with an emphasis on selected clinical applications that are either relevant, demonstrative or both.

Keywords: Multimodality imaging · hybrid modalities · advanced imaging · interventional guidance

1 Introduction

Biological complexity is such that it cannot be described through any single exploration technique. Accordingly, the fusion of various information that are all pertaining to the study of either the morphology or the function of the human body naturally became an endeavor of medical practice. Complementary information brought by various modalities (e.g., for imaging components) assist physicians for their diagnosis as well as for treatment management and delivery. In the research arena, such approaches help us in furthering our knowledge and grasp of what makes up a living organism and, therefore, the best ways to interact with it. Before the development

L. Bidaut[*]
Depart. of Imaging Physics, University of Texas – M. D. Anderson Cancer Center,
1515 Holcombe Blvd, Unit 1352 Houston, TX 77030, USA
e-mail: lbidaut@mdanderson.org

P. Jannin
Visages, U 746, Inserm-Inria-CNRS, Medical School – University of Rennes, CS 34317, F-35043 Rennes Cedex, France

Y. Lemoigne, A. Caner (eds.) *Molecular Imaging: Computer Reconstruction and Practice,* 249
and Experiments,
© Springer Science+Business Media B.V., 2008.

and advent of computer-based registration tools, most of the fusion process was achieved mentally whenever possible and with all the implicit limitations. The development of new and old modalities and of imaging systems that include networks, standards for communication and exchange and workstations that are able to visualize complex data sets have considerably eased the fusion process leading to new paradigms that could only be dreamt of in earlier days.[1–6]

This paper will first present the historical and present context and paradigms for multimodality imaging. There will then be a description of various methods for registering data sets, which will be followed by the rationale for validation of these methods and by the description of various approaches for this complex and necessary step. Finally, the fusion of previously registered datasets will be described along with a few selected examples that will be presented before the discussion and conclusion.

2 Registration of Biomedical Imaging Data Sets

2.1 The Diverse Nature of Biomedical Imaging Data Sets

Clinical applications of multimodality imaging bring together numerous types and sources of data such as measurements (e.g., images or signals), locations (e.g., the patient's, instruments) and preliminary knowledge (e.g., databases, atlas, models).

Nowadays, the major distinction is between morphological and functional information.

The modalities that provide morphological or anatomical information – e.g., computed tomography (CT), magnetic resonance (MR) or echography (also called ultrasound, US) imaging – do so through their intrinsic physical characteristics that provide either a measure of tissue (e.g., x-rays and CT) or of proton density in a given unit volume (e.g., MR), or a measure of acoustic transmission properties (e.g., US).

Contrast agents can be used with all of the morphological modalities to enhance the detection of specific structures (e.g., angiography for imaging vasculature). By using dynamic or multiframe acquisition after the injection of a contrast agent, morphological modalities can even complement or bridge the gap with functional modalities, by showing the timed accumulation and clearance of the agent in the structures of interest.

Functional modalities provide information that can be linked directly to the metabolism or the physiology of live organisms. Some information of great interest for cancer and other diseases is tissue perfusion that can be measured by MR perfusion (MRp), CT perfusion (CTp), single photon emission computed tomography (SPECT) or even positron emission tomography (PET). Although tissue perfusion is absolute, the surrogate measurements provided by different machines and protocols – that often do not have the same physical principles and may be selected

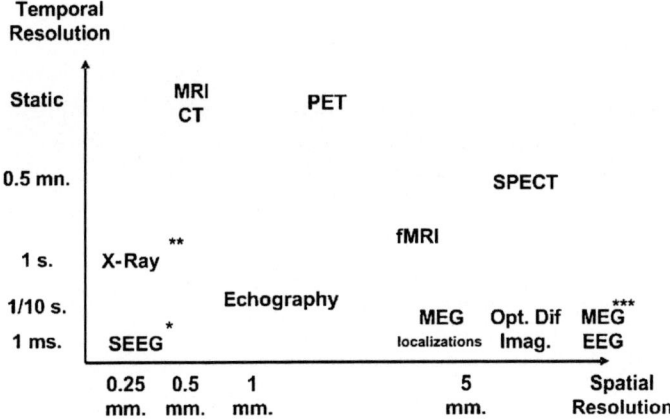

Fig. 1 Complementary aspect and diversity of some available medical modalities (images, signals). * in situ measurement, ** projection, *** surface measurement

for convenience, availability or consistency – do not necessarily provide the exact same value for all and therefore need to be cross-calibrated and validated to authorize any fair comparison. Other functional information of interest are glucose utilization via fluoro-deoxy-glucose with PET, or changes in blood oxygen concentration that can be assessed via functional MR. Optical diffusion imaging may also track oxy- and deoxyhemoglobine. Other functional modalities may explore electro-physiology (e.g., electro-cardio-graphy ECG/EKG, electro-myo-graphy EMG, electro-encephalo-graphy EEG, magneto-encephalo-graphy MEG).

For PET and SPECT, the functional information is provided through the detection of a radioactive compound – a tracer – that is incorporated into the patient's metabolism after injection or inhalation.

Because of the nature of the information, the need to restrict the concentration and radioactivity of the metabolized tracers, and also the physical limitations of the detection and reconstruction principles, functional modalities tend to be of lower spatial and temporal resolution than the morphological ones. Still, the different time characteristics, and the intrinsically complementary nature of both types of modality is what led to the clinical need for multimodality imaging through registration and fusion (see Fig. 1).

2.2 Biomedical Applications of Multimodality Imaging

An imaging study, whether with single or multiple modalities, is always aimed at addressing a clinical need. For multimodality studies, this goal is naturally expected to be reached only through the combined exploitation of the various datasets or of the complementary information that is derived from them. Such exploitation can for instance be based on the superposition of regions of interest (ROIs) extracted from various modalities, distances between such ROIs, statistical analyses of measurements derived from the content of these ROIs.

Clinical needs themselves are obviously linked to the actual clinical protocol that can for instance include comparisons within a given patient population, diagnosis assistance or interventional planning.

The development and implementation of a clinical imaging fusion application therefore requires a clear and precise definition of the clinical context and of the clinical end goal.

Once the modalities or data sets that are needed in a given context have been identified, two generic operations are required to combine the information in a manner suitable for their exploitation. The first operation is called "registration" and is the step at which a geometric transformation is estimated between a target and reference datasets. Once registration has been performed, "fusion" can take place that will combine the registered datasets – e.g., either via cross-processing or visualization – to provide the end user with the information that is expected from the multimodality process, which can often encompass more than two separate modalities/datasets.

The fusion process main issue stems from dissimilarities between the data to be matched. These "dissimilarities" are due to different acquisition conditions, different types of measured information, or different imaged subjects. The "dissimilarities" both include variations in the measures for which we want to compensate using registration methods: geometrical or intensity related variations and relevant information we want to highlight, using data fusion methods (e.g. complementary nature of measures).[5] The fusion paradigm stands in this duality between dissimilarities to be corrected and dissimilarities to be highlighted.

The initial assumption for fusion is as follows. The fusion process consists in matching various images – or measures of physical or physiological phenomena – concerning one or more physical entities considered as significantly similar. "Significantly similar" can mean that the different measures correspond to the same anatomical region. It could also mean that both measures conform to a common a priori model. For instance, the fusion between different brain MRI data sets of different subjects assumes that the overall shapes and contents of the individual brains are similar in spite of local anatomical dissimilarities.

Biomedical applications of multimodality imaging can generally be classified through their specific context. In the following, four major contexts are defined along with the corresponding characteristics of the multimodality process and selected examples. To remain general, "subject" might mean either patient, normal volunteer or even study animal for the case of pre-clinical imaging.

2.2.1 Intra-subject + Intra-modality

This context is related to the multimodality handling of datasets coming from the same modality for the same subject.

Such a context may, for instance, occur for tracking changes in a subject's anatomy – or function – through time, comparing various physiological states through time or subtraction imaging.

Tracking

Tracking anatomical changes through time may for instance be used to study the evolution of a target lesion in response to a given treatment regimen (including no treatment at all) or to understand the growth process for the lesion.

Comparison of States

This can be used to compare several different states in the subject physiology. For instance, functional MRI (fMRI) relies on a direct comparison between baseline/rest and activated states to ascertain what portions of the brain are activated through a specific stimulus. Similarly, comparing peri-ictal and baseline/rest SPECT datasets provide a clue for areas of the brain likely involved in an epileptic seizure.[7]

Subtraction Imaging

Once registered together, two datasets can be combined for instance through a simple arithmetic operation such as a subtraction. Among other possible applications, this is routinely used for clinical digital subtracted angiography (DSA) in which a baseline acquisition is subtracted from one in which contrast agent has been injected. Because of the properties of the contrast agent, the result of the subtraction shows the vasculature with all other anatomical structures removed. Because of live subject motion, a perfect result can seldom be obtained without requiring an additional registration step to get rid of motion artifacts.

2.2.2 Intra-subject + Inter-modality

This context is related to the fusion of various modalities for the same subject.

As summarized previously, no single modality provides all the information that can be acquired on any given subject. Consequently, in a modern clinical workup, subjects undergo different types of imaging acquisition that are then ideally managed in a multimodality fashion to extract and combine the complementary information of interest.

A classical example of such a context for multiple potential target organs or diseases is the multimodality handling of morphological modalities such as CT or MR in combination with functional modalities such as PET or SPECT.

A more refined instance of this context is when one of the modalities actually is not producing images but require a complex modeling in order to do so. Such "modalities" are the ones based on MR spectroscopy (MRS) or on electro-physiology (e.g., EEG and MEG). Others can be parametric maps derived from traditionally image prone modalities (e.g., pharmaco-kinetic modeling in PET, dynamic contrast enhancement – DCE – in MR or CT).

A complex clinical protocol for the surgery of epilepsy could for instance involve CT to show bones and vasculature, MRI to show soft tissues (including brain structures), fMRI to show brain functional areas, PET to show hypo-metabolic foci areas, EEG/MEG-derived information to show electro-physiology based foci delineation (see Fig. 2).[8–10] All this compound information can then be registered to the real-life O.R. environment to assist the intervention (see section 2.2.3).[11]

2.2.3 Real-Life Subject to Digital Imaging Space

This context relates to the multimodality handling of digital datasets to provide a direct link with a real-life setting such as a patient in an O.R. Such context allows Image Guided Surgery or Therapy.[12]

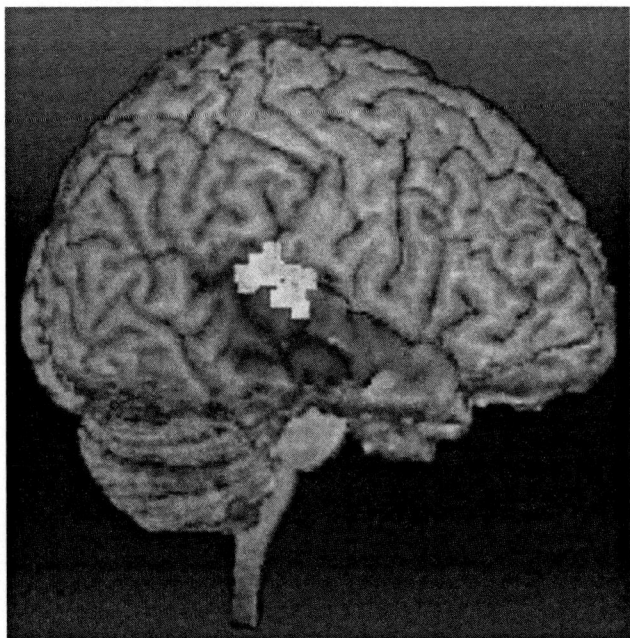

Fig. 2 3D display of 3 imaging modalities for a patient in epilepsy surgery. The brain was segmented from anatomical MR images. The colored texture was computed from registered peri-ictal SPECT. The grey dots show auditory activations computed from MEG. See Appendix for color version of the picture

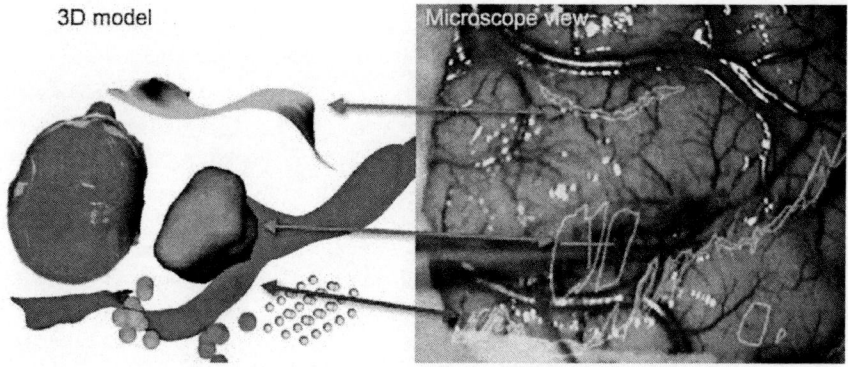

Fig. 3 Left: 3D structures segmented from multimodal images of the same patient for surgical planning. The sulci are segmented from MRI. The cavernoma is also segmented from MRI. The spheres are computed from registered functional MRI and MEG. Both functional information shows somato-sensory and motor areas. Right: these 3D preoperative 3D multimodal structures are displayed in the surgical microscope during the surgical procedure as 2D overlays. See Appendix for color version of the picture

To be complete and serve its main purpose, this context should not only include the real-life subject and the digital imaging datasets that pertain to his/her workup, but it should also be able to track instruments, such as surgical tools, surgical microscope (see Fig. 3) and intra operative imaging devices – and possibly persons – involved in the procedure.

Other than for the imaging modalities, the main components of a tracking process are so-called 3D digitizers that permit the real-life localization and tracking of subject, actors and instruments of many types. Such digitizers can be based on passive or active robotic technology, or on optical or electromagnetic tracking devices. Each technology has its strength and weaknesses but they all provide a way to relate a tracked subject or device to its spatial environment.

2.2.4 Inter-subject + Various Modalities

This context relates to pooling together datasets from various subjects in order to derive statistically significant information that pertains to a given population[13] (including sex and age range) or to a given pathology or state.

Compared to single subject approaches – for which it is often assumed that the various datasets are relatively close to each other and therefore do not require too advanced registration mechanisms – this context requires more advanced registration paradigms to be able to bring together information that is based on widely different anatomic-functional substrate as per the normal population's variance for such entities.

Because of this complexity, this context is still largely in the research domain. Still, computer assisted detection or diagnosis (CAD) approaches that may be based on statistical atlases [14] are slowly coming to the clinical side and probably one of the most demonstrative aspects for the potential of such approaches.

2.3 Registration Methods

Both multimodality context and clinical context are intrinsic components of any multimodality paradigm and will drive the choice of the technique(s) that will be used to reach the intended goal via the complementary and sequential steps of the fusion process. The global fusion process includes not only the actual registration process but also all the ancillaries – e.g., processing, visualization, interaction, and further analysis – that are necessary to accomplish this specific task.

The registration process itself is made up of a paradigm and of the processing methods that support and make up this paradigm. For registration, the paradigm is made up of homologous information – that will be used in the registration process and that will need to be present in both the target and the reference datasets – and of a spatial transformation model that needs to match whatever relevant information the incoming datasets will be able to provide.

To reach the optimal geometric transformation between target and reference datasets, a cost function similar to a distance is generally calculated. This cost is derived from the corresponding figure of merit (e.g., a straight Euclidean distance between similar landmarks or a "difference" between image-derived values) between the target dataset transformed by the current transformation and the reference dataset.

In some cases (e.g., when target and reference volumes are assumed to remain rigid, if there is no significant motion or other simplifying hypotheses), a simple analytical transformation can be readily derived from the datasets. In most real-life cases though, such simplifications do not apply and the registration process is slightly more convoluted and potentially open to errors.

The following sections will describe registration and fusion in more detail.

3 Registration

For the purpose of multimodality imaging, registration is principally the process that estimates a geometric transformation between a target and a reference dataset.

Any registration method requires the preliminary definition and setting of a search space, of a parameter space, of a similarity measure for the registration cost and of a search strategy.[5] Both search and parameter spaces are linked to the registration paradigm that has been summarized earlier. The type of geometric transformation that is used to define the dataset discrepancies defines the search space, and the parameter space is defined by the homologous information between the datasets to register. The similarity cost and the search strategy are linked to the computational methods. The similarity cost (or cost function) provides a surrogate to assess the "quality" of the registration for a given geometric transformation. The search strategy leads to the choice of an optimization method that will provide the best registration transformation for a given cost function. In summary, registration between two datasets is based on the optimization of a similarity cost measured on homologous information extracted from the datasets. The geometric transformation thus estimated can then be applied to the full extent of the datasets so that there is a unique and sometimes reversible spatial path from one point of a dataset to the similar point of the other dataset. It is important to note that this does not imply the uniqueness of either path.

When describing registration approaches, homologous information and cost function are seldom separated. In this paper we clearly differentiate between homologous information – that pertains to the datasets – and cost function – that pertains to the registration process.

3.1 Spatial Reference Systems

When performing registration, the first step is to define clearly the various reference systems that underlie the various geometric transformations.

Traditionally, three such reference systems are defined: one linked to the datasets, one linked to the subject, and a global one that is considered absolute. The reference system linked to the datasets is based on the grid of pixels (picture elements) or voxels (volume elements) that constitute the dataset volume. The system linked to the subject generally is centered on the dataset and expressed in millimeters (mm) that are derived from the size of the individual pixels or voxels. The global reference system is a combination of the subject's system and of information (e.g., stored with each clinical imaging dataset) that is linked to the "absolute" positioning of the subject during a specific dataset acquisition.

So defined, these reference systems already permit the initialization of the registration process but they can also be assisted through external devices (e.g., stereotaxy or other types of frames) or spatial relations to a common space (e.g., atlas such as Talairach's[15]) that may be derived from anatomical landmarks.

3.2 Registration Paradigms

3.2.1 Homologous Information

This information is the one that can be extracted from both datasets that need to be registered and that will be used for the search of the optimal geometric transformation for a given cost function.

Homologous information can be classified by its own dimension – D (point), 1D (line, contour), 2D (surface), 3D (volume) or nD (hypersurface) – and by its search space dimension – 2D (image, surface or projection), 3D (volume, hypersurface) or nD (hypersurface) with or without a time component. In addition, its type as either extrinsic or intrinsic often classifies homologous information.

Registration methods based on extrinsic information use artificial entities that are physically linked to the subject and are designed to be easily identifiable in the images. As a consequence, these methods require a prospective acquisition protocol to ensure that the extrinsic information will be available for subsequent registration. For clinical applications, extrinsic information may come from invasively implanted markers (e.g., screws in bony structures, stereotaxy frames, and the like) or from less invasive approaches (e.g., self-sticking fiducials on the subject's skin). In Image Guided Surgery, when patient to image registration is required, homologous structures on the patient's side may consist of points identified with a pointer tracked by a localizer. Points can be skin (see Fig. 4) or bone markers, or other relevant anatomical points. It may also consist of complete skin surface patch acquired with the same pointer or using stereovision or laser. From the image side, the corresponding (homologous) points and surfaces are manually or semi automatically identified in the imaging data sets.

Registration methods that are based on intrinsic information use information that is available from the datasets without requiring any artificial feature. Such information can be derived from anatomy (e.g., specific structures) or from the contents (e.g., voxel values) of the whole volume or from only selected portions thereof. Such information can for instance be significant – e.g., branching – points that are manually or (semi-) automatically selected, or knowledge-based features that are significant for specific anatomical structures. These features can be selected interactively or with the help of more objective pre-processing that might for instance select the points of maximum curvature as the most significant ones for the subsequent registration process. Extracting complex features – or subvolumes for the content-based approach – generally requires a preliminary step of image segmentation, a process

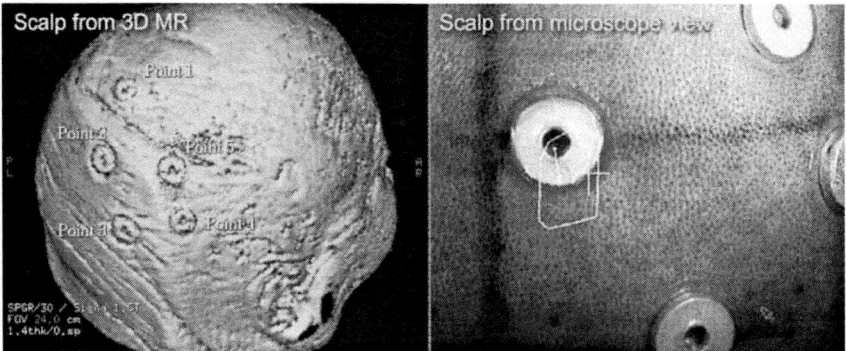

Fig. 4 Left: five skin markers identified on T1 MR images. Right: skin markers as seen in the operating room through the oculars of a surgical microscope

that is beyond the scope of this paper. Content-based intrinsic methods use the voxels values, either directly or through various arithmetic or statistical combinations. Such techniques are for instance relying on optical flow or on mutual information to estimate a cost function (see also section 3.3.1) that will lead to the optimal geometric transformation for this similarity measure. For mutual information – that has become popular recently because it requires relatively little pre-processing or supervision to provide good results in many cases - the homologous information is the empirical probability distribution of all the intensities in the image dataset, and the cost function (see also section 3.3.1) is the mutual information that characterizes the statistical dependency between the target and the reference datasets for a given geometric transformation.[16] As a rule of thumb, a registration method that uses intrinsic homologous information based on content should be preferred – or at least envisioned – every time it could be difficult to extract similar geometric features from both the target and the reference dataset. Such techniques also generally require less preparation and supervision than others, which make them better suited to the clinical environment.

3.2.2 Geometric Transformations

Their main goal is to represent the geometric discrepancies between datasets that the registration process intends to void. These transformations can be characterized both through their nature (e.g., rigid/linear, affine or non-rigid/non-linear) and by their application domain (e.g., local or global).

Linear Transformations

A transformation T is said to be linear if, and only if:

$$\forall x_1 \in E_1, \forall x_2 \in E_2, \forall c \in \Re$$
$$T(x_1 + x_2) = T(x_1) + T(x_2)$$
$$T(c \cdot x_1) = c \cdot T(x_1)$$

where $E1$ and $E2$ are the datasets to be registered and T is the geometric transformation from E1 to E2.

Because they simplify significantly the registration process, linear transformations should be used every time a-priori information about the data justify the hypothesis of linearity, e.g., through the conditions of the acquisitions or the known rigidity of the objects/structures of interest.

A transformation T is said to be affine if, and only if T(x)-T(0) is linear. An affine transformation between two spaces of dimension N is: T(x) = Ax + b where x and b are two vectors of dimension N and A is an NxN matrix. Affine transformations take into account a great variety of usual spatial distortions as they can be conceived as a combination of N translations (vector b), N rotations, N scale factors and N distortion factors. For volume registrations, N = 3 and affine transformations can be described through 12 independent parameters.

Under some conditions, the data to be registered may be considered as non-affected by – or already corrected for – spatial distortions. In such case, a so-called rigid transformation may be enough to register two volumes from the same subject. Rigid transformations are a special case of affine transformations for which only translations, rotations and – rarely – a global isotropic scale factor are considered. As the global scale factor can generally be derived from the individual voxel sizes in each volume, there are only three rotations and three translations, i.e., a total of six independent parameters left to identify.

A global linear transformation is applied to the full volume whereas a locally/piecewise linear transformation is only applied to a subset of the volume. An example of a piecewise linear transformation is the one that transforms a given brain into the Talairach referential system. [15]

Non Linear Transformations

Non-linear transformations are used to represent complex spatial deformations. Projection transformations where at least one spatial dimension is lost are a special case of non-linear transformations that is especially relevant to 3D/2D registration paradigms. [17]

Non-linear transformations are generally also called elastic, deformable or non-rigid[†] transformations. [18] Such transformations can be represented through a simple non-linear parametric function that can be defined on the complete volume (in the 3D registration case), or through local/piecewise non-linear deformations. The latter ones are generally better adapted to complex deformations and can be estimated through high-density deformations fields where there is one deformation vector for each node of a grid that is defined on the data. These transformations are generally defined via a similarity measure between corresponding nodes in the volumes to be registered together, and via a regularization measure to ensure some kind of spatial continuity for the final geometric transformation. From the deformations estimated on each individual nodes of the initial grid, interpolation techniques can be used to derive individual deformations in between grid nodes while ensuring global continuity. For homologous entities of the point type, continuity is obtained via radial interpolation. For homologous entities of a higher dimension (e.g., surfaces), continuity may be obtained via mechanical deformation models or via underlying super-quadric or spline-based deformations. For homologous entities based on local similarity measures, continuity may be ensured via visco-elastic or via fluid deformation models. Because of their intrinsic versatility, non-linear transformations are also suitable to account for intensity variations in the data to be registered together.

Non linear transformations are principally used in two main cases: (1) when the data sets to register do not correspond to the exact same anatomical substrate

---

[†] Non rigid transformations include affine transformations.

(e.g., between different subjects or between a subject and an atlas) and (2) when a common underlying substrate is deformed between the data sets to be registered together (e.g., different acquisitions, times for the same subject; distortions due to acquisition techniques, interventions or physiological variations).

3.3 Mathematical Entities Linked to the Registration Process

3.3.1 Cost Functions or Similarity Measures

A cost function is an objective criterion that is used to assess the goodness of a registration for a given spatial transformation between to data sets. This function is defined in relation to the homologous information used in the registration process as well as to the type of the spatial transformation to be assessed, i.e., to the number of parameters that relate to the type of the transformation. Through Brown's formalism,[5] the cost function defines a measurement of similarity through which the goodness of a current registration transformation can be estimated for a given set of transformation parameters within a given parameter search space. There are two main categories of cost functions: the ones that are based on Euclidean distances and the ones based on statistical similarity measures.

When the homologous information is extrinsic to the data, the cost function is often an estimate of an average Euclidean distance. In the case of external markers or anatomical landmarks, the cost function is often defined in a least square formalism. When homologous structures are more complicated, other distance criteria may be used such as the "hat-head" or the "chamfer" methods[19] that are either applied on continuous or discrete spatial reference system. These methods have proven their efficacy and robustness for multimodality registration of data from the same subject by using – for instance – the skin surface as homologous information to estimate an optimal rigid transformation between MRI, CT or PET data sets.[20]

When the homologous information is based on the distribution of signal intensity within each data set, the cost function then relies on measurement of statistical similarity. Such criteria are generally based on assumptions about the nature of the statistical dependency between the data to register together. For instance, the correlation coefficient measures a linear dependency between the intensity distributions, whereas the correlation ratio measures a functional dependency.[21] The so-called "Woods" criterion[22] is based on a uniformity assumption: a homogeneous intensity region in on of the data sets should correspond to a similarly homogeneous region in the other one. Finally, mutual information[16,23,24] measures a statistical dependency between data sets without any assumption as to what this dependency might actually be. Thus, the choice of a statistical similarity measure will depend on the a-priori knowledge that is available to describe the relationship between the data sets that are to be registered together.[25] For instance, correlation coefficient can be used for registering data sets from the same modality, such as sequential fMRI acquisitions. Mutual information on the other hand may be selected to register single and multi-modality data sets for which there is little a-priori information about the statistical dependency of the data sets (e.g., MRI/PET, various MRI sequences). Of special note, there are many more cost functions that exploit data content: variance ratio, histogram analysis, optical flow, entropy, etc. Any of these measures can be defined in either the spatial or the Fourier domain.[26]

Similarity measures are based on intensities within the data sets and are therefore sensitive to the interpolation techniques that are a component of any geometric transformation method.[27–29] This is principally due to the fact that the data to be registered together are almost exclusively discretely sampled (e.g., the nature of digital imaging) whereas geometric transformations are calculated on a continuous space. Most similarity measures previously mentioned are calculated via joint histograms that require data interpolation. The interpolation technique – e.g., nearest neighbor, trilinear, partial volume – should be carefully chosen so that the cost functions based on the joint histograms become continuous functions of the geometric transformation parameters, which then allows partial derivatives to be estimated.[30]

Among the most recent developments, various cost functions – that have different strengths and weaknesses – are being combined within the same registration process, or other information – such as anatomical models[31] – are being injected to supplement the theoretical cost function with more a-priori knowledge. A major limitation of most registration methods is that they have been designed and validated for normal anatomy and/or physiology. As such, they can fail on pathological data sets that may exhibit abnormal patterns in only one of the data sets to be registered (e.g., different tumor appearance in MRI, CT, SPECT and PET). In such cases, robust estimators[32] are combined with different similarity measures to identify and address differently regions with normal or too high a dissimilarity.[33]

3.3.2 Optimization Techniques

Whenever an exact geometric transformation cannot be directly calculated from the data at hand – i.e., the bulk of such cases – search mechanisms involve a method that permits reaching the optimal transformation for the retained cost function on the selected homologous information. Depending on the assumptions or constraints, the optimization process may have an analytical solution (for instance, least squares) or may necessitate an iterative approach. Many such approaches have been described in the literature and the choice of a method is based more on its core algorithm than on any given clinical context. For biomedical image registration, optimization methods – that are also chosen in relation to characteristics of the selected cost function, e.g., its derivability or continuity – can be classified as follows.[34]

Quadratic or semi-quadratic Approaches

These techniques are based on the assumption that the cost function is convex or quasi-convex near the optimal solution. Among them are the methods of gradient descent, Powell, Newton-Raphson or Simplex for convex solutions, and iterative closest point (ICP) for quasi-convex solutions. These methods are well adapted to reasonably constrained problems.

Stochastic or Statistical Approaches

Techniques such as simulated annealing and genetic algorithms are behaving relatively well in the presence of outliers. As such, they are best suited for problems with too few constraints and for avoiding local optima when searching for a global one. As a trade-off for their robustness, they are alas affected by high computational times.

Structural Approaches

These methods rely on tree or graph based optimization techniques. They imply an exhaustive exploration of the search space to find the optimal solution. A global optimum is always reached after lengthy calculations. These methods are well suited when similarity measures can be formalized through a hierarchical structure.

Heuristic Approaches

These methods generally rely on an interactive search for an optimum, for instance by trying to maximize visually the correspondence between two data sets.

In parallel to selecting a specific optimization method or subset thereof, it may also be relevant to select an optimization strategy to prevent classical issues such as local optima and detrimental calculation times. One such strategy is to use multi-scale[30] optimization that starts the process at a relatively coarse spatial resolution and then refines the results towards higher resolution once the optimization has completed at the previous level. Other possible strategies include initialization of the search near the solution, adapting the constraints (i.e., the homologous information) during the optimization process. Obviously, all strategies can be combined together to produce better results.

3.4 Hybrid Modalities

Hybrid modalities such as PET/CT, and more recently SPECT/CT, are a special case of multimodality imaging in which two modalities (a morphological one and a physiological/functional one) are acquired on a subject without much time or motion between the two.[35–38] While not simultaneous in the current designs, not moving the subject in between acquisitions dramatically reduces the discrepancies between the data sets. The match between data sets is generally considered good enough to use the CT data set to correct PET – or SPECT – for attenuation and to review data sets as they come from the machines.

Because the acquisitions are not simultaneous and do not span a similar time frame, though, it should be kept in mind that even hybrid modalities can benefit from post-registering data sets beyond what the machines are producing directly. Manufacturers have started noticing – and even addressing for some – this need. Of special note, because the subject stays on the same couch for both modalities and because there is not much time between the two acquisitions, the registration process starts much closer to the actual solution than if the data sets were acquired on distinct instruments.

4 Fusion

Once data sets have been registered through an optimal geometrical transformation, there is a direct spatial relationship between points/data in one data set and corresponding points/data in the other. In order to exploit further this correspondence, it is necessary to develop and implement visualization and interaction tools and paradigms. Such tools and paradigms may be designed for relatively simple qualitative evaluation or for generally more complex quantitative analysis that may help in extracting hidden characteristics from the data sets.

4.1 Visualization and Interaction

Before the advent of computer based multimodality imaging, physicians were merging such information mentally by looking at films and relying on their knowledge of the anatomy and of the principles underlying each modality.

More recently, computer based multimodality techniques permit a direct objective matching of the information that is coming from various sources. Two main approaches are used to display registered data sets: linked 3D cursor and visual superposition. Both approaches can be described in relation to the information that is shared between data sets (point, contour, image, surface, volume), to the graphical primitive used to display this information (graphics, texture, transparency, color encoding), and to the equipment used for the visualization (computer display, virtual reality glasses, microscope, etc.).[35,40]

4.1.1 Linked 3D Cursor

In this approach, a cursor interactively positioned in one of the data sets drives cursors in the other ones thereby permitting the identification of corresponding points/locations in all registered data sets.[40,41] The "slave" cursors are calculated

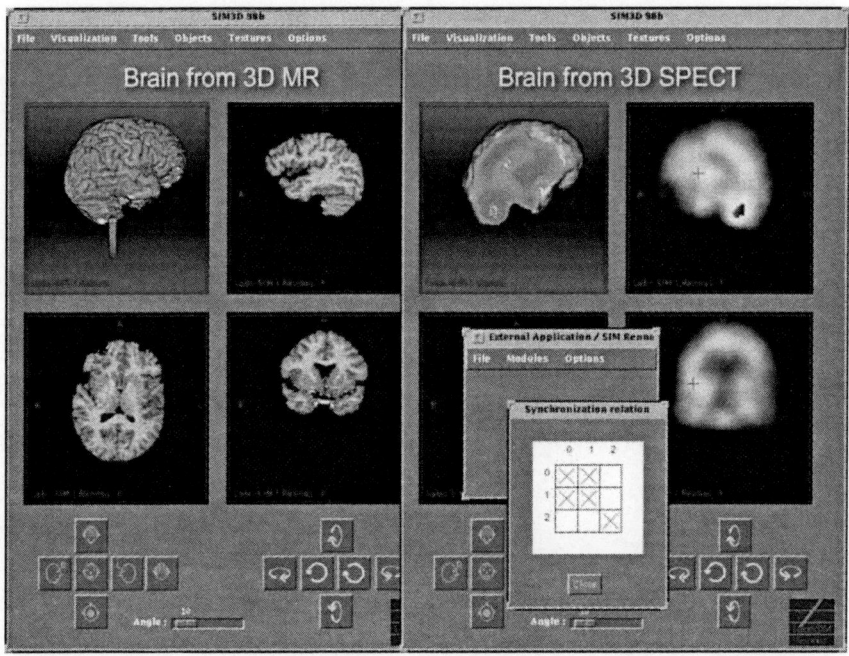

Fig. 5 Linked 3D cursor between 3D MR images and 3D peri-ictal SPECT images. Both volumes are not reformatted in the same coordinate system. The 3D cursor corresponds to the same 3D anatomical point in both imaging modalities

from the "master" one via the geometric transformations obtained during the registration process (see Fig. 5). When transformations can be inverted or combined, any data set can act as either the "master" or one (if more than two sets) of the "slaves".

4.1.2 Visual Superposition

Here, the principle is to calculate and to display a representation that combines the information of the registered data sets into a single entity. A standard paradigm is to combine the information point by point. This combination can for instance be performed via simple arithmetic operations on the registered voxels' intensities or colors (e.g., blending transparency), via more advanced color mapping (see Figs. 6–8) or spectral encoding, [42–44] or via fuzzy logic operators. [45] For 2D visualization, sliding 2D window or other patterns (chess board, etc.) can also be used that represent part of the image with data from one data set and the counterpart with registered data from the other data set. Relevant information from each data set can also be selectively extracted and then combined into a separate entity that can be presented to the physician for review. Whereas the latter approach permits representing information from more than two registered data sets, it can also confuse the physician by creating a representation that does not necessarily relate to any real modality.

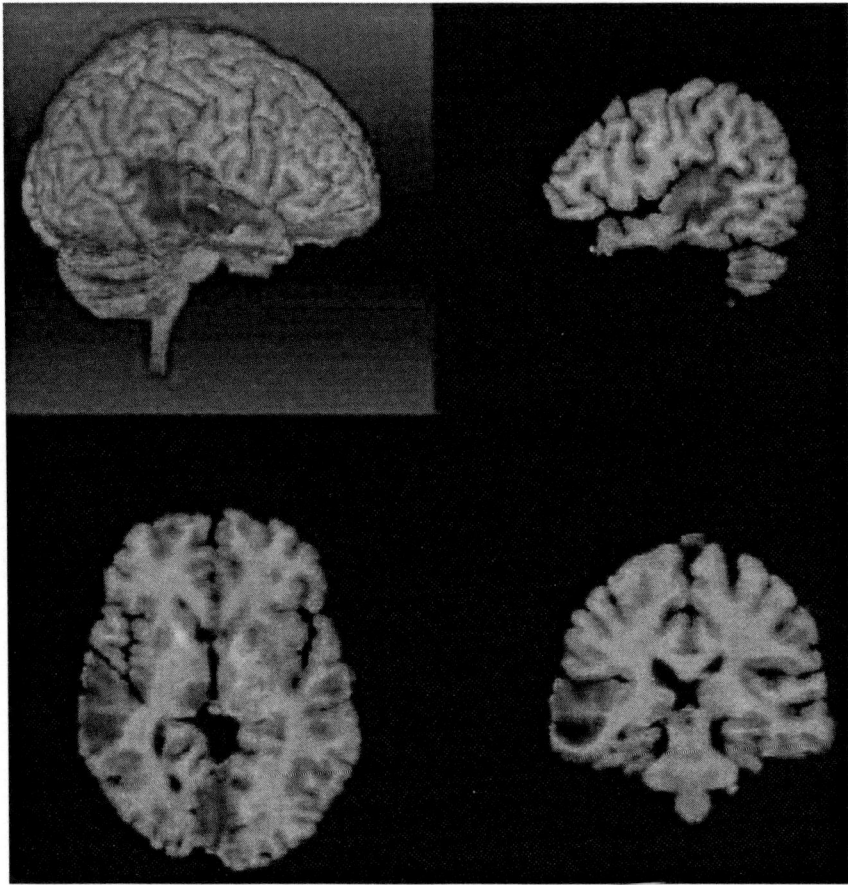

Fig. 6 Compared to Fig. 5, MR and SPECT imaging modalities are reformatted in the same coordinate system. The SPECT volume is then used to add color to MR-based objects. See Appendix for color version of the picture

Both of these approaches should be seen as complementary to each other and they should be chosen in relation to the clinical context or end goal.

Superposition may facilitate the understanding of complex multimodality entities by providing a more direct and global view of the relationship between relevant data from different data sets. Linked cursor also permits such analysis across many data sets without the need to transform any of them – as long as the cursor can be positioned on all of them via the corresponding geometrical transformations – but as such may only be consistent at the exact location of the cursor or within its immediate neighborhood. Whenever data sets need to be transformed through a registration transformation, there should be special attention toward the interpolation technique that is used to fill-up data at transformed locations and that may therefore affect the final representation.

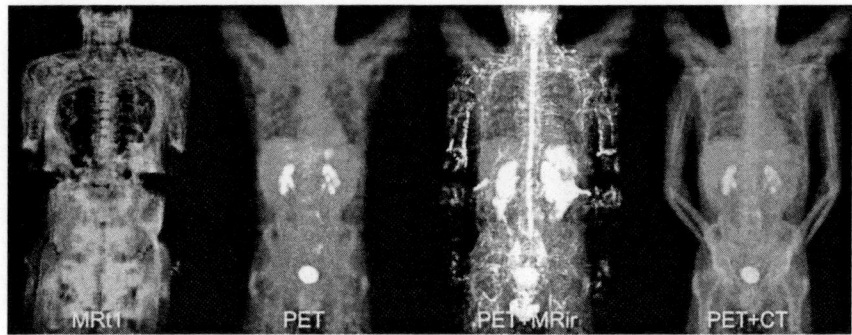

Fig. 7 Whole body multimodality imaging with discrete modalities. From left to right: 3D renderings of MR (T1 weighed), PET, MR (IR) and CT, the last two combined with PET. This figure also is an example of how simple variations in the acquisition protocols (here, arms lowered for MR and CT, arms raised for PET) can make the registration task much more complex. See Appendix for color version of the picture

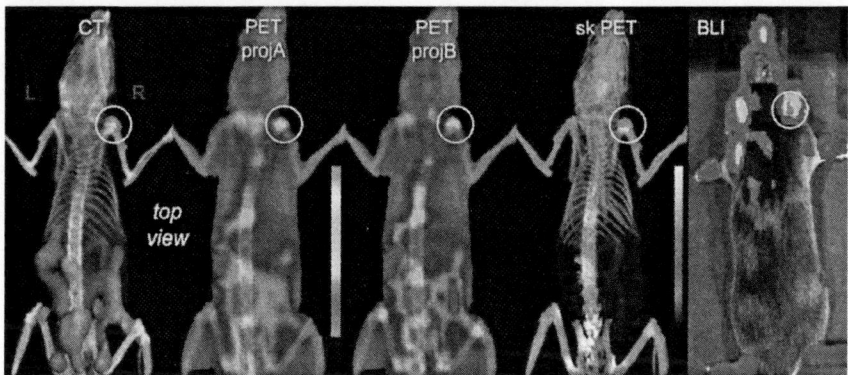

Fig. 8 Multimodality can also be performed with animal imaging. From left to right: 3D projection of CT with PET uptakes, whole body PET projections A and B, skeletal PET uptake and bioluminescence imaging for comparison (please note the corresponding signal in the right shoulder on all data sets). See Appendix for color version of the picture

4.2 Analysis Tools

Interactive visualization tools generally permit only a qualitative analysis of the data that are produced through a multimodality pipeline. Some applications and exploitation paradigms require going beyond qualitative aspects and closer to partial or fully quantitative analysis.

For instance, the registration process could lead to (1) an analysis of intensity variations between registered data sets (e.g., subtracted angiography, inter- and peri-ictal SPECT datasets to detect epilepsy foci[9,10,46,47]); (2) a statistical analysis to figure morphological variability among a subject population[48] or to study functional activation across subjects and/or modalities[49]; (3) the segmentation of images through an atlas or multispectral paradigms[50]; (4) the reconstruction of images through a-priori information or constraints introduced by at least one of the modalities/data sets that have been registered together[51] (hybrid modalities such as PET/CT and SPECT/CT actually rely on such a mechanism).

Another interesting approach is the analysis of sparse deformation fields computed in non-linear registration in order to understand where the differences between volumes are. For example, deformation based analysis is used to study longitudinal MR imaging examinations or to help the segmentation process.[52]

Whereas initially multimodality imaging mainly was focused on the registration process, i.e., the estimation of the geometrical transformation between two or more data sets, visualization and interaction tools soon became necessary to further the qualitative exploitation of registered data sets. More recently, quantitative analysis tools and paradigms were developed to tap into the real potential of multimodality data sets and go clearly beyond the simple sum of the parts. This move toward more quantification should eventually lead to real and exploitable computer assisted decision (CAD) systems that will bring together statistical analysis, a-priori knowledge, fuzzy logic classification,[53] distributed data bases[54] and artificial intelligence.[55]

5 Verification, Validation, and Evaluation

The role of image registration in medicine is proportional to the increasing importance of medical imaging in the medical workflow. Image registration may have an important influence on the medical decision making process and even on surgical actions. Therefore high quality and accuracy are expected. Sources of errors are numerous in image registration. Some errors are common to any image processing method, such as the ones related to the limited signal-to-noise ratio of the data, the limited spatial resolution of the images and the associated partial volume effect, the ones related to geometrical distortion in the images, or the ones related to the intrinsic data variability (e.g., patient movement during tomographic acquisition). Some others are specific to registration, such as the error associated to the identification of the homologous structures or to the optimization method when a local optimum is found, or the use of wrong rigidity hypothesis.

Assessment is a distinct and mandatory step of any multimodality method. The whole process of assessment is complex and includes a lot of different aspects. In software engineering, one distinguishes Verification, Validation and Evaluation (VVE) as following.[56] Verification consists in assessing that the system is built according to its specifications. Validation consists in assessing that the system actually fulfills the purpose for which it was intended. Evaluation consists in assessing that the system is accepted by the end users and well suited to its intended purpose. Verification, Validation and Evaluation can be performed all along the system life cycle in which the image fusion method is embedded: on the conceptual model representing the universe of discourse, on the requirements specification extracted from the model, on the design specification, on the executables modules, on the integrated application, on its results and finally on the results presented to the end-user.[57]

In particular, Verification and Validation permit the characteristics of the method, its performances, and its possible clinical interest in regard to other possible approaches to be formalized, demonstrated and analyzed. As such, it is a pre-requisite for any clinical implementation of a given method and it must be accomplished once and for all before any deployment. A validation protocol traditionally looks at precision, accuracy, robustness, complexity and computation time. As these parameters are obviously not independent, realistic compromises often need to be made. As a complement to the preliminary validation, strict quality control and verification must also be enforced at every use of the method. Evaluation is required in order to prove the clinical added value and interest of the method.

The following sections will mainly focus on introducing the main components of reference-based validation processes. The importance of strict quality control/assessment (QC/QA) and verification for any clinical use of these approaches will also be demonstrated.

The overall validation process starts by the specification the validation objective, which notably includes the clinical context in which the validation process has to be performed.[58] Validation objective also includes specification of the validation aim, such as study of the algorithmic performances of a method, verification of hypotheses in a registration context, comparison of various registration approaches or study of their relevance to a specific clinical context. The clinical context corresponds to the previously defined clinical context of multimodality imaging (section 2.2). The validation objective actually consists in specifying a hypothesis, relying on expected values (e.g. accuracy) required within the considered clinical context. The validation process will then aim at proposing an experiment to test such a hypothesis. The validation process itself starts by the definition of the validation data sets and the parameters that are designed to test some properties of the image registration method to be validated. Those data sets and parameters are thus applied to the image registration method, as well as another method, specifically chosen to provide a reference. Results computed by the image registration method and by the reference method are finally compared to be tested against the validation hypothesis in order to provide the validation result.[59] A complete validation process may necessitate the definition of several validation contexts and validation objectives.

5.1 Validation Data Sets

Three main types of validation data sets can be distinguished: numerical simulations, physical phantoms, and clinical data sets. One image modality can be simulated from another one. By acquiring images of physical phantoms one can control the geometry of the validation data sets and take into account the physical conditions of the image acquisition. Concerning clinical data sets as validation data sets, some may include an extrinsic system specifically used to estimate the Ground Truth or to control acquisition geometry (e.g., stereotactic frame, bone-implanted fiducial markers). Some validation studies on clinical data sets require a specific protocol for validation (e.g., intra-operative identification of anatomical landmarks or fiducial

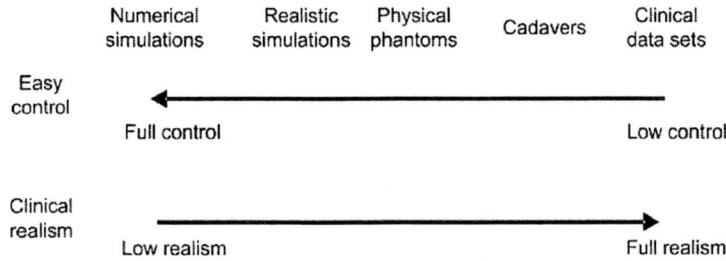

Fig. 9 Trade-off between easy control and clinical realism of validation data sets

markers using an optically tracked probe). Because there is always a compromise between realism and access to gold standard (see Fig. 9), some approaches actually combine various validation contexts and blur the line between them. For instance, simulated lesions can be added to normal clinical data sets and realistic data sets can be created through advanced computational simulation techniques that can take into account various real life characteristics of the acquisition process.

5.2 Validation Criteria

Accuracy is defined as the "degree to which a measurement is true or correct".[59] For each sample of experimental data, local accuracy is defined as the difference between observed values and theoretical ideal expected values. For a registration process, accuracy is related to the absolute difference between the computed and the true (i.e., theoretical) coordinates for every voxel within the volume of interest. Combining all such local errors leads to an estimate of a global registration error for the complete volume. In some cases, a spatial distribution of the accuracy can also be calculated from the local errors. *Precision* of a process is the resolution at which its results are repeatable, i.e., the value of the random fluctuation in the measurement made by the process. Precision is intrinsic to this process. The precision of a segmentation process – e.g., when identifying homologous structures as part of the registration process – may for instance be related either to the precision of the manual selection of the structures or to the segmentation method itself. The precision of a registration method is related to the minimum variation that can be detected when moving along the dimensions of its parameter space. The precision of a visualization method is a combination between the resolution of the image, of the display, and the characteristics of human vision. The main objective is to establish the minimum input variation that can be detected by the system and impact its output. Close to precision, *reliability* is defined as "the extent to which an observation that is repeated in the same, stable population yields the same result".[60]

For clinical applications, the global precision and accuracy of the complete registration process are at least as important to clinicians as their discrete values at every step. By compounding the whole process, global values provide realistic figures of merit for the quality of the process in relation to their clinical goals. Although most registration systems provide discrete information about every step of the process, clinicians are seldom informed about the global performance. For instance, most systems only provide data that relate to the spatial registration process, without much attempt to figure how the global multimodality process is actually impacted.

The *robustness* of a system refers to its performance in the presence of disruptive factors such as intrinsic data variability, data artifacts, pathology, or inter-individual anatomic or physiologic variability. Robustness can be established by studying the process' behavior when injecting simulated perturbations in the data. Additionally, robustness of a given method can also be studied by running it on a sample of data sets that is representative of its clinical application scope.

Complexity and *computation time* are intrinsic characteristics of any method and of its implementation. Algorithmic complexity may be estimated analytically and has a direct impact on the computation time for a given infrastructure. The functional complexity of any registration step may be either objective or subjective. Evaluating the complexity of a method therefore requires identifying and pooling together data that may relate to absolute processing times, to man-machine interactions and to implementation and exploitation in a clinical context. For instance, a single registration method may include many separate steps such as pre-processing of the data, definition/selection of homologous entities, initialization of the registration process, selection of the registration/transformation parameters (e.g., degrees of freedom, stopping criterion), etc. Even more so when many steps may be involved in a registration method, automating it may considerably reduce its functional complexity. Fully automatic methods only require that the user select the data sets to be registered. Semi-automatic methods on the other hand require more interactions such as the initialization of the geometrical transformation and/or the selection of homologous structures and/or manual verification and fine tuning of the results. Fully manual methods are a special case that requires all operations of the method to be accomplished by the user.

For most approaches, there is a compromise between automating, speed, accuracy and precision. For instance, many techniques could benefit from starting the registration process close enough to the optimal solution, which can often be performed easily by a trained user while eluding many advanced algorithms.

The true impact – and therefore feasibility – of a given level of complexity and/or computation time will obviously depend on the clinical application context. For instance, pre-surgical planning can generally comply with lengthy techniques while away from the intervention's pressure. On the other hand, interventional use will require speedy approaches that should not sacrifice accuracy, precision or robustness.

5.3 Validation Reference

As previously explained, the validation of a registration method requires to compare the results of the method with a reference that is assumed to be accurate, robust or precise. The terms of "Gold Standard" or "Ground Truth" are usually found in the literature. Ground truth and gold standard are very close concepts. Ground truth may be seen as a conceptual term relative to the knowledge of the truth concerning a specific question. It is the ideal expected result. Gold standard may be seen as the concrete realization of the ground truth. It is the practical expected result. The Gold standard can be computed or estimated from the validation data sets or from the parameters of the validation procedure. If the gold standard is estimated, it is called a bronze standard and will provide only an approximated value of the Ground Truth. The term of reference embraces these different notions.

The reference can be computed or estimated as follows. It can be the exact solution computed from a simulation. It can be an estimated solution derived from the results of a reference method, such as the solution computed from fiducials based registration. Finally, the reference may consist of an expert based solution that relies on assumptions or on a priori knowledge about the solution, such as the result of a manual registration performed by a radiologist.

Traditionally, any validation approach requires the definition of a reference that provides access to the ground truth onto which the validation relies. A key characteristic of the ground truth is that it should be derivable from the validation data. As such, the extent of the validation will necessarily be constrained by the intrinsic precision and accuracy of the method that is used to define the reference. This is why it is key to define first the validation context in order to choose a suitable reference. It also is essential to study the quality of the reference by studying, for instance, its accuracy. Additionally, the clinical realism of the reference and the way to access it have to be checked against the validation objective(s).

When the validation context requires true clinical data without access to an absolute reference, qualitative and semi-quantitative techniques (e.g., ROC curves[61]) are called upon that provide information that may be used in this context. For the special case of non-linear (also non-rigid) registration, data from realistic simulation can easily be used through the simulation of elastic deformations on high-resolution data sets. The precision and accuracy of a registration technique can then be derived – for instance – from the extent of overlap of homologous pre-segmented anatomical structures in the registered data sets. When comparing various approaches that are available as complete packages, the validation process should obviously rely on experimentation that relate to the clinical context of interest. As many such approaches are issued from professional grade developments that require thorough documentation, these specifications should also be part of the validation process and may even void the need for some of the focused experimentation altogether.

5.4 Transformation Before Comparison

Before comparing results of a method with reference, a transformation step is usually required. It has two main objectives. First, it aims to put results and reference into a same comparison system for allowing comparison. It therefore consists in a normalization step. Second, it allows performing comparison between results and reference on more clinically meaningful information. In image registration, it usually includes two sub-steps. First features are extracted (e.g. segmented, identified) from validation data sets to be registered. Second, these features are geometrically transformed into a same coordinate system. For example, anatomical points are manually identified in both the reference and the registered images. Brain or skin surfaces are segmented from both images. Then these points or surfaces are reoriented in the same coordinate system in order to be compared. Same approach can be followed with the whole volumes.

5.5 Comparison

Once relevant features for validation are identified and put into the same coordinate system. Operators are used for direct comparison. They are usually called *Validation Metrics*. In image registration, if the features consist of geometrical parameters, rotational and translational errors are computed. If the features consist of points, Target Registration Error provides relevant measures for accuracy.[62,63] If features consist of anatomical surfaces, Hausdorff distance and Iterative Closest Point are used to compare both surfaces. Finally, if features consist of the complete volumes, overlapping ratios computed with the Dice coefficient or ROC curves approach are relevant for comparison. Usually, such a comparison is performed on a set of features, or validation is performed on a set of experiments that each corresponds to different parameters or to different validation data sets. Consequently, the output of a comparison study consists of a set (or a distribution) of distances that corresponds to an error distribution. Quality indices are then computed on this error distribution such as median, minimal or maximal values, standard deviation or percentiles. A final statistical analysis is performed on these quality indices.

5.6 Others Validation Approaches

When Ground Truth and consequently reference may not be easily available from the validation data sets or from simulation, others approaches are available. This case is met when retrospective clinical data sets are used, or in case of non-linear registration validation. Statistical approaches were developed to estimate accuracy from the error performed in the identification of homologous structures in point-based registration. Accuracy is estimated via the estimation of TRE maps from the Fiducial Localization Error (FLE).[64] Repeatability studies may examine the intrinsic distribution error (e.g. mean value and standard deviation). Consistency of a registration method may be studied by checking differences of geometrical transformations computed in closed registration loops.[63,65] Such approaches provide information about the behavior of registration methods. But they are definitely not good indicator of accuracy. Additionally such approaches usually rely on strong assumptions about the data or the registration method that need to be checked. Finally, it is hard to associate these approaches with clinically relevant validation objectives.

5.7 Clinical Use

As previously described, validation only characterizes the experimental performances of a multimodality methodology. As such, this type of validation does not warrant that a multimodality method will behave well during a clinical application. Generally, registration algorithms provide the user with residual error figures. These figures are derived from the optimization process and do not necessarily represent well the quality of the registration. For instance, in the case of a local optimum, residual errors may be small with a registration result that is totally wrong.

Ideally, for any clinical procedure, users need (A) to verify results at every step of the registration and fusion process, and (B) to validate these results explicitly and separately. In practice though, because of the complexity and intricacy of such approaches it might not be easy for a physician to define and to verify objective criteria to assist in accepting or rejecting the results of a multimodality process.

For instance, in the case of neuro-navigation (e.g., for neurosurgical interventions), the surgeon may verify the goodness of the registration between digital data sets and real-life patient by pointing at easily identifiable anatomical landmarks on the real patient and verifying the match with the same landmarks on the digital 3D anatomy used by the navigation system. In order to ensure a relative independence between registration and verification, the latter needs to be performed via landmarks and/or fiducials that were not used in the registration process. The principal clinical aim of such verification is to make certain that a mathematical solution to the registration process is not anatomically incorrect.

Ideally, all along the multimodality process, any exploitation of the derived results or data – e.g., as combined visualizations – needs to include representations of the current/global level of confidence the user should always keep in mind. For instance, a statistical confidence representation of the fusion results could be used in place of a simpler – and possibly misleading – mode of combining the information.

6 Discussion

6.1 Implementing a Clinically Viable Multimodality Application

Many clinical approaches are inherently depending on many information sources, which explains why data fusion approaches were used even well before the advent of computer-based techniques. Since then, many approaches have been developed and implemented to assist with and to automate the fusion process while making it more accurate and objective.

In the previous sections of this chapter, the importance of the clinical context for the definition of a suitable fusion methodology has been underlined. It should now be clear to the reader that the design and implementation of a multimodality method needs to be preceded by a detailed study of the clinical objectives and paradigms (e.g., clinical data and applications): what are the information/data that need to be combined? What are the clinical needs related to the fusion process? Is the application for diagnosis, planning, therapy or surgery? Is the application inter- or intra-subject? Is the application inter- or intra-modality? Does the geometric registration need a rigid or a non-rigid transformation? What is the necessary accuracy level?

Additionally, the choice of a registration method will intrinsically depend on the homologous information and on the type of geometric transformation that will be used. The choice of the homologous information will depend on the information that is available in the data that will be collected to reach the clinical objectives. For instance, is it possible – or not – to use external markers in the protocol under scrutiny? Do easily identifiable common structures – e.g., surfaces or anatomical landmarks – exist in all data sets to be registered together? The choice of the geometric transformation is directly linked to the fusion context as well. While the implementation of the registration (e.g., cost function and optimization method) is intrinsically related to answers to the previous questions, the selection of the exploitation tools and paradigms (e.g., for analysis and visualization) will closely depend on the clinical objectives.

6.2 From Implementation to Clinical Use

Although research about multimodality methodology in biomedical imaging is certainly not new anymore, there are still relatively few commercially available products that can be used blindly in clinical routine. While initiated quite a while ago through the advanced exploitation (e.g., registration, processing and visualization) of then discrete modalities, the recent clinical advent of hybrid modalities has produced a significant shift in the medical community, imaging or otherwise. Through these new modalities, just about anyone now has direct access to – or a least knowledge of – instances of multimodality exploitation, which opens the door to even broader applications and support. While hybrid modalities can be credited for opening minds and making it easier to make a case about the potential of multimodality imaging in the biomedical area, their current design is still far from perfect and does not void the requirements for most of what has been presented herein.

For instance, validation remains important for the evaluation of any method and a prerequisite for any clinical application. As such, validation may be seen as an intrinsic component of the quality control process. Quality control is not an easy concept to define for biomedical imaging registration methods where both accuracy and precision may be hard to quantify. As a consequence, despite a few attempts that generally fall short of being comprehensive enough, there is still no standard procedure or data sets for the validation of such techniques.[58]

As a basic requirement, clinical implementation and use of multimodality approaches require a specific environment that includes digital imaging transfers and archiving, i.e., PACS (for Picture Archiving and Communication Systems). Advanced workstations and software tools are also necessary for the processing and visualization of multidimensional images (e.g., 2D, 3D and more). Standard image/data formats such as DICOM[66] (for Digital Imaging and Communications in Medicine) are essential to permit the integration of data sets that may come from various instrument and need to convey unambiguously all information that relates to the registration and fusion process (e.g., dimensions, sizes, location, orientation). In addition, the data registration and fusion process generates supplementary data – such as spatial reference systems, types and parameters of geometric transformations, of registration algorithms, geometric transforms, resampled volumes, segmentation results, etc. – that need to be stored with the original data sets. DICOM now offers the possibility to assign such information along with the images. Other than for relatively simple registration paradigms for which at least a few steps of the process can be automated, the registration and fusion process still are time and resource consuming. In a clinical setting, it is therefore recommended to arrange regular meetings with all parties – i.e., both makers and users – so that the complexity, requirements, end-goals and results of the process can be understood by all. As another consequence, the clinical benefit to the patient of such advanced procedures need to be objectively assessed v. their complexity and their cost.

6.3 Prospective Vision

Nowadays, most registration techniques that assume rigid or linear geometric transformations have reached a relative maturity but their underlying assumptions are not always satisfied.

To account for the most realistic deformations to compensate for distortions across modalities or various types of motion – e.g., intrinsic to the patient (e.g., physiology, breathing, different time-points) or due to mechanical causes (e.g., change of position between acquisitions on different machines, surgery) – more complex non-rigid registration is required to provide exploitable results. For clinical applications, non-rigid registration need to go well beyond simple "morphing". Other than refining the underlying transformations (that may be based on polynomial or other non-linear geometric primitives, as well as on complex mechanical deformation models), a priori information may also be taken into account to guide and/or constraint the range of adjustment for each parameter of the transformation.

Another field of active development is the optimization – both in speed and quality – of the convergence process to lead to the best registration transform. A usual strategy is to combine – at different scales that may intervene sequentially – various cost functions, optimization techniques and geometric transformations so that the process is accelerated throughout and refined only when relevant.

By performing from coarse to fine resolution, such an approach also warrants additional robustness, which is also much sought after in a clinical setting so that the process can be automated. Robustness can be further improved by developing appropriate strategies to reject outliers during registration.

All such approaches have already benefited immensely from the improvement in computers and visualization hardware and software. In addition, as mentioned earlier, another significant boost to these approaches has been the advent of hybrid modalities that have been addressing to some extent the major challenge of acquiring data sets on separate discrete modalities. The trend to integrate several modalities onto a single assembly does not seem ready to end and should therefore support fusion well beyond the current stage.

7 Conclusion

In this chapter, we have underlined at various occurrences the importance of the validation step so that data generated in a clinical workup can be trusted enough to be acted upon. For non-linear registration techniques – for which the ground truth is never easy to grasp – validation methods are still an open field. [67] In addition, we also clearly stated the need for standardized validation and QA/QC techniques and protocols to assess the quality of a particular data registration and fusion approach in a given context. While multimodality registration and fusion has already reached a definite maturity, moving beyond its current level would greatly benefit from such standards.

In the medium to longer term, multimodality tools and paradigms are likely to evolve toward multilayer systems for decision assistance that will further integrate multimodality approaches with quantification and advanced analysis capable of exploiting a-priori knowledge that may be either available locally or distributed globally. These technical improvements should then lead to the development of new applications in currently unexplored areas and could even foster entirely novel clinical approaches. The creation of multimodal patient-based anatomical or physiological models, and the modeling of therapy (e.g., surgery, radio-oncology, chemotherapy) planning will expand the possibilities for assistance both at the decisional and at the interventional levels. In essence, virtual clones will realistically – at least to the extent of their clinical intent – duplicate the anatomy and physiology of actual patients through the integration of individual multimodality data sets and parameters. These virtual clones may then be used for diagnosis, interventional planning or simulation, as well as for the quantitative evaluation of various strategies and of their possible outcome. [69]

Acknowledgements The authors would like to thank B. Gibaud from Visages, Rennes (France) and C. Grova from Montreal Neurological Institute, Montreal (Canada) for fruitful discussions and their help in formalizing concepts in these topics.

References

1. Maintz JA and Viergever MA, A survey of medical image registration, *Medical Image Analysis* 1, 1–36 (1998).
2. Maurer CR and Fitzpatrick JM, A review of medical image registration, *Interactive Image-Guided Neurosurgery*, edited by RJ Maciunas (American Association of Neurological Surgeons, Park Ridge, IL, 1993), pp. 17–44.
3. Van Den Elsen PA, Pol EJD and Viergever MA, Medical image matching - a review with classification, *IEEE Engineering in Medicine and Biology Magazine* 1, 26–39 (1993).
4. Hawkes DJ, Algorithms for radiological image registration and their clinical application, *Journal of Anatomy* 193, 347–361 (1998).
5. Brown LG, A survey of image registration techniques, *ACM Computing Surveys* 2, 325–376 (1992).
6. Crum WR, Griffin LD, Hill DLG and Hawkes DJ, Zen and the art of medical image registration : Correspondence, homology and quality, *NeuroImage* 20, 1425–1437 (2003).
7. Duncan R, SPECT imaging, in: *Focal Epilepsy* (Kluwer, Dordrecht, The Netherlands/ Boston, MA/ London, 1997), pp. 43–68.
8. Stefan H, Schneider S, Feistel H, Pawlik G, Schtiler P, Abraham-Fuchs K, Schelgel T, Neubauer U, Huk WJ, Ictal and interictal activity in partial epilepsy recorded with multichannel magnetoelectroencephalography: Correlation of Electroencephalography/Electrocorticography, magnetic resonance imaging, single photon emission computed tomography, and positron emission tomography findings, *Epilepsia* 3, 874–887 (1992).
9. Bidaut LM, Pascual-Marqui R, Delavelle J, et al., Three- to five-dimensional multisensor imaging for the assessment of neurological (dys)function, *Journal of Digital Imaging* 9, 185–198 (1996).
10. Bidaut L, Model-based multi-constrained integration of invasive electrophysiology with other modalities, *SPIE-Medical Imaging 2001* 4319, 681–692 (2001).
11. Papademetris X, Vives KP, DiStasio M, Staib LH, Neff M, Flossman S, Frielinghaus N, Zaveri H, Novotny EJ, Blumenfeld H, Constable RT, Hetherington HP, Duckrow RB, Spencer SS, Spencer DD and Duncan JS, Development of a research interface for image guided intervention: Initial application to epilepsy neurosurgery, *IEEE International Symposium on Biomedical Imaging (ISBI)*, 490–493 (2006).
12. Peters TM, Review: Images for guidance for surgical procedures, *Physics in Medicine and Biology* 51(14), R505–R540 (2006).
13. Gholipou A, Kehtarnavaz N, Briggs R, Devous M and Gopinath K, Brain functional localization: A survey of image registration techniques, *IEEE Transactions on Medical Imaging* 26(4), 427–451 (2007).
14. Mazziotta JC, Toga AW, Evans AC, Fox P and Lancaster JL, A probabilistic atlas of human brain: Theory and rationales for its development, *Neuroimage* 2, 89–101 (1995).
15. Talairach J and Toumoux P, *Co-Planar Stereotactic Atlas of the Human Brain* (Elsevier, Stuttgart, 1999).
16. Pluim JPW, Maintz JBA and Viergever MA, Mutual-information-based registration of medical images: A survey, *IEEE Transactions on Medical Imaging* 22(8), 986–1004 (2003).
17. Faugeras O, *Three-Dimensional Computer Vision: A Geometric Viewpoint* (MIT Press, Cambridge, MA, 1993).
18. Crum WR, Hartkens T and Hill DLG, Non-rigid image registration: Theory and practice, *British Journal of Radiology* 77, S140–S153 (2007).

19. Borgefors G, Hierarchical chamfer matching: A parametric edge matching algorithm, *IEEE Transactions on Pattern Analysis and Machine Intelligence* 1, 849–865 (1998).
20. Pelizzari CA, Chan GTY, Spelbring DR, Weichselbaum EE and Chen CT, Accurate three-dimensional registration of CT, PET and/or MR images of the brain, *Journal of Computed Assisted Tomography* 13, 20-26 (1989).
21. Roche A, Malandain G, Pennec X and Ayache N, The correlation ratio as a new similarity measure for multimodal image registration, in: *Proceedings of Medical Image Computing and Computer-Assisted Interventions, Boston*, edited by C Wells, A Colchester and S Delp (Lecture Notes in Computer Science, Springer, Londres, 1998), pp. 1115–1124.
22. Woods RP, Mazziotta JC and Cherry SR, MRI-PET registration with automated algorithm, *Journal of Computed Assisted Tomography* 1, 536–546 (1993).
23. Maes F, Collignon A, Vandermeulen D, Marchal G and Suetens P, Multimodality image registration by maximization of mutual information, *IEEE Transactions on Medical Imaging* 1, 187–198 (1997).
24. Wells III WM, Viola P, Atsumi H, Nakajima S and Kikinis R, Multi-modal volume registration by maximization of mutual information, *Medical Image Analysis* 1, 35–51 (1996).
25. Roche A, Malandain G and Ayache N, Unifying maximum likelihood approaches in medical image registration, INRIA Technical Report RR-3741 (1999).
26. De Castro E, Morandi C, Registration of translated and rotated images using finite Fourier transforms, *IEEE Transactions on Pattern Analysis and Machine Intelligence* 5, 700–703 (1987).
27. Pluim JPW, Maintz JBA and Viergever MA, Interpolation artefacts in mutual information-based image registration, *Computer Vision and Image Understanding* 7, 211–232 (2000).
28. Tsao J, Interpolation artifacts in multimodality image registration based on maximization of mutual information, *IEEE Transactions on Medical Imaging* 22(7), 854–864 (2003).
29. Xiuquan J, Pan H and Liang ZP, Further analysis of interpolation effects in mutual information-based image registration, *IEEE Transactions on Medical Imaging* 22(9), 1131–1140 (2003).
30. Maes F, Vendermeulen D and Suetens P, Comparative evaluation of multiresolution optimization strategies for multimodality image registration by maximisation of mutual information, *Medical Image Analysis* 4, 373–386 (1999).
31. Friston KJ, Ashbumer J, Poline JB, Frith CD, Heather JD and Frackowiak RSJ, Spatial registration and normalization of images, *Human Brain Mapping* 2, 165–189 (1995).
32. Black MJ and Rangarajan A, On the unification of line processes, outlier rejection, and robust statistics with applications in early vision, *International Journal of Computer Vision* 1, 57–91 (1996).
33. Nikou C, Heitz F, Armspach JP, Namer IJ and Grucker D, Registration of MR/MR and MRiSPECT brain images by fast stochastic optimization of robust voxel similarity measures, *Neuroimage* 8, 30–43 (1998).
34. Barillot C. Fusion de donnees et imagerie 3d en medecine [habilitation a diriger des recherches] (IRISA, Universite de Rennes I, Rennes, 1999).
35. Beyer T, Townsend DW, Brun T, et al., A combined PET/CT scanner for clinical oncology, *Journal of Nuclear Medicine* 41, 1369–1379 (2000).
36. Townsend DW and Beyer T, A combined PET/CT scanner: The path to true image fusion, *British Journal of Radiology* 75(suppl), 24–30 (2002).
37. Vogel WV, Oyen WJG, Barentsz JO, et al., PET/CT: Panacea, redundancy, or something in between? *The Journal of Nuclear Medicine* 45(1 suppl), 15S–24S (2004).
38. Hasegawa BH, Wong KH, Iwata K, et al., Dual-modality imaging of cancer with SPECT/CT, *Technology in Cancer Research and Treatment* 1(6), 449–458 (2002).
39. Jannin P, Bouliou A and Scarabin JM, Visual matching between real and virtual images in image guided neurosurgery, *Proceedings of SPIE Medical Imaging* (SPIE Proceeding Series, Newport Beach, CA/Bellingham, WA, 1997), pp. 518–526.

40. Jannin P, Grova C, Schwartz D, Barillot C and Gibaud B, Visual qualitative comparison between functional neuro-imaging (MEG, fMRI, SPECT), in: *Proceedings of Computed Assisted Radiology and Surgery*, edited by HU Lemke (Elsevier, Amsterdam, 1999), pp. 238–243.

41. Hawkes DJ, Hill DLG, Lehmann ED, Robinson GP, Maisey MN and Colchester ACF, Preliminary work on the interpretation of SPECT images with the aid of registered MR images and an MR derived 3D nemo-anatomical atlas, in: *NATO ASI Series* (Springer Berlin Heidelberg, 1990), pp. 241–251.

42. Viergever MA, Maintz JBA and Stokking R, Integration of functional and anatomical brain images, *Biophysical Chemistry* 68, 207–219 (1997).

43. Socoloinsky DA and Wolff LB, Image fusion for enhanced visualization of brain imaging, *Proceedings of SPIE Medical Imaging* (SPIE Proceeding Series, San Diego, CA/Bellingham, WA, 1999), pp. 352–362.

44. Noordmans HJ, Van der Voort HTM, Rutten GJM and Viergever MA. Physically realistic visualization of embedded volume structures for medical image data, *Proceedings of SPIE Medical Imaging* (SPIE Proceeding Series, San Diego, CA/Bellingham, WA, 1999), pp. 613–620.

45. Colin A and Boire JY, MRI-SPECT fusion for the synthesis of high resolution 3d functional brain images : A preliminary study, *Computer Methods and Programs in Biomedicine* 60, 107–116 (1999).

46. O'Brien TJ, O'Connor MK, Mullan BP, Brinkmann BH, Hanson D, Jack CR, et al., Subtraction ictal SPET co-registered to MRI in partial epilepsy : Description and technical validation of the method with phantom and patient studies, *Nuclear Medecine Communications* 19, 31–45 (1998).

47. Brinkmann BH, O'Brien TJ, Aharon S, O'Connor MK, Mullan BP, Hanson DP, et al., Quantitative and clinical analysis of SPECT image registration for epilepsy studies, *Journal of Nuclear Medecine* 4, 1098–1105 (1999).

48. Le Goualher G, Procyk E, Collins DL, Venugopal R, Barillot C and Evans AC, Automated extraction and variability analysis of sulcal neuroanatomy, *IEEE Transactions on Medical Imaging* 1, 206–217 (1999).

49. Friston KJ, Holmes AP, Worsley KJ, Poline JB, Frith CD and Frackowiak RSJ, Statistical parametric maps in functional imaging: A general linear approach, *Human Brain Mapping* 2, 189–210 (1995).

50. Lundervold A and Storvik G, Segmentation of brain parenchyma and cerebrospinal fluid in multispectral magnetic resonance images, *IEEE Transactions on Medical Imaging* 1, 339–349 (1995).

51. Liu AK, Belliveau JW and Dale AM, Spatiotemporal imaging of human brain activity using functional MRI constrained magnetoencephalography data: Monte carlo simulations, *Proceedings of the National Academy of Science [USA]* 95, 8945–8950 (1998).

52. Duchesne S and Collins DL, Analysis of deformation fields for appearance-based segmentation, in: *Proceedings of the 2001 Medical Image Computing and Computer-Assisted Intervention Conference* (Utrecht, Netherlands, 2001), vol. 2208 LNCS, pp. 1189–1190.

53. Barra V and Boire JY, A general framework for the fusion of anatomical and functional medical images, *Neuroimage* 13, 410–424 (2001).

54. Holtz B and Thirion JP, HeartPerfect: data mining in a large database of myocardial perfusion scintigraphy, in: *Proceedings of Medical Image Computing and Computer-Assisted Intervention*, edited by S Del, A DiGioia and B Jaramaz (Lecture Notes in Computer Science, Springer, Berlin, 2000), pp. 367–374.

55. Taylor CA, Draney MT, Ku JP, Parker D, Steele BN, Wang K and Zarins CK, Predictive medicine: Computational techniques in therapeutic decision-making, *Computer Aided Surgery* 4, 231–247 (1999).

56. General Principles of Software Validation; Final Guidance for Industry and FDA Staff v2.0 (2002); http://www.fda.gov/cdrh/comp/guidance/938.html

57. Balci O, Verification, validation and certification of modeling and simulation applications, in: *Proceedings of the 2003 Winter Simulation Conference* (2003), pp. 150–158.

58. Jannin P, Fitzpatrick JM, Hawkes DJ, Pennec X, Shahidi R and Vannier MW, Editorial: Validation of medical image processing in image-guided therapy, *IEEE Transactions on Medical Imaging* 21(11), 1445–1449 (2002).
59. Jannin P, Grova C and Maurer C, Model for designing and reporting reference based validation procedures in medical image processing, *International Journal of Computer Assisted Radiology and Surgery* 1(2), 1001–2115 (2006).
60. Goodman CS, Introduction to Health Care Technology Assessment, National Library of Medicine/NICHSR (2004); http://www.nlm.nih.gov/nichsr/hta101/ta101_c1.html
61. Metz CE, ROC Methodology in radiologic imaging, *Investigative Radiology* 21, 720–722 (1986).
62. West J, Fitzpatrick JM, Wang MY, Dawant BM, Maurer CR, Kessler RM, et al., Comparison and evaluation of retrospective intermodality brain image registration techniques, *Journal of Computer Assisted Tomography* 2, 554–566 (1997).
63. Fitzpatrick JM, Hill DLG and Maurer CR, Jr, Image registration, in: *Handbook of Medical Imaging: Medical Image Processing and Analysis*, edited by M Sonka and JM Fitzpatrick (SPIE Press, Bellingham, WA, 2000), vol. 2, pp. 447–513.
64. Fitzpatrick JM, West JB and Maurer CR Jr, Predicting error in rigid-body, point-based registration, *IEEE Transactions on Medical Imaging* 17(5), 694–702 (1998).
65. Pennec X and Thirion JP, A framework for uncertainty and validation of 3D registration methods based on points and frames, *International Journal of Computer Vision* 25(3), 203–229 (1997).
66. DICOM Committee, Digital Imaging and Communications in Medecine (DICOM). ACR-NEMA Standard PS3, 1–9 (1993).
67. Jannin P, Krupinski E and Warfield S, Validation in medical image processing, *IEEE Transactions on Medical Imaging* 25(11), 1405–1409 (2006).
68. Jannin P and Morandi X, Surgical models for computer-assisted neurosurgery, *Neuroimage* 37(3), 783–791 (2007).
69. Satava RM and Jones SB, Current and future applications of virtual reality for medicine, *Proceedings of the IEEE* 86, 484–489 (1998).

Appendix

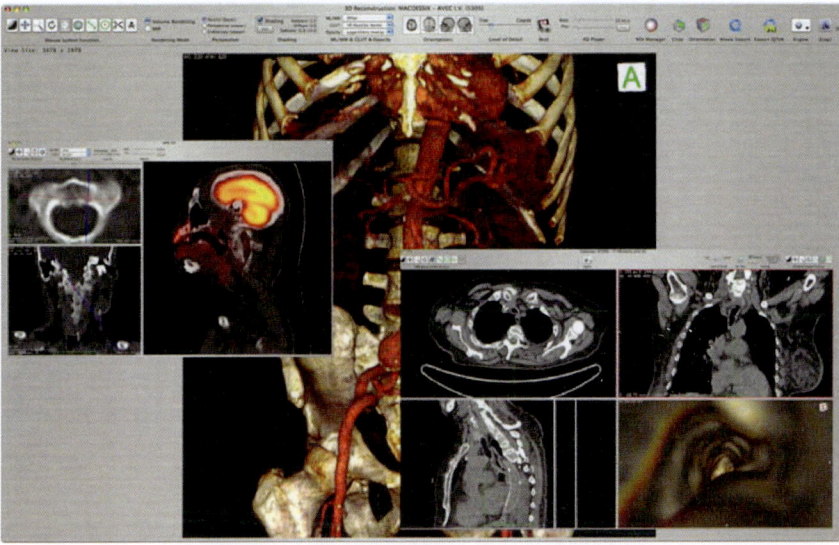

Fig. 1 Example of reading software with multidimensional image visualization capabilities (OsiriX software[5]): volume rendering, virtual endoscopy, multiplanar reconstruction. See Appendix for color version of the picture

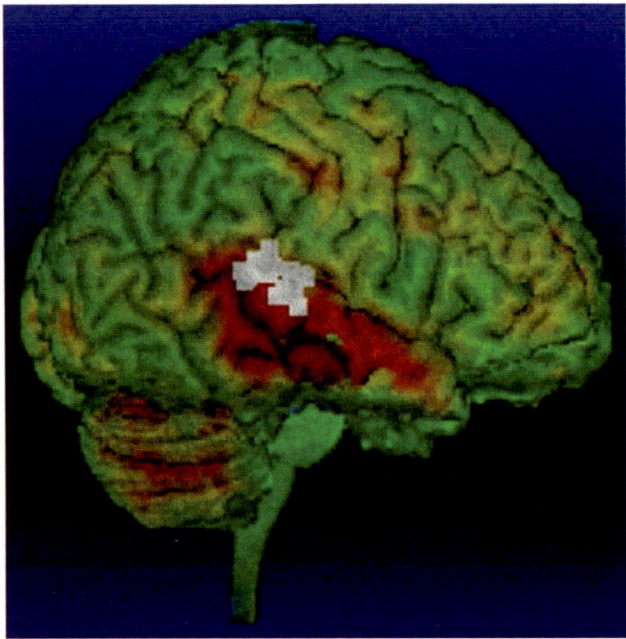

Fig. 2 3D display of 3 imaging modalities for a patient in epilepsy surgery. The brain was segmented from anatomical MR images. The colored texture was computed from registered periictal SPECT. The grey dots show auditory activations computed from MEG. See Appendix for color version of the picture

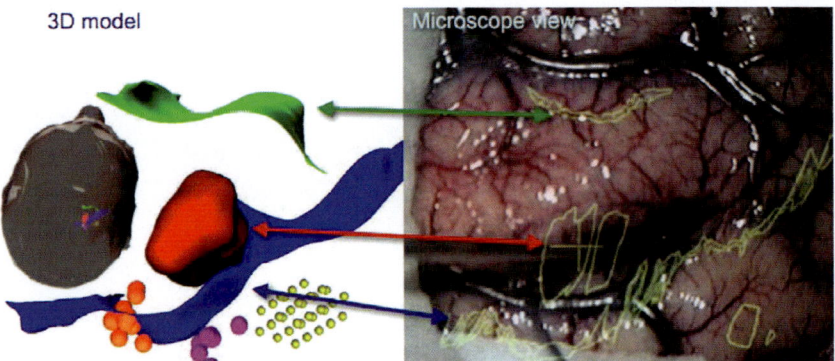

Fig. 3 Left: 3D structures segmented from multimodal images of the same patient for surgical planning. The sulci are segmented from MRI. The cavernoma is also segmented from MRI. The spheres are computed from registered functional MRI and MEG. Both functional information shows somato-sensory and motor areas. Right: these 3D preoperative 3D multimodal structures are displayed in the surgical microscope during the surgical procedure as 2D overlays. See Appendix for color version of the picture

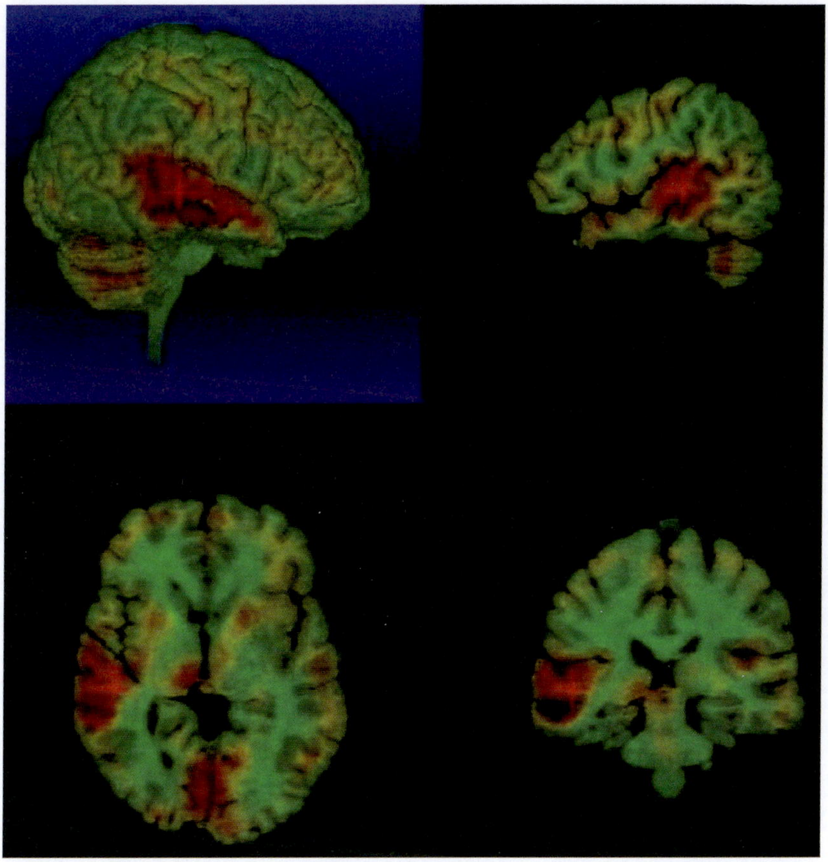

Fig. 4 Compared to Fig. 5, MR and SPECT imaging modalities are reformatted in the same coordinate system. The SPECT volume is then used to add color to MR-based objects. See Appendix for color version of the picture

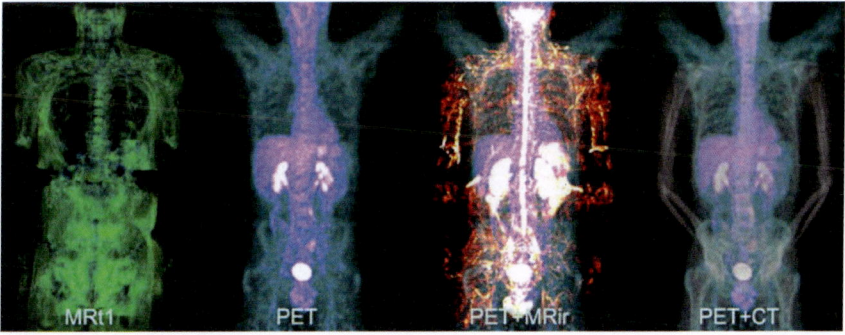

Fig. 5 Whole body multimodality imaging with discrete modalities. From left to right: 3D renderings of MR (T1 weighed), PET, MR (IR) and CT, the last two combined with PET. This figure also is an example of how simple variations in the acquisition protocols (here, arms lowered for MR and CT, arms raised for PET) can make the registration task much more complex. See Appendix for color version of the picture

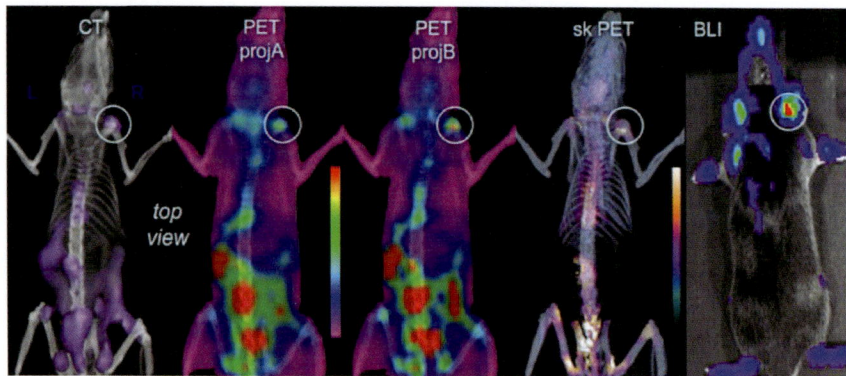

Fig. 6 Multimodality can also be performed with animal imaging. From left to right: 3D projection of CT with PET uptakes, whole body PET projections A and B, skeletal PET uptake and bioluminescence imaging for comparison (please note the corresponding signal in the right shoulder on all data sets). See Appendix for color version of the picture

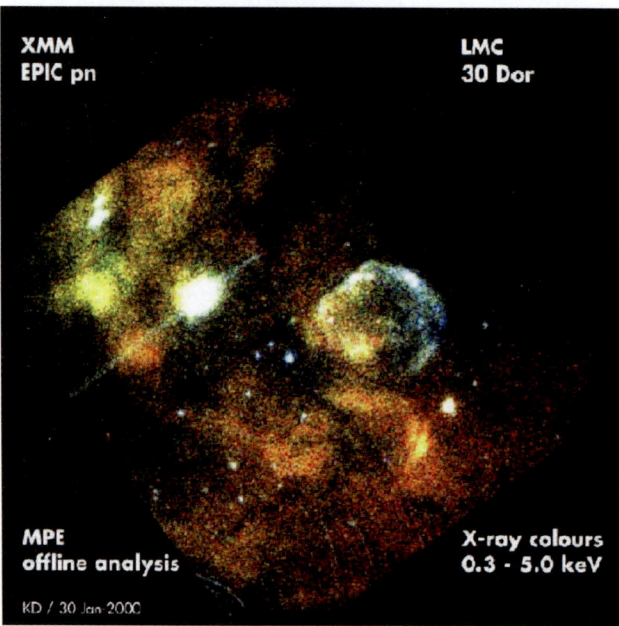

Fig. 7 The Large Magellanic Cloud in the X-ray range observed by the XMM EPIC. See Appendix I for color version of the picture